MW00583884

THE
EXPLORERS

THE
EXPLORERS

A NEW HISTORY OF AMERICA
IN TEN EXPEDITIONS

AMANDA BELLOWS

wm

WILLIAM MORROW
An Imprint of HarperCollinsPublishers

HarperCollins books may be purchased for educational, business, or sales promotional use. For information, please email the Special Markets Department at SPsales@harpercollins.com.

FIRST EDITION

Designed by Elina Cohen
Title page art courtesy of Shutterstock / Petrov Stanislav
Map by Nick Springer

Chapter opener photograph credits: Prologue: Amanda Bellows; Chapter 1: Wikimedia Commons/Rickmouser45; Chapter 2: Miriam Matthews Photograph Collection/UCLA Library; Chapter 3: Public Domain; Chapter 4: Library of Congress; Chapter 5: Notman Photographic Company; Chapter 6: Public Domain; Chapter 7: Library of Congress; Chapter 8: Library of Congress; Chapter 9: Public Domain; Chapter 10: NASA; Epilogue: NASA.

Photograph credits for insert: page 8, top right: photograph of Florence Merriam Bailey in camp near Queens, New Mexico, 1901, Box 19, Negative Number 14084, Coll. 554, Vernon Bailey Papers, American Heritage Center, University of Wyoming. Page 11, center right: Matthew Henson Collection, Box 5-1. Beulah M. Davis Special Collections, Morgan State University, Baltimore, Maryland.

Library of Congress Cataloging-in-Publication Data has been applied for.

ISBN 978-0-06-322740-8

24 25 26 27 28 LBC 5 4 3 2 1

CONTENTS

PART II

THE
EXPLORERS

CANADA

Seattle

WASHINGTON

Fort Clatsop

Columbia R.

Portland

OREGON

Salmon

IDAHO

Corps of Discovery Trail

Missouri River

Fort Ma

Knife R.

MONTANA

NORTH
DAKOTA

Fort Manuel

De S

SOUTH
DAKOTA

Yankton Reservation

WYOMING

Beckwourth Trail

American Valley

Bidwell's Bar

Beckwourth Pass

Salt Lake City

NEBRASKA

Coloma

Reno

Stockton

NEVADA

UTAH

Denver

KANS

COLORADO

San Francisco

Yosemite Valley

CALIFORNIA

The Santa F Trail

Santa Monica

Los Angeles

ARIZONA

OKLAH

Santa Fe

NEW MEXICO

PACIFIC OCEAN

TEX

MEXICO

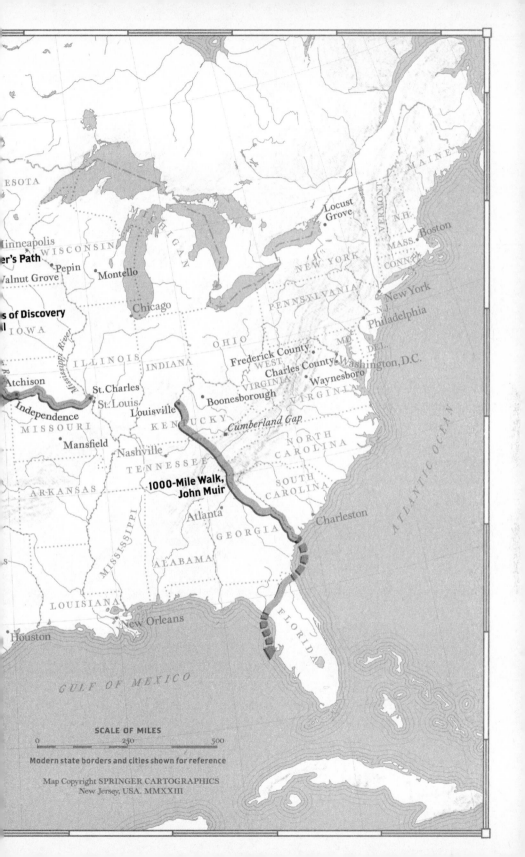

MAINE

ESOTA

MINNEAPOLIS

WISCONSIN

er's Path

Walnut Grove · Pepin

Montello

MICHIGAN

Locust
Grove

VERMONT

N.H.

MASS. · Boston

CONN.

NEW YORK

s of Discovery

Chicago

PENNSYLVANIA

New York

IOWA

Philadelphia

Mississippi River

OHIO

MD. · DEL.

Atchison

ILLINOIS

INDIANA

Frederick County

WEST

Charles County · Washington, D.C.

St. Charles
St. Louis

Louisville

VIRGINIA

Waynesboro

Independence

KENTUCKY

Boonesborough

VIRGINIA

MISSOURI

· Mansfield

· Nashville

Cumberland Gap

NORTH
CAROLINA

TENNESSEE

ARKANSAS

1000-Mile Walk,
John Muir

SOUTH
CAROLINA

· Atlanta

GEORGIA

Charleston

ATLANTIC OCEAN

MISSISSIPPI

ALABAMA

LOUISIANA

· New Orleans

FLORIDA

· Houston

GULF OF MEXICO

SCALE OF MILES

0   250   500

Modern state borders and cities shown for reference

Map Copyright SPRINGER CARTOGRAPHICS
New Jersey, USA. MMXXIII

AUTHOR'S NOTE

In places, the spelling and grammar of original source quotations have been modernized for readability.

PROLOGUE

The edge of America's western frontier was once a small wooden fort overlooking the Kentucky River. Beyond it stretched dense forests of walnut, ash, hickory, and oak. The stronghold was built near a salt lick fed by two sulfur springs, where bison gathered to taste the rich earth. Daniel Boone constructed Fort Boonesborough in Kentucky's Bluegrass region because it was a place of abundance, with fertile land and plentiful game. But it was also a disputed area, one prized by the Shawnee and Cherokee as a hunting ground.

In 1775, just nineteen days before the first shots of the Revolutionary War were fired in Massachusetts, Boone arrived at the future site of Fort Boonesborough after an exhausting journey. To reach the Bluegrass region, then located within an "Indian Reserve" established by the British after the Seven Years' War, he blazed a route through the Cumberland Gap. With an expedition of thirty men possessing critical survival skills, Boone followed historic buffalo traces and Indian trails, then headed north into Kentucky territory.

Although Native Americans and the colony of Virginia claimed the land where Fort Boonesborough stood, Boone's employer, Richard Henderson, asserted control over the territory through an illegal transaction. Violent conflict erupted in 1778, when Shawnee warriors initiated a failed siege of the fort and its occupants. Afterward, thousands of settlers heading west followed in Boone's footsteps through the cave-studded, rocky opening in the Appalachian Mountains. The flow of people who chose to live in Kentucky was so significant that, in 1792, it joined the newly formed nation as its fifteenth state.

In the years that followed, Daniel Boone's name became synonymous with exploration. He symbolized the quintessential adventurer of the American imagination, a heroic man who forged a path through what he described as the "howling wilderness" of Kentucky territory. Today many consider him to be, in the words of the historian Michael Lofaro, "the prototype and epitome of the American frontiersman, a near ideal representative of the westward movement of the nation." But why Daniel Boone?

Boone gained widespread fame through self-promotion during his own lifetime. After his autobiographical manuscript was swept away during a canoe accident, he retold his thrilling life story to the historian John Filson. Boone spoke of bitter winters, imprisonment by the Shawnee, and other dangers in Kentucky territory. On one memorable occasion, Boone's own daughter was abducted. Just ten days after the Continental Congress adopted the Declaration of Independence in Philadelphia, a group of Shawnee and Cherokee warriors kidnapped thirteen-year-old Jemima and her two friends. Daniel Boone spearheaded a dramatic rescue operation and brought the girls back home to Boonesborough.

Upon hearing Boone's biography, Filson knew that readers would be enthralled by his compelling tales. He subsequently condensed Boone's narrative into an appendix for his new book on Kentucky, which began circulating in 1784. Four years later, Filson, a frontiersman in his own right, perished during a suspected attack by the Shawnee while he surveyed the future site of Cincinnati. But his account of Boone's experiences quickly gained a life of its own.

A heavily edited version of Filson's *The Adventures of Col. Daniel Boone* was subsequently published by John Trumbull in 1786, three years after the Revolutionary War ended. The story was so popular that it would be reprinted twelve times for eager American audiences over the next thirty-seven years. European readers also gobbled up tales of Boone's exploits, as evidenced by multiple runs of Filson's original edition in France and Germany. The English Romantic poet Lord Byron even paid Boone homage in his book-length poem *Don Juan* (1823), writing admiringly of the "great . . . back-woodsman of Kentucky" who "was happiest amongst mortals anywhere" because he lived free in the wilderness. By sharing his experiences in books that reached an international reading public who wanted to know about life on the western frontier, Boone secured his place as one of the earliest heroes of American exploration.

As the eighteenth century came to a close, settlers pushed the frontier line farther west, and Fort Boonesborough faded into historical memory. Its residents, no longer seeking protection from Native Americans, moved away. By 1830, the settlement had fallen into disrepair. Daniel Boone himself had died ten years earlier, in Defiance, Missouri. But as the decades went on, Boone's status as America's favorite explorer only grew. Nearly one hundred years later, the historian Clarence Alvord then observed, "The fame of Boone was so universally accepted that few, if any, raised a question, when his name was inscribed among the greatest of the land in the Hall of Fame."

Boone's celebrity reached new heights as the twentieth century continued. His name and visage adorned ceramic mugs, puzzles, and books: mass-oriented products that appealed to consumers in an increasingly modern world, where Kentucky's wilderness was a distant dream. As the nation confronted the Soviet Union during the Cold War era, children channeled Boone's rugged, patriotic spirit by wearing faux coonskin caps and carrying popguns. Teenagers went to the cinema to see films like *Daniel Boone: Trail Blazer* (1956), which depicted an armed Boone, towering over a group of attacking Native Americans, as a "conqueror of the savage frontier!" By the century's end, the explorer had fully captured the imagination of those who viewed him, in the words of the historian Stephen Aron, as "the most celebrated of American frontiersmen," a man who "long personified the territorial expansion of the United States" and stood as "the foremost symbol of intrepid pioneering."

Today the time has come to revisit the history of American exploration. For too long, we have focused on adventurers like Boone, whom we elevated to mythical status in our collective imagination. By doing so, we have overlooked other important explorers—male and female, Black and white, Indigenous and immigrant—whose discoveries also helped make the United States the country that it is today.

In addition, we have embraced a limited definition of exploration, which emphasizes the acquisition of land and understates the consequences of settlement for Native peoples. What if we turned our attention to the contributions of a broader group of explorers: American Indians and African Americans who charted routes to acquire knowledge, immigrants who fought to preserve the wilderness, and women who tested new technologies to travel around the world? What might their pioneering stories tell us about exploration and the United States over the course of the past two centuries? How might their experiences help us rethink the nation's history, the reasons for exploration, and what it means to be an American?

THE FIRST EXPLORERS of the continental United States arrived about twenty thousand years ago. They came from the west and the east, likely traipsing across the Bering Land Bridge that once linked Asia and North America or sailing primitive vessels over the Atlantic Ocean. Then they spread out across thousands of miles of land, forming communities in dense forests and arid deserts, on windswept coasts and grassy plains. Much remains undiscovered about these brave early explorers, who left behind mere traces of their existence in stone tools and weaponry.

When the Norseman Leif Erikson became the first known European to walk upon North American shores around the year 1000, millennia had passed since the last transcontinental migrations. To medieval Europeans like Erikson, the continent was shrouded in mystery. Five centuries later, the first European colonists established settlements along the Atlantic Seaboard. But they knew little about the Indigenous people who already populated the continent. Over the next two centuries, Europeans arrived by the hundreds of thousands—some free and some indentured, some wealthy and some poor, but all searching for opportunity in what they called the "New World." Slave traders also brought people by force via ships transporting captive Africans who were compelled to help settle North America under the lash of the whip.

In 1776, revolution set the continent afire, when fifty-six men signed the Declaration of Independence. After seven bloody years of war, Great Britain relinquished much of its territorial claim to the new democratic republic, the United States of America. The country now stretched across 800,000 square miles, from the Eastern Seaboard to the Mississippi River. Nearly four million people called the United States home, but the vast majority of them resided in states on the East Coast.

Most Americans had never traveled to the continent's vast interior and knew not what, or whom, they would find there. They yearned to learn about the types of animals populating its dense forests, the height of the soaring mountains that waited to be summited, and whether stores of minerals lay undisturbed in its rocks. They considered whether fertile land, where food or cash crops could be grown, awaited them. They wondered about the kinds of people who lived there. Finally, they questioned how to journey most efficiently across the continent.

And so they began to penetrate deeper into the country. Just who were the men and women who ventured into undeveloped regions during the nation's earliest years? What motivated them to leave behind familiar lives and risk their fortunes, their health, and their safety in the American West? Did they possess particular personality traits or survival skills that made them uniquely suited for the rugged, dangerous conditions that explorers often confront? Finally, did they make discoveries that would influence the course of US history?

IN PART I of *The Explorers*, we will meet five intrepid men and women who crisscrossed the western frontier during the nineteenth century. They came from diverse backgrounds, escaping from lives circumscribed by racism, sexism, poverty, or discrimination as they took on great risk to exercise personal freedom and seek opportunity in a country brimming with possibility. Some possessed an individualistic spirit like that of Daniel Boone, while others brought loved ones on their journeys: husbands and wives, parents and children.

In a century of changing territorial borders and conflict over slavery's future in the United States, some explorers also faced uncertainty about their legal status. While some of the five explorers enjoyed the full privileges of citizenship, others lacked particular protections due to their sex, the color of their skin, or their tribal status. Most remained undaunted by these challenges and the varied dangers of life on the frontier. For them, exploration offered independence and personal autonomy. Far from their homes, these explorers sought physical freedom in the West, as they searched for new paths through treacherous terrain, mined for gold, set up humble homesteads, or even took on new identities.

They may have been some of the first Americans to reach the most remote regions of the western frontier, but they would not be the last. Their discoveries helped open up the continent for further human settlement. Millions of people followed in search of new lives, but these settlers also left a wake of destruction. Nineteenth-century westward expansion resulted in the violent displacement of Native Americans, ecological disruption, and environmental devastation.

Adventurers in the late nineteenth century began to recognize the negative effects of unbridled expansion. After the closing of the western frontier in 1890, two of the five explorers in Part I confronted the powerful entities whose practices destroyed natural landscapes. They sought to conserve America's remaining areas of natural beauty by creating national parks and preventing the decimation of wildlife through overhunting. Their actions set the stage for the modern environmental movement.

PART II OF *The Explorers* brings us into the twentieth century, when five new American adventurers set their sights beyond the limits of the continental United States. In the 1900s and 1910s, men overcame the obstacles of the Jim Crow era and women defied gender conventions to create careers for themselves as daring explorers. During this era of imperialism, when the world's most powerful empires sought to exploit colonized peoples for material gain, these adventurers went abroad, traveling to the deepest jungles of the Congo, the greatest mountains of Peru, and even the North Pole, seeking scientific and anthropological knowledge.

Technological advances in aviation and rocketry, accelerated by global conflicts early in the century, made it possible for American adventurers to redefine the notion of the frontier after the 1920s. Pilots pushed the limits of travel as they set speed records and flew across terrain where no person had ever set foot. The onset of the Cold War in the late 1940s catalyzed the space race, as American astronauts and Soviet cosmonauts competed to send people past the earth's atmosphere. Reaching the moon in 1969 was but one milestone in America's effort to secure geopolitical dominance and make new discoveries about the universe.

The wilderness—be it the frozen tundra of the Arctic or the airless environment of outer space—demanded much of the men and women of *The Explorers*. To prepare for unspeakable hardships, these ten adventurers turned inward and focused on readying the body and the mind. Most prioritized their objectives as explorers above all else; it was difficult to successfully balance the challenges of adventuring with family life. All of the explorers were married at some point in their lives, but not always happily. Six of the adventurers were women, but only two of them had children. By and large, they were a serious lot. They carried the weight of others' hopes and expectations, the fear of facing death, and the strong desire to discover the unknown.

When faced with extreme circumstances, these ten American explorers also learned about themselves. They tested their abilities to withstand the intolerable, from subzero temperatures in the Arctic to the violent effects of malarial infection in the jungle. When they experienced the aching loneliness of an excursion across new terrain, be it on a boat traveling up the Congo River or on horseback in the Andes, they often drew strength from their religious faith or from making connections with the people of different lands. Their journeys produced not only new insights about the wide world. The explorers' sense of self, their relationship with humankind, and their connection to the divine also expanded.

THE EXPLORERS AIMS to tell the story of the United States across two centuries. Each of the book's chapters brings us into a different key period of US history: the Louisiana Purchase, the Gold Rush, the settlement of the Great Plains, the Gilded Age, the Progressive Era, the rise of imperialism, the Great Depression, World War II, and the Cold War. We will travel across America's vast prairies and venture skyward into its mountains. We will also join American explorers on their journeys beyond the boundaries of the continental United States. We will cross the frozen Arctic Ocean and descend into the jungles of South America and Africa; we will journey by canoe and horseback, dogsled and steamship, airplane and space shuttle. We will experience the exhilarating history of American exploration alongside the men and women who shared a deep drive to discover the unknown.

Relying on sources including journals, autobiographies, correspondence, newspaper articles, drawings, and photographs, *The Explorers* seeks to reorient our understanding of who these American explorers were, what motivated them, and what contributions they made. It will show how a diverse group of Americans, including Native Americans, African Americans, immigrants, and women, went well beyond the boundaries of the western frontier once marked by Daniel Boone, then traversed the farthest reaches of the globe, and eventually penetrated outer space to explore the unknown.

We will see how, at different historical moments, exploration resulted in the destruction of the natural world *and* the birth of the conservation movement; the displacement of Native Americans *and* increased international immigration; scientific discovery *and* economic growth. By considering the lives of ten individuals, their particular aspirations and distinct moments in history, we will see how the goals of exploration changed dramatically over the course of two dynamic centuries.

These ten American adventurers may, at this point, be less celebrated than others. *The Explorers* will reveal their true and remarkable impact and legacy, while broadening our sense of the possibilities of exploration. Along the way, we will gain a fuller understanding of the American story—something richer, more complex, and more diverse than we ever imagined.

PART I

Sacagawea

THE NAVIGATOR

A violent kidnapping set off a chain of events that led to a history-changing exploration of the American West. In 1800, a twelve-year-old girl named Sacagawea was encamped with her people, the Agaidika Shoshones, at the headwaters of the Missouri River near present-day Three Forks, Montana. Her homeland was in the Lemhi River Valley of today's eastern Idaho, where pronghorn grazed on tall grasses and bighorn sheep stepped nimbly among the rocky hillsides. There, bitterroots, serviceberries, and elderberries provided nourishment for the Shoshones. They also fished for the silvery spotted salmon that filled the flowing river. Food was scarce during wintertime, so the Shoshones traveled to western Montana that year to procure bison meat to sustain them in the months ahead.

They made camp at a site where three smaller tributaries converged to form the powerful Missouri River. It was a place of staggering beauty, with soaring mountains that towered above the wide plains.

There, the Shoshones likely constructed brush lodges to provide shelter for the duration of the bison hunt. While the women remained close to their temporary home, the men ventured onto the prairies to hunt bison. Sacagawea must have hoped for a good hunt, providing the Shoshone women with enough bison hides to sew together and stretch around wooden poles to form tipis, structures to protect them from winter's chill.

Perhaps Sacagawea was assisting her mother with chores or playing with the other children when she first sensed danger. Without warning, the serene verdant landscape was pierced by loud cries. A group of Hidatsa warriors, members of an enemy nation, had emerged from hiding. Their homeland was farther east, in present-day North Dakota. But they had been weakened by smallpox outbreaks over the prior decade. By assaulting the Shoshones, they could get horses, a valuable commodity. The Hidatsa also wanted to enslave Shoshone women and force them to give birth to children, to repopulate their nation. Now the Hidatsa were executing a vicious surprise attack.

Running for cover, Sacagawea likely hid among thick groves of cottonwood trees along the riverbank with other terrified Shoshones. But it was too late. As the Hidatsa men chased them, Sacagawea headed toward the river and plunged in. Midway across, she was overtaken by her pursuers, who, in an act of war, killed at least ten Shoshone men, women, and children that day. For unknown reasons, they spared Sacagawea and several others. But tragedy nonetheless awaited her: the warriors enslaved Sacagawea and forced her to travel hundreds of miles east to their village, Metaharta. Though lucky to be alive, she grieved the loss of her family and her homeland.

In the years ahead, it became clear that there was little possibility of escape from Metaharta, one of five villages that stood near the intersection of the Knife River and the Missouri River in North Dakota. Slowly, Sacagawea began to adapt to her new life among strangers. In Metaharta and the four adjacent villages, which together formed a crucial trading center of the Northern Plains region, more than three thousand Hidatsa people and their neighbors, the Mandans, grew vegetables and exchanged goods, such as flint and corn, with Indigenous, European, and American visitors.

Living in one of Metaharta's fifty earthen lodges, Sacagawea learned the Hidatsa language, new customs and rituals, and strategies for survival that were suited to agrarian life near the Upper Missouri River. Although she had never farmed among the semi-nomadic Sho-shones, Sacagawea likely helped prepare the ground each spring for planting corn, beans, squash, tobacco, and sunflowers, using a sharp stick or a hoe made from the shoulder blade of a buffalo. In sum-mer, the women of her village may have taught her to shape pots from the brown clay of the Knife River, which flowed past their dwellings. When Hidatsa men brought wild game from their hunting excursions, she would have worked alongside the other women to scrape the flesh from the hides and prepare them for use as clothing and blankets. Because she was enslaved, Sacagawea's captors considered her their human property, a person owned by others. How she was treated as a prisoner of war is unknown, but she had little power to determine her future.

After about four years of enslavement, Sacagawea found herself in another kind of prison. Her captors likely sold her into marriage to a French Canadian trader named Toussaint Charbonneau, who was about thirty-seven years of age. Sacagawea was only sixteen. Char-bonneau was then living among the Hidatsa people with his first wife, Otter Woman, who may also have been a Shoshone prisoner of war. Charbonneau spent much of his life abusing Indigenous women; he raped a Saulteau girl in Canada in 1795 and purportedly trafficked Arapaho women in 1814.

Compelled to submit to Charbonneau's sexual desires as his en-slaved bride, Sacagawea found herself pregnant in the fall of 1804. Separated from her native country, she lacked the support of the Shoshone women who would have prepared her for the transition into motherhood. This was not the life she had imagined during her child-hood, when she was promised in marriage to an older Shoshone man. As Sacagawea prepared to deliver her child during North Dakota's most dangerous season, when winter blizzards flattened the frozen prairie grasses and temperatures plunged below zero degrees Fahrenheit, she could not have anticipated that an opportunity to escape her current situation would soon present itself.

ON OCTOBER 27, 1804, word traveled quickly among the Hidatsa and Mandan people about a large group of foreign men who had appeared near the Mandan village of Ruptáre, one of the five Knife River villages. The local people were curious about these strangers, who had invited them to a meeting the following day. When the Mandans approached, some brought boiled hominy as a gift to the newcomers. They later offered goods in return, presenting medals, flags, and clothes to the village leaders as well as a corn mill to the Mandans. In a display of technological and military power, they also fired their guns. The noisy spectacle astounded the attentive observers. One week after their arrival, the strangers signaled their intention to stay. Preparations for the construction of a winter fort began on November 2, 1804, with selection of a campsite near a wooded area downstream from Ruptáre on the Missouri River.

Charbonneau wasted no time getting acquainted with the foreigners. Leaving Sacagawea and Otter Woman behind, he approached their campsite on November 4 and introduced himself. Learning that the men comprised a US government–funded expedition called the Corps of Discovery, Charbonneau quickly offered his services as a paid interpreter. He would be able to translate Hidatsa into French during the corps' intended journey to the Pacific Ocean. One of the corps' men, the interpreter George Drouillard, who was of Shawnee and French Canadian descent, would then translate Charbonneau's French into English. To sweeten the deal, Charbonneau added that his Shoshone wives could also assist as interpreters in the western part of the country, where they had been captured and enslaved by the Hidatsa during different raids only a few years before.

The expedition's co-commanders immediately agreed to bring

along Charbonneau and just one of his wives. When Charbonneau returned to the lodge and told Sacagawea and Otter Woman this news, the women must have wondered if this trek might provide the chance to see the place of their birth once again. But which woman would join the expedition? And just who were these strangers asking for help to navigate the immense territory of the American West on their quest to reach the Pacific Ocean?

THE FORTY-FIVE MEN who composed the Corps of Discovery harbored grand ambitions to cross North America in search of a water passage to the Pacific Ocean. They were led by the Virginians Meriwether Lewis, President Thomas Jefferson's twenty-nine-year-old personal secretary, and William Clark, a thirty-three-year-old former Army lieutenant and experienced frontiersman. The corps sought to traverse the 828,000 square miles of largely unknown terrain that the US government had recently acquired from France in 1803, in the Louisiana Purchase. A momentous transaction spearheaded by President Jefferson, the Louisiana Purchase doubled the geographic scope of the young nation and secured the potential for its rapid expansion. The US government paid $15 million to France for the territory, but most of it remained unceded by the Native Americans who lived there.

At the turn of the nineteenth century, Americans craved land. By 1800, the US population exceeded 5.3 million people and had grown by 35 percent in the last decade alone. Most of these Americans lived on the East Coast, but settlement was increasing in the Ohio River Valley, then populated by Native Americans including the Miami, Shawnee, and Delaware Indians. European merchants and African Americans also were living there. At the time of the Louisiana Purchase, American migrants increasingly looked westward in search of land and trading opportunities.

Little was known to the US government, however, about the territory that stretched west from the winding Mississippi River to the towering Rocky Mountains. Did a water passage exist via the Missouri

River to a great river leading to the West Coast? What kinds of animals populated the continent's western forests, plains, and deserts? Which Indigenous peoples resided there, what were their claims on the land, and what were their customs? To answer these questions, President Jefferson launched the Corps of Discovery.

In his instructions to Meriwether Lewis in June 1803, President Jefferson outlined the expedition's goals. "The object of your mission is to explore the Missouri River, and such principal stream of it, as, by its course and communication with the water of the Pacific Ocean may offer the most direct and practicable water communication across this continent, for the purposes of commerce," he wrote. The idea of a "Northwest Passage" intrigued Jefferson because he sought to secure supremacy over the northwestern coast of the continent, where Great Britain also wanted to gain an economic foothold.

During the prior decade, Great Britain had made strides in surveying the Pacific Northwest, where dealers hoped to obtain luxurious furs to trade with Cantonese merchants for Chinese products like tea and porcelain. In 1792, the British naval captain George Vancouver and his team of men successfully mapped the portion of the Pacific coast stretching from present-day San Francisco to present-day British Columbia. Vancouver's lieutenant, William Broughton, even penetrated the interior of what became known in 1848 as Oregon Territory, by voyaging a hundred miles up the Columbia River. Jefferson hoped that the discovery of a water route between the Missouri River and the Columbia River would foster the dramatic growth of international trade and bolster American power.

But almost nothing was known to US and European political leaders about the continent's western interior. Increased knowledge about the West would fuel settlement, drawing American migrants who desired to farm, mine, hunt, or trap. To that end, Jefferson also asked Lewis to gather information relating to soil, wild animals, minerals, and weather conditions of different regions. Finally, he expressed curiosity about the different Indigenous peoples in the western part of the continent, with an eye to establishing lucrative commercial arrangements and what the scholar James Ronda describes as "the vital matter of announcing American sovereignty over lands of the Louisiana Purchase."

When Lewis and Clark launched their expedition in the spring of 1804, they knew the stakes were great: their mission had not only scientific and economic goals, but also geopolitical ones. The potential for US westward expansion would be determined in part by their findings and by the relationships they would establish with Native American tribes. It remained to be seen, however, whether the Corps of Discovery could successfully navigate the roiling waters of the Missouri River and the fraught dynamics they might encounter with the Indigenous people living in the West.

ON MONDAY, MAY 21, 1804, Lewis, Clark, three sergeants, more than two dozen privates, one interpreter, and an enslaved African American man named York departed from Saint Charles, Missouri, a town northwest of Saint Louis. These men formed what Lewis called "the party destined for the discovery of the interior of the continent of North America." Aboard the fifty-five-foot-long keeled boat, and two smaller boats named the white pirogue and the red pirogue, were soldiers and hunters, boatmen and traders. Each man possessed skills that would help him survive in the wilderness and ensure the expedition's success. There were blacksmiths and gunsmiths who could keep their equipment in shape, sharpshooters whose precise aim could provide food and offer protection from grizzly bears. Lewis's Newfoundland dog, Seaman, would help guard their campsite at night. The keeled boat's precious cargo, weighing twelve tons, included ink powder and notebooks for recording observations of people, material culture, wildlife, and geography. The vessels that made their way up the Missouri River lugged rifles, tomahawks, blankets, cups, utensils, medicines—and eager men who nonetheless wondered if they would ever see home again.

Lewis and Clark assumed they could travel by keeled boat only as far north as the Hidatsa and Mandan villages where Sacagawea was taken as a prisoner four years earlier. They would not be the first non-Native people to reach this region; Spanish and British merchants had traded up the Missouri River in the 1790s while competing to establish trade dominance. But beyond that point, Lewis and Clark knew little about the river's course. The two men brought crude maps with the most current information about the location of rivers, lakes, mountains, and Native American settlements in the West, but these were imperfect surveys at best. The delicate script of one map that Lewis carried identified much of the unknown territory between the Hidatsa and Mandan villages and the Pacific coast as "Conjectural." It provided few clues about the vast plains and the soaring, snow-capped Rocky Mountains that separated the villages and the ocean. During the journey, Lewis and Clark intended to make their own maps, using a chronometer to tell the time at a specific location; a sextant to ascertain latitude and longitude by calculating the angle between the line of the horizon and the sun, the stars, or the moon; and dead reckoning, an estimation of distance traveled without the use of celestial navigational aids.

Covering ten to fifteen miles each day, the men made their way up the Missouri River through present-day Kansas, Nebraska, Iowa, and South Dakota. Herds of woolly buffalo roamed the plains, stately elk bathed in the cool waters of the river, and furtive coyotes prowled the yellow clay bluffs. Along their route, the members of the corps met the Oto, the Omaha, and the Yankton Sioux. Lewis and Clark exchanged goods with tribal leaders, smoked peace pipes, and studied the people's dress, habits, and intertribal relationships. Clark wrote admiringly of *akicita*, a special group of warriors within the Yankton Sioux nation: "brave active young men who [took] a vow never to give back let the danger be what it may" during battle. To further the goals of the US government, Lewis and Clark also delivered speeches asserting American dominion over the Louisiana Purchase territory and expressing a desire to establish trade relations with Native Americans.

At the ruins of a former Omaha tribal settlement in present-day

Nebraska, the men witnessed the tragic consequences of past contact between Indigenous people and settlers. Just four years earlier, a
smallpox outbreak had killed four hundred Omaha: men, women,
and children, more than a fifth of the nation. Since the earliest days
of European contact, smallpox and other viruses brought by colonists
wreaked havoc among Native Americans, who were susceptible to
diseases to which they had never been exposed. Up to 95 percent of
Indigenous people in the Americas may have died from exposure to
European diseases between 1492 and 1700, a shocking figure that at-
tests to the devastating effects of colonization.

In the Omaha village, the catastrophic smallpox outbreak in 1800
produced great terror among the people, resulting in what Clark described as "their frenzy to very extraordinary length, not only of burning their village, but they put their *wives* and children to *death* with a
view of their all going together to some better country." The epidemic
also sapped the nation of its military strength, which "left them to
the insults of their weaker neighbors which before was glad to be on
friendly terms with them." With heavy hearts, the men of the Corps
of Discovery left behind the barren fields and dilapidated lodges that
now barely sustained the once robust population.

Days later, as the group passed along the banks of today's Iowa,
Sergeant Charles Floyd suddenly took ill with a case of what Clark
described in his journal as "Beliose Cholick," or "biliary colic," developing severe diarrhea and a weak pulse. Lewis, Clark, and York
tried to relieve his discomfort, but his condition, today believed to be
appendicitis, an aneurysm, or a gastrointestinal infection, confounded
them all. Within a day, the man who had impressed his peers with
what Clark called "firmness and determined resolution to do service
to his country and honor to himself" was dead. On August 20, the
men buried Floyd beneath a red cedar post on a bluff overlooking a
small river. They named both the bluff and the river after him, then
resumed their journey in grief. The hazards of the expedition were
clearer than ever.

Throughout autumn, the men pushed northward up the Missouri

River, finally making their way into present-day North Dakota. On a cloudy day in late October, with a dusting of snow on the ground, Clark recorded that they entered a "beautiful country on both sides [of the river], bottoms covered with wood." Above the riverbanks he could see tall, circular earthen lodges, packed closely together. The expedition had reached the villages where Sacagawea now lived, settlements that appeared on one of Lewis's maps. But as far as the corps knew, white men had journeyed no farther west than this. What they might encounter going forward was a mystery.

Bitterly cold winds signaled winter's rapid approach and the need to find a safe place to build a campsite. First, however, the corps wanted to establish positive diplomatic relations with the local people. The residents of five adjacent villages were the Mandans and Hidatsa. These two distinct tribes had become closer in the aftermath of a devastating smallpox epidemic in the 1780s. Lewis and Clark hosted curious visitors, exchanged gifts, and spoke with influential village leaders including Posecopsahe, a prominent Mandan chief. Within a few days, they determined it was safe to stay. "I went in the pirogue to the island seven miles above to look out a proper place for to winter, it being near the time the ice begins to run at this place," Clark recorded on October 30. Not enough trees grew there to support the construction of a stronghold, however, so the men selected a location further downstream. Three days later, work began on Fort Mandan, the site of their humble lodgings during the frigid winter of 1804–5.

Built from the timber of cottonwood trees, Fort Mandan slowly took shape over weeks. Tall logs with sharpened points formed a palisade eighteen feet high, surrounding seven interior dwelling rooms, storage rooms, and a workspace for the corps' blacksmith. The residences contained fireplaces for cooking and keeping warm in sub-zero temperatures. Construction proceeded quickly as the men raced against time to finish before winter arrived. While Fort Mandan materialized, Lewis and Clark turned their attention to planning the next, more dangerous phase of their journey into the western part of the continent, where they hoped to find a water route to the Pacific Ocean.

THE CORPS OF Discovery's commanders must have been relieved to have the interpretive services of Charbonneau and one of his wives on November 4, just days after the expedition's arrival at the Mandan and Hidatsa villages. Should the corps require horses to make their way across rocky terrain, or other supplies, they could now communicate with the Shoshone people for trading purposes. One week after Lewis and Clark accepted the French Canadian trapper's offer to accompany them on their excursion, Charbonneau returned to Fort Mandan, this time accompanied by Sacagawea and Otter Woman. They presented four warm buffalo robes as gifts to the Corps of Discovery's officers in an act of goodwill.

Sacagawea, then in the third trimester of her pregnancy, surely wondered at the sight of the strangely dressed men, their supplies, and the wooden huts within their fort, which differed so much from the earthen lodge in which she lived. Soon she and Otter Woman would move into Fort Mandan with Charbonneau for the long winter. They likely shared a large room with René Jusseaume, another French-Canadian man who was employed by the North West Company and served as an interpreter for the Corps of Discovery during their time at Fort Mandan, and his Native American wife. Like Sacagawea, the other residents of the Mandan and Hidatsa villages became increasingly familiar with members of the corps during the winter months that followed. They dropped by the fort, exchanged goods, and formed relationships.

What little we know about Sacagawea's life comes primarily from the journal entries of the members of the Corps of Discovery. In addition, William Clark's observations of life among the Mandan and Hidatsa people during the long winter shed light on how Indigenous women like Sacagawea sometimes lacked agency, security, or control over their own bodies. In January 1805, Clark described a ceremony in which younger Mandan men offered their wives to senior men for sexual relations, which they believed would ensure the success of an upcoming buffalo hunt. "A Buffalo Dance (or Medicine) for three nights passed in the first village, a curious custom," Clark wrote in his journal. "The young men who have their wives . . . go to one of the old men with a whining tone and [request] the old man to take his wife (who presents naked except a robe) and—(or sleep with him) the girl then takes the old man (who very often can scarcely walk) and leads him to a convenient place for the business," he continued. An unnamed member of the corps opportunistically volunteered to participate in the Buffalo Dance that winter. "We sent a man to this Medicine [Dance] last night, they gave him four girls," Clark remarked.

Troubling instances of abuse of young Native American women also occurred near Fort Mandan not long after the Corps of Discovery's arrival. On November 22, Clark recounted, one of the men standing guard alerted him that "an Indian was about to kill his wife," an acquaintance of Sacagawea and Otter Woman, after she had sex with Sergeant John Ordway. The unnamed woman fled to the hut where Sacagawea, Otter Woman, Charbonneau, and Jusseaume and his wife lived, likely seeking protection from them and the soldiers in the fort. Clark viewed the activity between Ordway and the woman as a coercive act, suggesting that the husband gave Ordway "use of her for a night, in his own bed." After Clark learned that the woman's husband had "much beat, and stabbed [her] in three places" during the preceding days, he ordered Ordway to make amends and forbade his men to "touch the [woman], or the wife of any Indian." Weeks later, in early January, Clark also recorded the visit to the camp of a Hidatsa woman who "had been much abused [by her husband], & came here for protection." During the Corps of Discovery's journey the following year, Clark witnessed Charbonneau assaulting his young wife Sacagawea, an event that prompted him to issue what Lewis called "a severe reprimand."

Despite the social conflict that could ensue when members of the

Corps of Discovery engaged in sexual relations with Mandan and Hidatsa women, desire overwhelmed the better judgment of Lewis and Clark's men, many of whom contracted sexually transmitted diseases and spread them to their partners. Clark noted that the men remained in good health for the entire winter, with the exception of those who expressed frequent "venereal complaints."

But the Corps of Discovery's commanders had come prepared. Lewis brought pewter syringes and two pounds of potassium nitrate to treat the painful symptoms of gonorrhea, as well as mercury ointment as a remedy for syphilis. He would make regular use of these treatments during the winter, when numerous men came to him for relief. Lewis himself may have contracted syphilis during the journey, a pervasive infection that may have contributed to his death by suicide years later.

IN FEBRUARY 1805, just weeks before the Corps of Discovery's planned departure from Fort Mandan, Sacagawea went into labor. Indigenous women of the Great Plains typically delivered their babies in a special structure; mothers clung to wooden stakes during the grueling, sometimes days-long process. But as a resident of Fort Mandan, living among predominantly English-speaking men, Sacagawea found herself in an unusual position. In the days leading up to her labor, she may have sought comfort or advice from Otter Woman or the Indigenous wife of Jusseaume. When the time came to deliver her baby within the fort's walls, Sacagawea's "labor was tedious and the pain violent," Lewis recorded, perhaps because "this was the first child which this woman had borne." Remarkably, the other residents of Fort Mandan joined together to help Sacagawea deliver her baby.

To alleviate her pain, Jusseaume proposed a remedy. Describing a traditional Mandan practice used during childbirth, the French Canadian interpreter told Lewis that he "had frequently administered a small portion of the rattle of the rattle-snake" to speed up a delivery. Checking his medical supplies, Lewis located "the rattle of a snake," and Jusseaume quickly took "two rings of it to the woman broken in small pieces with the fingers and added to a small quantity of water." Ten minutes later, Sacagawea gave one final, mighty push and delivered her son, who was given a French Canadian name: Jean Baptiste (John the Baptist). Clark, who quickly grew attached to the infant, called him Pomp, or "firstborn" in the Shoshone language. Everyone surely rejoiced that mother and baby had survived the perilous labor. During the days ahead, Sacagawea began to recover and prepare for her westward journey with the Corps of Discovery.

In early April, Sacagawea packed up her belongings in the shared room at Fort Mandan. Perhaps she returned to the earthen lodge where she had lived with Charbonneau and Otter Woman, walking carefully around its circular interior to ensure she had left nothing behind. Only two months postpartum, she was likely exhausted from sleepless nights spent nursing her infant son. Although Indigenous women from the Great Plains typically had little time to rest after childbirth, Sacagawea's health would likely still have been delicate. Her pelvic floor would have been weak, her ligaments loose, and her muscle tone diminished, conditions that would make the rough journey ahead all the more difficult. Nonetheless, the Corps of Discovery was ready to leave Fort Mandan and, for reasons lost to history, Sacagawea had been chosen to join them. Perhaps Sacagawea persuaded Lewis and Clark that she, not Otter Woman, possessed the skills necessary for the journey.

On the afternoon of April 7, 1805, Sacagawea stepped into one of the eight vessels now comprising what Lewis called his "little fleet." On her back she carried Jean Baptiste, his drowsy eyes opening up to the world, in a "bier" likely made of netting. The central contingent of the Corps of Discovery, composed of thirty-three people, would journey westward aboard these vessels, which included the two pirogues and six dugout canoes, while the keeled boat returned to Saint Louis with about a dozen members of the expedition tasked with transporting the commanders' reports and assorted scientific objects to Washington, DC. For the men of the corps, the western territory that lay before them represented a terra incognita that Lewis described as "two-thousand miles in width, on which the foot of civilized man had never trodden." But for Sacagawea, who had crossed much of the land following her enslavement by the Hidatsa warriors, the journey was a return to her birthplace.

THE SUN SPARKLED upon the muddy blue waters of the Missouri River as the Corps of Discovery made its way upstream. In good spirits, the men took pleasure in spotting blooming flowers, flocks of black and white geese, and even a nimble gopher burrowing under the prairie grasses. Sacagawea quickly impressed Lewis and Clark with her knowledge of edible plants, using a stick to sift through a pile of driftwood to locate wild artichokes for the party's consumption on April 9. That week, the members of the expedition also foraged for onions on the plains and hunted for beaver and venison.

But the pleasant weather did not last long. On April 13, a fierce gust of wind violently shook the convoy of boats. The white pirogue piloted by Charbonneau, which held the corps' most precious supplies, tipped precariously and nearly flipped on its side, almost spilling its cargo into the river. In addition, Lewis later noted, this vessel carried the "three men who could not swim and [Sacagawea] with the young child, all of whom, had the pirogue overset, would most probably have perished, as the waves were high, and the pirogue upwards of two hundred yards from the nearest shore." An unimpressive boatman, Charbonneau panicked and nearly capsized the vessel, but Lewis ordered George Drouillard to take control of the helm in the nick of time. Disaster was averted.

But one month later, on a cold May evening when the men were paddling through rapids in present-day Montana, a squall again upset the white pirogue carrying Sacagawea. This time, the situation became dire. Frigid water flooded the boat and swept away some of the corps' supplies. Charbonneau, whom Lewis described as "perhaps the most timid waterman in the world," was again "unfortunately . . . at

the helm" of the white pirogue and began to panic. For thirty agoniz-
ing seconds, Lewis and Clark watched in horror from the shore, where
they had been hunting. The boat tipped on its side, and waves began
crashing over it. Too far away to be heard by the pirogue's crew, even
after firing their guns into the air, Lewis and Clark remained helpless
observers to the events that followed.

As the boat began to sink, Charbonneau became overwhelmed by
fear of drowning and loudly called out to God for help. A crew member, Pierre Cruzatte, demanded that Charbonneau come to his senses,
but his warnings were of no use. Cruzatte then "threatened to shoot
him instantly if he did not take hold of the rudder and do his duty."
Taking charge, Cruzatte "ordered two of the men to throw out the
water with some kettles that fortunately were convenient," Lewis remembered.

Although Sacagawea did not know how to swim, she suppressed
her fears amid the chaos. When icy water inundated the boat and ir-
replaceable supplies floated away, she ignored the danger to her baby
and herself. Hurriedly, she began yanking precious papers, books, and
surveying instruments from the raging waters. Finally, the crew successfully rowed the swamped pirogue to the riverbank, where Lewis
and Clark helped dry out the drenched equipment. Thanks to Sacagawea's "fortitude and resolution," Lewis later recounted, "the loss we
sustained was not so great as we had at first apprehended." Only weeks
after leaving Fort Mandan, Sacagawea had proven herself an essential
member of the expedition.

AFTER RECOVERING FROM the nearly catastrophic events of
May 14, the Corps of Discovery proceeded upstream with renewed
vigor. Soon Lewis noticed that the Missouri River was beginning to
narrow, and the number of cottonwood trees growing along the banks
was diminishing. On the morning of May 26, William Clark and one
of the expedition's men went ashore and climbed to a high plain. From
there they caught a glimpse of the Rockies in the distance. Later that

day, Lewis took a look for himself, recording that he had finally "beheld the Rocky Mountains for the first time."

The sun brightened the lofty mountains' white peaks, a sight that filled Lewis with both delight and apprehension. "While I viewed these mountains I felt a secret pleasure in finding myself so near the head of the heretofore conceived boundless Missouri; but when I reflected on the difficulties which this snowy barrier would most probably throw in my way to the Pacific, and the sufferings and hardships of myself and party in them, it in some measure counterbalanced the joy I had felt in the first moments in which I gazed on them," Lewis wrote in his journal.

The members of the Corps of Discovery now suspected that passage through the mountains would be far more difficult than they had hoped. On June 3, they were forced to stop at a junction in the river. They could not proceed before discerning which of the two streams was the Missouri. Different parties of men explored possible routes in the days ahead, but they failed to come to an agreement.

To make matters worse, Sacagawea fell seriously ill on June 10 from what may have been a gonorrheal pelvic infection; she likely contracted the disease from Charbonneau. At first, Clark sought to nurse her back to health through periodic bleedings and by placing Peruvian bark on her pelvis. Both Lewis and Clark wrote of their alarm about her illness in their journals, taking pains to describe her ailments and their attempts to alleviate her suffering. They carefully spelled out Sacagawea's name in some of these entries rather than referring to her merely as the "Indian woman" or the "interpreter's wife." In the months that had passed since their departure from Fort Mandan, Sacagawea had become more than just the wife of Toussaint Charbonneau: she was now an increasingly valued member of the expedition.

When Lewis went off to survey their surroundings, Sacagawea's condition worsened. After he returned to camp five days later, having identified which of the two rivers was the Missouri, Lewis was distressed to find Sacagawea so unwell. He began to consider the expedition's fate, should she perish. "[She was] our only dependence for a friendly negotiation with the Snake [Shoshone] Indians on whom we depend for horses to assist us in our portage from the Missouri to the Columbia River," Lewis wrote, conceding that the Corps of Discovery would require the aid of the Shoshones to pass between the two rivers—a land route of yet unknown difficulty. He immediately directed his attention to Sacagawea, applying poultices and giving her mineral water to drink. "Her pulse was scarcely perceptible, very quick, frequently irregular and attended with strong nervous symptoms," Lewis remembered.

So dire was her condition that Lewis made a commitment to stay at the camp until her health returned. That evening, he applied additional poultices containing Peruvian bark and opium to ease her pain. The next day, the worst seemed to be behind them. Sacagawea "is free from pain, clear of fever, her pulse regular, and eats as heartily as I am willing to permit her of broiled buffalo well seasoned with pepper and salt and rich soup of the same meat," he recorded in his journal. With relief, Lewis concluded, "I think therefore that there is every rational hope of her recovery." The following day, Sacagawea had regained enough strength to take her first steps since falling ill on June 10. She was ready to join the members of the expedition when they departed from the river junction on June 21.

During the next three weeks, the members of the Corps of Discovery accomplished a difficult portage by lugging much of their gear eighteen miles overland to get around the Great Falls of the Missouri in present-day Montana. Keenly aware of the time lost during the portage and their need to cross the Rocky Mountains before winter arrived, they pressed onward. "We all believe that we are now about to enter on the most perilous and difficult part of our voyage, yet I see no one repining; all appear ready to meet those difficulties which wait us with resolution and becoming fortitude," mused Lewis. The territory through which they passed remained unfamiliar to all until July 22, when Sacagawea recognized the landscape around her, "assur[ing the men] that this is the river on which her relations live, and that the [Three Forks of the Missouri River] are of no great distance," Lewis wrote. "This piece of information has cheered the spirits of the party," he added, "who now begin to console themselves with the anticipation of shortly seeing the head of the Missouri yet unknown to the civilized world."

The expedition arrived at the Three Forks of the Missouri River just five days later. While the men celebrated reaching such a significant landmark after surviving severe weather and bouts of illness, Sacagawea was less jubilant. For her, the Three Forks represented the painful break between her childhood as a Shoshone girl and her forced transition into adulthood among the Hidatsa. In a calm voice, she told the men that their campsite was "precisely on the spot" of the Hidatsa raid four years earlier. Sacagawea described for the first time how her people had fled their enemies, running three miles up what Lewis and Clark would name the Jefferson River, before they were either killed or enslaved. As Sacagawea recounted her traumatic story, Lewis expressed his surprise at her apparent ambivalence. "I cannot discover that she shows any emotion of sorrow in recollecting these events, or of joy in being again restored to her native country; if she has enough to eat and a few trinkets to wear I believe she would be perfectly content anywhere." More likely, Sacagawea's response that day reflected her remarkable stoicism rather than a lack of feeling.

As summer's evening light dwindled at the Three Forks campsite, Lewis began contemplating their next move. He knew that they had to find the Shoshones.

We begin to feel considerable anxiety with respect to the Snake [Shoshone] Indians. If we do not find them or some other nation who have horses, I fear the successful issue of our voyage will be very doubtful or at all events much more difficult in its accomplishment. We are now several hundred miles within the bosom of this wild and mountainous country, where game may rationally be expected shortly to become scarce and subsistence precarious without any information with respect to the country not knowing how far these mountains continue, or where to direct our course to pass them to advantage or intercept a navigable branch of the Columbia, or even were we on such a one the probability is that we should not find any timber within these mountains large enough for canoes if we judge from the portion of them through which we have passed.

Thus, Lewis concluded, it was essential to go out in search of the Shoshones, who could provide them with horses and directions to the Columbia River.

ON AUGUST 1 AT 8:30 A.M., Lewis and a contingent of three men bade farewell to the remaining members of the corps, who would be led by Clark in Lewis's absence, and set off into the mountains. The sun scorched Lewis's back as he tramped across barren terrain and into deep valleys with his men. Although they failed to locate any members of the Shoshone nation, on August 3, Clark, exploring different terrain, saw "a track which he supposed to be that of an Indian" who had "thus discovered them and ran off," Lewis recorded. On August 6, the parties reunited and continued to make their way up Beaverhead River together the following day.

Distressed by their inability to locate the Shoshones, Lewis and Clark surely rejoiced on August 8, when Sacagawea again recognized her surroundings. Near Beaverhead Rock, on a high plain near present-day Dillon, Montana, the Shoshones typically established their summer dwellings. Sacagawea expressed her confidence to the members of the expedition that they would "either find her people on this river or on the river immediately west of its source." Sacagawea's knowledge galvanized Lewis, who sprang into action and left the following day with a small party of men.

But hope was in short supply. Four days later, Lewis and his men crossed the Continental Divide for the first time via Lemhi Pass and found the headwaters of the Missouri River, a moment that helped confirm no Northwest Passage existed. It was a question that had consumed Lewis for years and he was relieved to have finally found the Missouri River's source. But the discovery of "immense ranges of high mountains still to the West of us with their tops partially covered with snow," he wrote, signaled the perils ahead. As autumn approached, the chance of reaching the Pacific Ocean by winter seemed slim.

Then, the Corps of Discovery had a stroke of luck. After days of searching, Lewis and his men finally found the Agaidika Shoshones at their summer campsite on the Lemhi River near the Salmon-Challis National Forest of present-day Idaho. Their chief, Cameahwait, facilitated introductions between the expedition members and the Shoshone men and women. As they became acquainted, the Shoshones offered Lewis and his men berries, fish, and pronghorn before performing elaborate dances that stretched into the night. The next day, Cameahwait and an elderly member of the tribe provided Lewis with long-desired information about the local geography, advising him of the difficulties of crossing the Rocky Mountains. It became clear to Lewis that accomplishing an overland trek on foot would be impossible.

But the Shoshones had hundreds of horses. Lewis hoped Cameahwait would be willing to share them. He asked Cameahwait to bring thirty horses to the campsite at the fork of the Beaverhead River, where Lewis anticipated that the rest of the Corps of Discovery would be waiting. Cameahwait agreed and together, a large group of Shoshones set off with Lewis and his men the following day. When they reached the Beaverhead River, however, there was no sign of Clark, Sacagawea, or the other members of the expedition. Lewis urged the confused Shoshones to wait, explaining multiple times that Sacagawea, a "woman of his nation who had been taken prisoner by the [Hidatsa]," would soon arrive and assist as a translator between the two groups.

When they woke the next day, the rest of the Corps of Discovery were still nowhere to be found. Then, at last, Sacagawea and the other members of the expedition appeared in the distance. Scanning the faces of the Agaidika Shoshones, Sacagawea danced with joy. There was Jumping Fish, a friend who had been caught by the Hidatsa during the same raid in which Sacagawea was imprisoned, but later escaped. And incredibly, Cameahwait, the Shoshone chief, turned out to be her own brother. Even Lewis was moved, recording that "the meeting of those people was really affecting," though his eyes remained on the prize: Sacagawea's potential facilitation of the exchange of horses and goods that would help them reach the Pacific Ocean.

At 4 p.m. that day, negotiations for horses and a Shoshone guide began with the help of a team of translators including Sacagawea, Charbonneau, and François Labiche, a corps member with French Canadian and Omaha heritage. Lewis recalled their collective efforts to convince the Shoshones of their friendly intentions as well as the strength of the US government. As a sign of goodwill, the corps gave gifts of peace medals, clothing, tobacco, knives, beads, mirrors, and corn. "Every article about us appeared to excite astonishment in their minds," Lewis remembered, including "the appearance of the men, their arms, the canoes, our manner of working them, the black man York and the sagacity of my dog were equally objects of admiration." Negotiations continued in the days ahead, with Sacagawea and Charbonneau playing a critical role. Thanks to Sacagawea's skills in translation and diplomacy, the Corps of Discovery began their westward overland trek on August 30. Twenty-nine horses assisted in carrying their baggage and an elderly Shoshone guide, Toby, would lead them over the steep mountain passes by following a series of established traces.

Might Sacagawea have stayed behind with the members of her nation? For a moment, her future in the corps seemed uncertain. Upon her return, a Shoshone man reminded Sacagawea that her parents had once promised her in marriage to him once she reached thirteen or fourteen years of age. "He was more than double her age and had two other wives," Lewis recalled. The man publicly asserted his authority over Sacagawea but ultimately relinquished his claim on the grounds that "as she had had a child by another man, who was Charbonneau, he did not want her," Lewis wrote. Thus Sacagawea continued on the Corps of Discovery's journey to reach the Pacific Ocean.

The days ahead would test their stamina like no other part of the journey. The seemingly endless Bitterroot Mountains—desolate, steep, and rocky—fatigued the members of the corps and depleted their morale. Wild game was exceedingly scarce and conditions were treacherous. Falling snow made the slippery mountain passes even more challenging, and hunger enfeebled the group when they needed nourishment most. "I find myself growing weak for the want of food and most of the men complain of a similar deficiency and have fallen off very much," Lewis gloomily recorded in his journal on September 21, 1805.

Teetering on the edge of starvation, the Corps of Discovery noticed an unmistakable change in the landscape when they approached Weippe Prairie in Idaho. The very next day, Sacagawea and the men "descend[ed] once more to a level and fertile country where there was every rational hope of finding a comfortable subsistence," Lewis wrote. He cheered at the sight of the changing topography, adding, "The pleasure I now felt in having triumphed over the Rocky Mountains . . . can be more readily conceived than expressed, nor was the flattering prospect of the final success of the expedition less pleasing." They had traversed about a hundred miles of rugged, sometimes snowy terrain along an Indian trace that wove through the mountains.

Before long, the Corps of Discovery reached the Nez Perce tribe, who shared with them dried fish, roots, and berries. They made camp at the Clearwater River, and, after regaining some of their strength, the men set to work felling fragrant ponderosa pine trees, which seemed to scrape the sky with their long, spiny branches. Then they hollowed them out to form five dugout canoes. On October 7, the men loaded their vessels with supplies and pushed off. Moving swiftly down the Clearwater River, the Snake River, and then the Columbia River, they encountered members of Pacific Northwest nations who lived along the riverbanks. After more than a month of travel, Sacagawea and the men glimpsed in the distance what they believed to be the ocean. "Great joy in camp we are in view of the ocean, this great Pacific Ocean which we been so long anxious to see," Clark wrote on November 7.

What they spotted, however, was only the Columbia River estuary.

Twenty miles still separated the Corps of Discovery and the Pacific. The next day, sheets of rain drenched the members of the expedition, so they could not complete their journey. They were forced to take shelter on the shore at various points, including one spot Clark named the "dismal nitch." Finally, on November 15, they sighted the Pacific Ocean in the distance. Corps member Patrick Gass took a moment to reflect on the occasion in his journal, writing: "We are now at the end of our voyage, which has been completely accomplished according to the intention of the expedition, the object of which was to discover a passage by the way of the Missouri and Columbia Rivers to the Pacific Ocean; notwithstanding the difficulties, privations and dangers, which we had to encounter, endure and surmount."

On a sandy beach, the men set up camp, listening in awe to what Joseph Whitehouse described as the sounds of the "waves rolling and the surf roaring very loud." Three days later, on November 18, a contingent of ten men went with Clark to what they would name Cape Disappointment, where they viewed the Pacific Ocean in full. But Sacagawea, whose work in negotiating with the Shoshones for horses had helped secure their passage to the sea, was left behind. As the men gazed upon the deep blue surf, they celebrated reaching the Western Seaboard, confirming for President Jefferson that no Northwest Passage existed. The Corps of Discovery's journey to the edge of the continent had come to an end, but no one back east yet knew of their success. They would have to make the long return trip across the continent to share their findings with their countrymen. But first, they needed to endure another winter.

INCESSANT RAIN WAS an essential part of life on the coast of Oregon Country during much of the year. Fierce winds and wet conditions troubled the members of the Corps of Discovery. Massive waves flooded their camp, causing great discomfort. "Oh! How horrible is the day," Clark moaned, "we are all confined to our camp and wet."

The men began to consider building their winter lodgings farther inland, near the Chinook and Clatsop tribes who lived at the mouth of the Columbia River. To determine the site of the future Fort Clatsop, Lewis and Clark decided to hold a vote.

Although women and enslaved African Americans could not legally exercise the franchise in the United States, Sacagawea and York made history by casting their votes alongside the other members of the expedition. Taking into account the views of the group, Lewis and Clark chose to build Fort Clatsop on the south bank of the Columbia River, where elk were plentiful, where the men could extract salt from seawater for curing meat, and, Sacagawea had argued, where wapato roots abounded.

Before long, the members of the expedition had constructed Fort Clatsop inland from the powerful ocean, which Clark joked was hardly "pacific," given that "its waters [were] forming and perpetually break with immense waves on the sands and rocky coasts, tempestuous and horrible."

Prevented from visiting the Pacific Ocean on November 18, Sacagawea later insisted on joining an excursion to see a blue whale's massive carcass on the seashore in January 1806. She made a compelling case, Lewis remembered, arguing "that she had traveled a long way with us to see the great waters, and that now that monstrous fish was also to be seen, she thought it very hard she could not be permitted to see either." For a woman who had grown up amidst the Rocky Mountains and spent her teenage years in the Great Plains, it was a majestic sight.

At Fort Clatsop, the Corps of Discovery endured four months of unceasing rain, vicious fleas, and dampened spirits, and so they were relieved when the time finally came to depart. On March 23, 1806, they bid farewell to the fort and began their long journey back from Oregon Country to the United States. Sacagawea provided one final act of assistance to the corps by locating a passage between the steep Bridger and Gallatin Mountains in July 1806. "The Indian woman who has been of great service to me as a pilot through this country recommends a gap in the mountain more south which I shall cross," Clark wrote of her significant navigational contribution.

When Sacagawea, Charbonneau, and Jean Baptiste reached the Knife River villages in August, they decided to return to their former life and say goodbye to the friends with whom they had crossed a continent. Lewis and Clark paid Charbonneau for his services as their interpreter but neglected to remunerate Sacagawea for her contributions to the success of the expedition. Clark quickly regretted this, writing to Charbonneau several days later that Sacagawea "deserved a greater reward for her attention and services on that route than we had in our power to give her at the Mandans." He also repeated an offer made on the day of his departure: to care for Jean Baptiste, "a beautiful, promising child" who had captured his heart, and raise him back east. But Sacagawea and Charbonneau chose to keep little Pomp with them for another year.

On September 23, 1806, the Corps of Discovery reached Saint Louis, where enthusiastic villagers turned out in droves to celebrate their arrival. Their days of foraging for food, searching for the Northwest Passage, and wondering if they would make it home alive were over. Local festivities would fill the days ahead, but the true scale of the corps' achievements was hard to fathom by those who greeted them.

The expedition's findings revealed to the world the vast expanse of the North American continent and the impossibility of crossing it by water alone. The journals of the Corps of Discovery's commanders advanced the frontiers of knowledge, thanks to their documentation of flora, fauna, and Native American culture. To the delight of botanists and zoologists, Lewis and Clark described about 178 new plants and 122 new animals, such as striped skunks, prairie dogs, and yellow-bellied marmots, which were unknown to Americans at the time. The pages of their journals overflowed with drawings of Native American material culture, including carved wooden canoes and delicate fishing nets. In a major cartographic achievement, William Clark subsequently created "Clark's Map of 1810." The map would be widely used by explorers of the West for the next twenty-five years, even though it was not perfectly accurate.

The coming years would cement Lewis and Clark's fame. But without Sacagawea's guidance the Corps of Discovery might have perished from starvation before crossing the Rocky Mountains and reaching the Pacific Ocean. In fact, her contributions were so great that the historian Alvin Josephy believed the corps' decision to hire Sacagawea and Charbonneau was "the most important decision in western American history, since Sacagawea proved to be the deciding factor between success and failure of the expedition."

What happened to Sacagawea later seems even more tragic, considering her skillful leadership during the journey. Although the Corps of Discovery's men obtained 320 acres of land each and money from Congress, Sacagawea received nothing. She and Charbonneau moved to Saint Ferdinand Township in Missouri in 1810, where Clark sold them property, but they did not stay for long. Charbonneau signed on as an interpreter for the fur trader Manuel Lisa's westward excursion in 1811, despite Sacagawea's ill health at the time. Leaving Jean Baptiste with Clark, the two departed from Saint Louis in April of that year. Soon Sacagawea was pregnant again and gave birth to a daughter, Lisette, in August 1812.

Some historians believe that Sacagawea died just four months later, on December 20, from a fever at Fort Manuel Lisa in present-day South Dakota. She would have been just twenty-four years old. William Clark, who became the legal guardian of Jean Baptiste and Lisette in 1813, recorded Sacagawea's death on a list written between 1825 and 1828. Other scholars propose that she passed away much later, in 1884, on the Wind River Indian Reservation in Wyoming. As evidence, they cite the statements of Native Americans and white people describing a Shoshone woman residing there during the late nineteenth century. A memorial site bearing Sacagawea's name stands at Fort Washakie on the Wind River Indian Reservation today.

We will never know what Sacagawea might have thought about the rapid transformation of the United States, which brought fortune to some and disaster to others. As settlers moved west in search of economic opportunity, Indigenous people were increasingly displaced from their homelands, forced onto reservations, exposed to deadly diseases, or killed in battles with federal troops. The fate of Sacagawea's daughter, Lisette, remains a mystery, but Jean Baptiste, the youngest member of the Corps of Discovery, who traveled across the continent on his mother's back, continued her legacy of exploration. As an adult, he worked as a guide, fur trader, and gold prospector out west. Jean Baptiste and his fellow "mountain men" made up the next group of adventurers to reach the frontier. With extensive language and survival skills, these mountain men were uniquely positioned to make exciting discoveries during the transformative century that followed.

James Beckwourth

THE MOUNTAIN MAN

On a frigid morning in January 1848, a contractor named James Marshall approached a water-powered sawmill situated on the South Fork of the American River near Coloma, California. The sawmill's construction was almost complete, a process Marshall managed for John Sutter, a Swiss immigrant and landowner. Above him rose the dusty foothills of California's Sierra Nevada, a landscape studded with emerald-green trees and shrubs that the local Nisenan Indians had named Cullumah, or "Beautiful Valley." Marshall heard the sound of the water tumbling over the smooth tan stones and breathed in the fragrant scent of the lean pines. When he reached the river, he began inspecting the shallow tailrace of the sawmill. The machine had been giving Marshall and his crew some trouble, and his workers had recently delved into the bedrock to improve the waterflow.

As Marshall surveyed the waters that flowed through the tailrace, he noticed something sparkling amid the gravel and sand. Plunging his hand into the stream, he retrieved a small, gleaming nugget from the riverbed. "It made my heart thump," Marshall later recalled, "for I felt certain it was gold." Suddenly, he spotted a second glimmering rock and quickly snatched it up. Holding the wet nuggets in his hand, he examined them as they glittered softly in the dim winter sunlight. Could he truly have found gold? Then, Marshall likely stuffed the pieces into a rag and rushed to show them to his men. Later, he hurried to John Sutter's private offices.

Bursting with excitement, Marshall revealed his findings to Sutter, who concurred that further investigation was warranted. The two eager men soon "made a prospecting promenade," Sutter remembered, then headed "to the tail-race of the mill, through which the water was running during the night, to clean out the gravel which had been made loose . . . to search for gold." To their delight, more pieces were visible. At first, the men tried to keep their discovery secret, but the news rapidly spread. Within weeks, Sutter reported that his laborers and members of the local community had come down with "gold fever." Afraid of missing the chance to strike it rich, they abandoned their jobs and began searching for gold. Over the next four years, more than 300,000 people from around the world came to California: miners and merchants, trappers and traders, all in search of opportunity.

JAMES BECKWOURTH WAS one of those men. Born into slavery in Virginia at the turn of the nineteenth century, Beckwourth found an unlikely path out of bondage. Most enslaved men, women, and children in Virginia, treated as human property by slaveholders, spent their lives enriching their owners through unpaid labor. Working from dawn until sunset on farms or plantations, slaves planted and harvested crops like tobacco and cotton. Enslaved men and women knew that their owners could whip, rape, or even kill them; they had few rights and were treated like chattel. As the child of an enslaved woman and a man who may have been her white owner, Beckwourth faced a bleak future.

Many of the details of Beckwourth's early life at the turn of the nineteenth century remain uncertain. Born between 1798 and 1800, James Pierson Beckwith, who later became known as Jim Beckwourth, was the son of Jennings Beckwith, a landowner and slaveholder in Frederick County, Virginia, and an enslaved woman possibly named Miss Kill. Legally a slave, Beckwourth was raised by his father, who moved their family to Saint Louis, Missouri, in about 1810, just seven years after the US government acquired the town through the Louisiana Purchase. According to one observer from his later life, Beckwourth possessed a "copper complexion" that enabled him to pass in societies ranging from brigades of seasoned mountain men to different Native American tribes.

He was a charismatic man, one whom the poet Ina Coolbrith described as a "famous scout . . . one of the most beautiful creatures that ever lived." As a mature explorer, he "wore his hair in two long braids, twisted with colored cord. . . . a leather coat and moccasins and rode a horse without a saddle." Wherever he traveled, Beckwourth made a strong impression, both with his striking appearance and the enthralling stories he told of life as a mountain man.

When Beckwourth arrived in Saint Louis, it was a thriving town with a diverse population that served as a gateway to the western frontier. Established in 1764 by the French to facilitate the growing trade in bear, beaver, fox, otter, and deer pelts, by 1810 it had become an economic hub. Located near the confluence of the Missouri River and the Mississippi River, Saint Louis benefited from its central position on waterways that created a trade network stretching from Canada to New Orleans. Local residents including Osage, French, Spanish, British, and American traders swapped tomahawks and muskets, beads and jewelry, cloth and cooking utensils. Cross-cultural exchange occurred as well, leading to the development of blended identities among Saint Louisans who came from ethnically diverse families, spoke multiple languages, and adopted ways of dress that combined elements of Native American and Euro-American fashions.

Jim Beckwourth spent his formative years there, in a place of possibility wholly unlike Frederick County. In his memoir, which he dictated to journalist Thomas D. Bonner, Beckwourth marveled at the diversity of the citizens of the "grand trading depôt for the regions of the West and Northwest," seemingly aware at a young age of the outsized "profits derived from the intercourse" between trappers and merchants. Although most slaveholders prevented enslaved children from learning to read, Jim was able to attend school for a few years in Saint Louis. Afterward, he began an apprenticeship with a local blacksmith. Before long, however, Beckwourth began to chafe under the man's rules, and in 1819, he abruptly left. Jim, then about nineteen years old, decided to make his own way as a frontiersman.

At that time, the western frontier of the United States was changing rapidly. The Corps of Discovery's expedition (1804–6) through the newly purchased Louisiana Territory, extending from the Mississippi River to the Rocky Mountains, paved the way for subsequent exploration, trade, and human migration. According to the historian Bernard DeVoto, the Corps of Discovery's success had a dramatic effect: "It satisfied desire and it created desire: the desire of a westering nation." Subsequent Jeffersonian expeditions included those of William Hunter and George Dunbar, explorers who journeyed through present-day Louisiana and Arkansas in 1804; Zebulon Pike, a US soldier who surveyed territory in the Southwest, which included present-day Colorado in 1806–7; and Thomas Freeman and Peter Custis, a surveyor and a naturalist who traveled along Arkansas's Red River in 1806. With the completion of each trip, the US government gained more information about the varied geographies within Louisiana Purchase territory, the best routes for migration, local wildlife and botany, and the different Native American nations.

And some intrepid travelers set out to make a fortune. The German immigrant John Jacob Astor, the French American members of Saint Louis's Chouteau family, and the Spanish-American merchant Manuel Lisa capitalized on the opening of the West by vastly expanding the lucrative fur trade. Competition to control supply chains was intense, especially in the Pacific Northwest, where British, Canadian, and Russian traders vied for dominance. Global demand for pelts, particularly for making hats from felted beaver fur, fueled a rush among suppliers. Furs from the region were traded as far away as China and Hawaii, where some gentlemen wore black beaver hats to shield their heads from the sun. During the War of 1812, British blockades and disquiet among Native Americans temporarily disrupted the fur trade, but after the ratification of the Treaty of Ghent in 1815, the industry began to regain its footing.

The war's end also prompted Americans to migrate further westward, as Jim Beckwourth's family had done five years earlier, in search of new opportunities. Many people headed to the Ohio River Valley, where Indiana and Illinois were admitted to the Union as new states in 1816 and 1818, respectively. Settlers to the region included yeomen farmers who planned to cultivate land for planting corn, to raise hogs, and to expand their families. Thousands of people also moved to the Mississippi River Valley between 1814 and 1819; many were enslaved African Americans and white slaveholders who established farms and plantations to grow cotton.

Life was also changing for Indigenous people as contact with white settlers increased. When Native Americans accessed new material goods, such as manufactured tools and guns, they often adapted their ways of living. For instance, the traditional practice of constructing pottery declined with the increased availability of metal cooking equipment, but Native Americans continued to use bows and arrows as well as guns during hunts. Contact with Europeans and Americans also brought continued exposure to deadly viruses to which Indigenous people had no immunity. Smallpox and measles swept through Native American villages between 1800 and 1850, particularly in the Great Plains region, where tribes faced mortality rates of up to 70 percent. Finally, settlers often dramatically disrupted the environment through logging, hunting, farming, and mining, activities that depleted natural resources, reshaped the terrain, and imperiled traditional ways of life.

IT WAS DURING this period of national upheaval and transformation that Beckwourth departed Saint Louis for the western frontier.

At some point before 1824, Beckwourth's father legally emancipated him, an act that would help ensure his ability to travel freely, without fear of re-enslavement in the years ahead. He would become a man constantly on the move, an explorer who was, according to his biographer Elinor Wilson, "restless, longing to travel, eager for adventure and 'renown,' [who] never stayed long in one place, never settled down to one occupation." Leaving behind the town of his childhood, he joined a new generation of explorers: the so-called mountain men. Coming from all walks of life, they were trappers and traders, miners and scouts. With myriad skills, they sought careers on the western frontier where they could escape poverty, slavery, or prejudice. Most of all, mountain men possessed the rare kind of courage that prepared them to face a grizzly bear, survive in a blizzard, or find their way through the wilderness.

In 1822, Beckwourth journeyed to the frontier in today's southwestern Wisconsin as a member of an expedition led by James Johnson, a War of 1812 veteran. There, eager migrants hoped to make a fortune off a mineral used to make everything from bullets to pewter plates, from paint to cosmetics: lead. Native Americans in Wisconsin had mined it for centuries. Guided by Johnson, who had been granted a three-year lease from the US government to mine at Galena River, a group of approximately one hundred men traveled up the tributary of the Mississippi. They were greeted by a stern armed group of Sauk and Meskwaki Indians who, Beckwourth noted, "were already acquainted with the object of our expedition." Under pressure from US soldiers from nearby forts, who were present during negotiations, the Native Americans eventually permitted Johnson's men to mine for lead.

The next year, Beckwourth worked to quickly extract the shiny gray mineral from the banks of the river. The frenzy among miners like Beckwourth was so great that many migrants to present-day Wisconsin dug large holes in the ground for homes instead of constructing permanent housing, prompting others to call them "badgers." But Beckwourth preferred to live more sustainably, seeking guidance from local Native Americans who showed him "their choicest hunting-grounds," where "there was abundance of game, including deer, bears, wild turkey, raccoons, and numerous other wild animals." The wilderness survival skills Beckwourth acquired and the knowledge he gained about Native American culture would serve him well on future expeditions.

Before long, Beckwourth found himself craving new adventures. This time, he wanted to traverse the western interior of the continent, where a growing number of trappers and traders hoped to make a fortune. "Being possessed with a strong desire to see the celebrated Rocky Mountains, and the great Western wilderness so much talked about," Beckwourth wrote, he joined the fur trading partnership of frontiersmen Jedediah Smith and William Ashley. A former miner, War of 1812 veteran, and lieutenant governor of Missouri, Ashley assembled an expedition of men to trap for pelts. In autumn 1824, Beckwourth set off for the Rocky Mountains.

For months, Beckwourth honed his hunting skills, subsisting on "deer, wild turkeys . . . bear-meat . . . flour, sugar, and coffee" in good times and "dead horses" in periods of scarcity. He endured harsh weather, encountered Native Americans, learned how to navigate rocky rivers and steep passes, and gained confidence in his abilities as a trapper and hunter. Jim became so acclimated to life in the West that he decided in 1829 to live among the Crow people after convincing them that he was a long-lost member of their nation.

With his biracial heritage, long hair, and familiarity with Native American customs, Beckwourth became successfully integrated into the Crow tribe in the Rocky Mountains. He drew upon his skills to trade goods on behalf of the American Fur Company, exchanging pelts and furs procured by members of the Crow nation for the weapons and fabric they desired. During his time on the western frontier, Beckwourth took as wives no fewer than seven Indigenous women and girls, one of whom he physically assaulted, and fathered at least one child.

In time, Beckwourth reported rising to the level of chief. Although he had a tendency to embellish his personal history, it is possible that he occupied this position because of his stated advantage in straddling two different worlds as a trader. He claimed that this role enabled him to sell pelts procured by the Crow hunters for "twice as much . . . as [they] ever got before." Among outsiders, Beckwourth passed for a native Crow because, he reflected, he could speak the Crow language, "dressed like a Crow, [wore his hair] as long a Crow's, and [was] as black as a crow."

After seven years, however, Beckwourth began to tire of life among the Crow. Perhaps, he reminisced in his autobiography, his sense of longing for change came from having stayed in one place for too long. "But what had been my career?" he asked himself. "I had just visited the Indian territory to gratify a youthful search for adventure . . . I had traversed the fastnesses of the far Rocky Mountains in summer heats and frosts; I had encountered savage beasts and wild men, until my deliverance was a prevailing miracle." Despite his success in assimilating into the Crow tribe, he felt deep shame for having "accompanied them in their mutual slaughters, and dyed [his] hand crimson with the blood of victims who had never injured me." In the end, Beckwourth believed that he had "wasted [his] time," spending years in the Rocky Mountains that resulted in his accumulation of "a catalogue of ruthless deeds." The moment had come to return to Saint Louis. In the summer of 1836, he took to the Yellowstone River by boat, heading eastward toward his childhood hometown. His wives did not accompany him. Beckwourth abandoned them when he left the Crow nation.

Beckwourth claimed to crave stability and a more sedentary life, but it quickly become clear that he could not stay in one place for long. He soon left Saint Louis for Florida, then a US territory acquired from Spain through the Transcontinental Treaty of 1819. After a brief stint serving in the US Army during the Second Seminole War, he turned his attention westward once again, this time to the Santa Fe Trail, a newly constructed trading road connecting Independence, Missouri, and Santa Fe, New Mexico. Established in 1821 by a trader named William Becknell, the Santa Fe Trail provided a route for suppliers bearing goods some twelve hundred miles on a road that wound through the uncultivated grasslands of present-day Kansas into Colorado and New Mexico. Upon reaching Santa Fe, merchants who did not wish to return home could proceed south to Mexico City or west to Los Angeles. Beckwourth could have returned to the Rocky Mountains at this critical juncture in his career, but he likely recognized that the most lucrative opportunities were now in the Southwest. Overhunting, the substitution of cheaper nutria pelts for beaver pelts, and declining consumer interest in the late 1830s and early 1840s had decimated the beaver fur trade in Beckwourth's former stomping grounds. By the 1850s, journalists observed, silk top hats had become de rigueur, replacing the dated beaver hats once worn by the suitors of "Grandmothers in their maiden days."

The Santa Fe Trail was a world apart from the treacherous trails that looped through the Rockies. The settler James A. Little, who journeyed on the Santa Fe Trail in 1854 by wagon, described much of the landscape as exceptionally flat, ensuring that "no bridges were needed for all streams were easily forded." He marveled at how changes in elevation came gradually, remarking, "In its entire length there was not a hill. Even in crossing the mountains before reaching Sante Fe or Albuquerque, nature seemed to have arranged the great canyons in a way that our trains were never obstructed."

But dangers nonetheless awaited travelers on what many believed to be a straightforward path to prosperity. Miles from established towns, disease or traumatic accidents left travelers without access to adequate medical care. Inclement weather appeared suddenly, drenching surprised travelers with rain or pelting them with hailstones. Travelers on the Santa Fe Trail also faced the prospect of attack from Indian nations, including the Comanche and the Apache. In 1849, one newspaper recorded an assault on "a party of emigrants destined for California." Two hundred Apache warriors attacked the group after they departed from Santa Fe, killing seven of the nine men.

Beckwourth embarked upon his southwestern journey as a trader for Louis Vasquez and Andrew Sublette, two men with experience trapping in the Rocky Mountains. In 1838, Beckwourth headed west on the Santa Fe Trail to what would become known as Fort Vasquez, a dusty trading post in present-day Platteville, Colorado, where he exchanged bison robes, alcohol, and other goods with the Cheyenne, the Sioux, and others. Beckwourth also traded for Bent, St. Vrain, and Company in 1840 and moved to present-day New Mexico around 1841, building a trading post in the town of Taos and marrying a Spanish woman named Louise Sandeville.

During his regional travels in 1842, Beckwourth spent time with Jean Baptiste Charbonneau, the son of the Shoshone woman Saca-gawea and the French Canadian trader Toussaint Charbonneau. Now a fellow mountain man, Jean Baptiste had been an infant while travel-ing with the Corps of Discovery in 1805. He likely befriended Beck-wourth when the two men trapped in the Rocky Mountains in the early 1830s. They had a great deal in common: both men were biracial, grew up without their mothers, and, through mountaineering, traveled freely and extensively throughout the United States and its western territories. Their relationship would grow stronger over the following decade, when Beckwourth and Charbonneau headed to the Mother Lode region of California in hopes of making a fortune, one way or another.

WHEN JOHN MARSHALL discovered gold in January 1848, California was not yet part of the United States. Mere weeks after he plucked two nuggets from the cold waters of the South Fork of the American River, Mexico ceded present-day California, which it had claimed since 1821, to the United States as part of the Treaty of Guadalupe Hidalgo, ending the Mexican-American War. Califor-nia was a land of remarkable variety, with arid mountains, lush red-wood forests, and abundant Pacific coastlines. Indigenous people had populated the state for thousands of years, numbering approximately 300,000 in 1492. The arrival of Spanish explorers and colonists in the 1540s, however, heralded disaster for these diverse tribes, who spoke 135 languages and lived peacefully in a land characterized by plentiful resources. Diseases like smallpox, measles, and syphilis, introduced by Spanish settlers who established Catholic missions throughout the state, killed Native Americans by the tens of thousands. In 1848, when migrants from all over the world rushed to California in search of gold, only about 150,000 Indigenous people remained. Twenty years later, their number had been cut down to 30,000.

The first non-Indigenous residents of California included the descendants of Hispanic settlers who had resided there for centuries by the time a broader mix of European and American settlers arrived in the first half of the nineteenth century. By the decade immediately preceding the Gold Rush, approximately thirteen thousand non-Indigenous people, mostly ranchers, farmers, and tradesmen, lived in California. Beckwourth first traveled to California in 1844, just two years before the outbreak of the Mexican-American War. Drawn by trading opportunities, he likely traversed the Old Spanish Trail, a road that linked Santa Fe with Los Angeles.

It was a fruitful journey. On the way, Beckwourth found eager buyers for his wares among the Indigenous population, swapping his "merchandise for elk, deer, and antelope skins, very beautifully dressed." After arriving in Los Angeles, a city whose late eighteenth-century founders were primarily of African descent, Beckwourth "indulged [his] new passion for trade, and did a very profitable business for several months." But when war came to California two years later, Beckwourth hastily returned to the safety of New Mexico, where he discovered that his wife had been unfaithful to him. Breaking off his relationship with Sandeville, he resumed his itinerant lifestyle, taking a position as a wartime mail courier for the US government from Santa Fe to Fort Leavenworth, Kansas.

The Mexican-American War, which lasted from April 1846 to February 1848, led to the deaths of 12,535 US soldiers, 88 percent of whom succumbed to diseases from unsanitary conditions in military camps. Victory over Mexico, a young nation that had gained independence from Spain in 1821, added some 525,000 square miles to the United States. Acquired land included the state of Texas, which had been annexed and admitted as a state in 1845, as well as additional territory encompassing part or all of the present-day states of Utah, New Mexico, Arizona, Nevada, California, Colorado, Kansas, and Wyoming.

President James Polk touted the value of New Mexico and California in particular, declaring in his message to Congress shortly after the war's conclusion: "They constitute of themselves a country large enough for a great empire; and their acquisition is second only in importance to that of Louisiana in 1803." Keenly aware of Marshall's exciting discovery of gold in Coloma just months earlier, Polk emphasized the region's prospects as a state "rich in mineral and agricultural resources" to be "developed by American energy and enterprise."

But the president also recognized that regional settlement would fuel long-term economic growth, announcing to Congress in a subsequent message that "Upper California, irrespective of the vast mineral wealth recently developed there, holds at this day, in point of value and importance, to the rest of the Union the same relation that Louisiana did when that fine territory was acquired from France forty-five years ago." California's coastal cities would serve as central hubs for global trade, he argued, while its "temperate climate and an extensive interior of fertile lands" offered an ideal environment for farming and settlement.

Although Beckwourth likely would not have read Polk's speech, he too recognized the promise of California. After completing his final mission for the US government, delivering Army dispatches, this time from Fort Leavenworth to Los Angeles, he decided to remain in California. Beckwourth was not alone in this decision.

It was a heady moment in California history, one that saw explosive growth in a sliver of time. Marshall's discovery of gold in 1848 initiated a migratory frenzy that would become, in the words of the historian Malcolm Rohrbough, "the most significant event in the first half of the nineteenth century, from Thomas Jefferson's purchase of Louisiana in the autumn of 1803 to South Carolina's secession from the Union in the winter of 1860." The consequences of the Gold Rush were myriad; Rohrbough argues that "no other series of events produced so much movement among peoples; called into question so many basic values—marriage, family, work, wealth, and leisure; led to so many varied consequences, and left such vivid memories among its participants."

As news of the discovery of gold in California reached distant lands, migrants of all backgrounds quickly packed their belongings, bid farewell to their families, and hurried to the West Coast. The Gold Rush had begun. By 1849, 100,000 people called themselves residents of California and the population more than doubled over the next two years. The new inhabitants, who hailed from faraway places such as China, Chile, and New Zealand, became known as "Argonauts" or "Forty-Niners."

Gold-seekers came from across the North American continent too, crossing hundreds of miles of arid deserts, windswept plains, and treacherous mountain passes. One miner who reached California in November 1849 described how, despite the taxing journey, he and the members of his party were "hardy, healthy, and in good spirits, buoyant with hope." For thousands of men and women like him, the possibility of striking it rich in Mother Lode country justified the risks of travel and the price of leaving their old lives behind.

AS A MAN who saw more than his share of violence while living in the Rocky Mountains, Beckwourth was inured to the harshness and brutality of life on the western frontier by the time the Gold Rush began. But the lawlessness of California during this period was so extreme, even Beckwourth was taken aback. "At this time society in California was in the worst condition to be found, probably, in any part of the world, to call it civilized," he wrote in his autobiography.

As frantic Argonauts flooded the state, they jostled for the best land and resources. One merchant noted how each person "fear[ed] that his neighbor [would] get more than himself." Brutal competition led to extreme violence in the 1850s. Duels were commonplace. Thieves held up stagecoaches carrying valuable goods, and gangs of Mexican men called bandidos robbed and killed Chinese, Mexican, and American miners working in the hills of rural California.

Government control over the region was severely lacking. Instead, Beckwourth remembered, "the rankest excesses were familiar occurrences, and men were butchered under the very eyes of the officers of justice, and no action was taken in the matter." Emboldened individuals formed extrajudicial "Vigilance Committees," with "mock officers" who seized purported criminals and lynched them.

The rest of the country began to take notice. In 1849, the *Baltimore Patriot* published a gruesome letter from someone in San Francisco, describing the phenomenon of lynching in California: "Crimes are punishable with dreadful penalties, hanging, shooting and whipping, cutting off the ears, etc." The author went on to recount the seemingly medieval punishment meted out to a man accused of stealing eight thousand dollars' worth of gold: "He was sentenced to have his head shaved, both ears cut off close to his head, to receive 100 lashes on the bare back," and to leave California or be hanged.

As a man of color, Jim Beckwourth faced serious personal risks in this anarchic environment. A white man's or woman's accusation of wrongdoing could end in his murder. In California, African Americans—enslaved or free—possessed few legal rights during the tumultuous Gold Rush era. Approximately a thousand African Americans were living in northern California by 1852, working as miners, farmers, and business owners. By then, California had entered the Union as a free state under the Compromise of 1850. But life was still fraught for fugitives from slavery, who feared the prospect of legal recapture under California's Fugitive Slave Law of 1852.

De facto segregation prevented African Americans from entering most public establishments, and the Supreme Court decision in *Dred Scott v. Sandford* (1857) denied all African Americans citizenship and its accompanying privileges and immunities. Black Californians could not vote or testify in court. The battle to survive during the Gold Rush era was difficult enough for most travelers to the state, but in the words of the historian Sucheng Chan, "Blacks in gold-rush California [also] had to fight larger battles for equality, social justice, and basic human dignity." But Beckwourth remained.

When the Gold Rush began and Forty-Niners started pouring into the state, Beckwourth decided to make his way to the Mother Lode country. Already in the hilly coastal town of Monterey, where he had taken a job delivering dispatches for the local commissariat, Beckwourth had a head-start advantage over those coming from the East or abroad. Journey by water up the coast of California was more efficient than an overland trek, so he made his way to Monterey's bustling docks.

As luck would have it, the first US steamship soon arrived. The *California* had traveled from New York City around the Strait of Magellan to Panama, where desperate gold-seekers begged to board. Having burned up most of its fuel, the *California* stopped in Monterey to refresh its stores of wood. There, Beckwourth boarded the vessel and rode along until the ship reached its final destination: San Francisco.

So eager were the *California*'s passengers and crew to strike it rich, they immediately "started for the mines," one newspaper reported in May 1849. "Great apprehension was felt that the steamer would be unable to go to sea," the journalist concluded.

From San Francisco, a town with a population of just 812 non-Indigenous residents in the spring of 1848, Beckwourth made his way inland to Stockton, a town with a growing Black population mostly made up of men and women from the US South. Stockton was where prospectors stocked up on supplies before heading into the fields. They had visions of imminent gold, but Beckwourth had other ideas.

After purchasing clothing in Stockton to resell to prospectors, he and a friend transported the merchandise sixty miles to Sonora, a small mining town where gold had been discovered for the first time just weeks before his arrival. Rather than establish himself as a prospector, he sold his valuable goods to the influx of eager miners at "fabulous prices." Business was good. In the months ahead, Beckwourth and his friend would continue their partnership, with the friend making round trips to Stockton for provisions while Beckwourth stayed in Sonora.

In northern California, Indigenous Californians joined miners from around the world, comprising the single most populous group in the first wave of prospectors. Many, however, searched for gold under the control of American or European settlers, who often swindled them out of their earnings by trading food or clothing of a lower value for the gold they had discovered. It became tragically commonplace for Native Americans to work as forced laborers; one newspaper even cruelly urged European settlers who reached California "to make the Indians work for them. . . . [because California Indians], being most of them docile, can be made to be of great service, after they are once trained into submission."

Although he had been born enslaved, Beckwourth used California Indian labor to his own advantage during the early months of the Gold Rush. He gave Native Americans tools and housing in return for half of their profits in order to maximize his earnings. For the most part, Beckwourth remained unconcerned with the welfare of Indigenous Californians, who faced growing violence as settlers poured into the region. By 1851, the situation prompted a journalist from the *Daily Alta California* to write with alarm, "[Settlers] have abused and outraged the confidence and friendship of the trusting Indians, robbed and murdered them without compunction, and, in short, perpetuated all those outrages against humanity" that began during the nation's earliest days.

For a time, Beckwourth was satisfied with his life in Sonora as a merchant and manager. But, ever the itinerant mountain man, Beckwourth eventually found that "inactivity fatigued [him] to death." It was time to move on in search of new adventures. Auburn, a thriving mining town, attracted his attention. His old friend Jean Baptiste Charbonneau had moved there and welcomed him into a temporary business partnership. Together, the men ran an inn near Murderer's Bar, a mining site on the middle fork of the American River. Murderer's Bar was the site of an industrial prospecting operation, described as consisting of "deep shafts sunk" into the earth at the river's fork. As the rushing waters flowed past, miners worked to enrich themselves with each load of dirt, which yielded "fifty cents to the pan." In the early days of running the inn, Beckwourth and Charbonneau surely welcomed an eclectic group of visitors, providing them with much-needed sustenance and swapping stories about life on the frontier.

Winter's arrival prompted Beckwourth to relocate again, this time to nearby Greenwood Valley. It was situated in El Dorado County, named in 1850 after the mythical city of gold that sixteenth-century explorers supposed lay hidden somewhere in the so-called New World. When spring came, Beckwourth decided to try his hand as a miner. He may have packed these supplies: a prospecting hammer and a pickax for cutting into the rocky ground, a large pan for sifting wet silt, scales for weighing the precious metal, and a deerskin "poke" that served as a makeshift wallet for gold dust. Then Beckwourth departed Greenwood for the American Valley. His hope of finding gold likely buoyed his spirits as he journeyed north more than a hundred miles. Once again, Beckwourth would draw upon his experiences in the Rockies as he navigated unfamiliar territory in the Sierra Nevada of northern California. And then he made a startling discovery.

IT WAS THE spring of 1850, and Beckwourth was searching for gold with a male companion. The odds of striking it rich were against them: most prospectors during the Gold Rush era made far less money than those who focused on selling goods or acquiring real estate amid the frenzy. Nonetheless, they reached the American Valley and headed north toward the Pitt River, where, in the distance, Beckwourth glimpsed "a place far away to the southward that seemed lower than any other." He kept his thoughts to himself, but decided to explore the area further on a later trip.

Weeks later, Beckwourth returned with a group of twelve other prospectors intent on finding gold. But Jim had other plans: he "had come to discover what [he] suspected to be a pass." From the American Valley, they climbed up into the mountains, scrambling among snow-covered rocks until they reached what Beckwourth described as "an extensive valley at the northwest extremity of the Sierra range." A multitude of colorful flowers dotted the hills, and deer, antelope, geese, and ducks filled its meadows. Although the region had been historically inhabited by California's Maidu people, much of the region remained unexplored by settlers; Beckwourth believed that his steps and those of his companions "were the first that ever marked the spot."

As he and the other prospectors climbed the steep hills, they gazed upon rows of mountains that seemed to stretch endlessly into the distance. Navigating the terrain was no easy task, and the men were likely wary of attacks from the massive grizzly bears that Beckwourth said were "very plenty" in the Sierra Nevada. But as they trekked deeper into the wilderness, Beckwourth was thrilled to find a hidden path through the treacherous mountain range. "We struck across this beautiful valley to the waters of the Yuba, from thence to the waters of the Truchy, which latter flowed in an easterly direction, telling us we were on the eastern slope of the mountain range," Beckwourth wrote. This pass, he predicted, would "afford the best wagon-road into the American Valley approaching from the eastward," a route that would significantly accelerate settlement in the region and increase its prosperity. In Beckwourth's mind, his discovery was as good as gold.

Beckwourth proposed a theory to three trusted prospectors: they could make greater profits by bringing eastern settlers through the Sierra Nevada than they could by seeking gold. His companions concurred. The route, potentially spanning two hundred miles, was viable. Beckwourth began to plot his next move. Even when the group detected a small quantity of gold, Beckwourth remained focused on the even better profits to be made by guiding settlers through the mountains.

He descended to the American Valley, where he proposed the creation of the trail to a local investor named Mr. Turner. If Beckwourth could build a road from his pass point to the mining camp of Bidwell's Bar, the two men agreed, Beckwourth would be "a made-man for life." From Bidwell's Bar, settlers could take an existing route to the town of Marysville. Turner wrote a subscription list and put down $200 to support the endeavor; other investors already living in the region quickly followed suit. The mayor of Marysville estimated that "profits resulting from the speculation could not be less than from six to ten thousand dollars." Placing his trust in the mayor and the citizens of Marysville to compensate him for diverting travel to the town, Beckwourth returned to the Sierra Nevada to begin work.

In the months that followed, Beckwourth and his partners began leveling a road through the mountains upon which wagons could travel. They used axes to fell trees and cut a swath, then removed heavy rocks from the path using shovels. It was hard work, but the group made quick progress. It must have seemed to Jim that his dream would soon be realized. Beckwourth left the men behind and headed to the banks of the Truckee River, where he hoped to meet settlers traveling on the California Trail from the east. There, he sought to "turn emigration into [his] newly-discovered route." One hundred miles from civilization, however, disaster struck.

It started with a strange rash. Before long, it spread quickly across his body, creating a painful burning sensation that was difficult to ignore. Then Beckwourth likely became feverish. Extreme fatigue overcame him, and he "abandoned all hopes of recovery," given his location in the wilderness where his "only shelter . . . was a brush tent." Although he had survived frigid winters in the Rocky Mountains and violent encounters with Native Americans, it seemed as if this invisible enemy—a mysterious affliction of the skin—would lead to his demise.

The cause of Beckwourth's illness was a bacterial infection called erysipelas, a condition successfully treated by penicillin in the twentieth century. No antibiotics existed during the Gold Rush, however, and Beckwourth prepared himself for death, even writing his last will and testament. As he lay alone by the side of the road, half conscious and severely weakened, he suddenly heard the sound of wheels rumbling in the distance. It was a wagon train carrying a group of settlers.

Taking pity on Beckwourth, they climbed down from their wagons to attend to him. In the hours ahead, the women paid special attention to the charismatic mountain man, gently nursing him back to health. They may have tried nineteenth-century remedies like bleeding him, giving him monkshood, a native but poisonous plant of the Sierra Nevada, or providing him with food and water. Miraculously, Beckwourth regained his strength. Finally, Jim remembered, he was able to "mount [his] horse, and lead the first train, consisting of seventeen wagons, through 'Beckwourth Pass.'" The party climbed to an elevation of just 5,221 feet, the lowest route through the Sierra Nevada in its day. The route saved the settlers miles of extra travel and, at this crossing, spared their animals the extra strain of hauling the wagons even higher into the mountains. From the top of Beckwourth Pass, the curved outlines of distant surrounding mountains stood out against the bright blue sky.

Upon arriving in Marysville with his first party of travelers in late summer of 1851, Beckwourth recalled feeling "proud of [his] achievement, and was foolish enough to promise [himself] a substantial recognition of [his] labors." He knew that the construction of the trail would transform the region through future settlers' establishment of mining operations, farms, and ranches. But the residents of Marysville who had agreed to fund Beckwourth's endeavor failed to make good on their promises following two devastating fires that summer.

Beckwourth had spent $1,600 of his own money to build the road but received only $200 in compensation from a small group of individuals living in Marysville. Thus, the pride Beckwourth took in his discovery was overshadowed by the bitterness he felt when reflecting upon the number of men and women who subsequently crossed the Sierra Nevada using his route. The city of Marysville, Beckwourth remarked sourly in his autobiography, "greatly benefited" in the years that followed from the construction of a pass that funneled prospectors and settlers who "would otherwise have gone to Sacramento." But what could Beckwourth have done differently? It was impossible, he acknowledged with regret, to "roll a mountain into the pass and shut it up." In the end, Beckwourth slowly came to terms with the injustice, balancing his "love of country" and desire to "advance her interests" with his feelings of betrayal.

Despite his disappointment, Beckwourth set down roots in the area, in what became known as Beckwourth Valley. A verdant place filled with abundant grasses, the expansive valley spanned thirty-five miles in width. Streams of clear fresh water that flowed down the mountainside abounded with trout and otter, and grizzly bears clambered through the meadows in search of serviceberries. Pine and cedar trees provided Beckwourth with wood for his modest cabin and tinder for his fireplace, while his garden, which he described to visitors as his "plantation," supplied him with root vegetables and cabbages. The area was sparsely populated during Beckwourth's time there: in summer, the closest person lived four miles away from his cabin.

Beckwourth's house was situated about fourteen miles west of the pass; there he lived for about six years, working as an innkeeper and trader. During that time, a growing number of wagons carrying fatigued travelers descended from Beckwourth Pass into the valley. Many of them stopped at Beckwourth's log cabin to purchase supplies, rest, and learn about life on the California frontier from one of its earliest American settlers. His home became known as "the emigrant's landing-place," Beckwourth recalled fondly, "as it [was] the first ranch he arrives at in the golden state."

Month after month, Beckwourth welcomed settlers drawn by the prospect of mining, raising cattle, or growing crops. "When the weary, toil-worn emigrant reaches this valley," he assured his nineteenth-century readers, "he feels himself secure; he can lay himself down and taste refreshing repose." The travelers who reached Beckwourth's home were often in dire straits: their resources depleted from the long journey and their health failing. "Numbers have put up at my ranch without a morsel of food," Beckwourth remembered, "and without a dollar to procure any." Having endured harsh conditions as a mountain man, Beckwourth empathized with the settlers and generously provided them with food and a place to recuperate. Most settlers could never hope to repay Beckwourth, but Jim always found a way to help them.

To generate additional funds, Beckwourth agreed to dictate his autobiography to the journalist Thomas D. Bonner in 1854. Published just two years later by Harper and Brothers, *The Life and Adventures of James P. Beckwourth* immediately attracted readers and provoked debate in response to Beckwourth's sensational tales, some of which likely contained inaccuracies and exaggerations. One reviewer for the *National Era* described it as a "long yarn," a "fish story," and "half fiction," criticizing the "incongruities of the story, and the unbounded self-glorification which is the staple of the book." Nonetheless, the critic acknowledged ruefully, "Numerous readers will swallow down its marvels." Already a well-known mountain man, Beckwourth gained fame through his book.

California remained Beckwourth's home until 1858, when, for reasons lost to history, he returned to Missouri, probably via the California Trail. During the last eight years of his life, he remained characteristically active. After leaving California, Beckwourth participated in trading expeditions, married a woman named Elizabeth Ledbetter, fathered a daughter who did not survive childhood, separated from Ledbetter, and married Sue, a Crow woman. During the Civil War (1861–65), he served as a military guide in Colorado, where he led federal troops to the Cheyenne nation in advance of the tragic Sand Creek Massacre and provided crucial testimony to the government in its aftermath.

Much remains unknown about the final days of Beckwourth's life, but it is likely he perished in 1866 among the Crow people whom he once had served as a chief. Multiple newspapers reported on the death of Jim Beckwourth, who remained an outsized figure in the popular imagination of the American West until the end. As the *Gold Hill Daily News* put it, "There are few men in the world who have passed through a more exciting or eventful history than he, and there are but few men who had a bigger heart than he."

The pass that bears Beckwourth's name welcomed countless settlers to California as part of the nation's largest human migration at the time. Shortly after Beckwourth died, the state's population reached 560,000. For California's first inhabitants, life would never again be the same. The Gold Rush had reshaped the environment, wreaking havoc on the landscape in mining regions and leading to rapid urbanization in other parts of the state. Violence and disease devastated Indigenous communities, who made up no more than 6 percent of California's population by 1870. When the frenzy to strike it rich finally ended, California had been permanently transformed. Soon, however, a new rush would transform the yet-unsettled interior of the United States. This time, the migrants who arrived in great numbers were driven not by gold but by a different tantalizing prospect: cheap fertile land.

3

Laura Ingalls Wilder

THE HOMESTEADER

It was over in an instant. With the stroke of a pen, eleven million acres of land were gone forever. Thirteen Yankton men signed the Treaty of Washington on a bleak day in April 1858. Their names evoked the wildlife of eastern South Dakota: Walking Elk, Owl Man, Grabbing Hawk, Smutty Bear. But most of the men wrote their signature as an X as US government agents and officials looked on. Then they walked out onto the damp streets of the District of Columbia, contemplating how months of negotiations over the fate of their homeland had led to this moment.

Acting as representatives of their people, the Yankton men agreed to cede their territory in eastern South Dakota and move their people to a reservation consisting of a mere 430,405 acres. As compensation, they would receive annual government payments of $32,000 for the next half century. Newspapers hailed the completion of negotiations, reporting that "a very important treaty has just been signed. . . . [which]

opens to settlement a very valuable portion of Dakota Territory on the Missouri and Sioux rivers." But what of the men and women of the Yankton nation?

Outrage and disappointment soon swept through the lodges of eastern South Dakota, where families and friends told one another about the loss of their land. One white traveler to the region in the weeks that preceded the forced removal of the Yankton reported that "the Indians in the upper country [were] in a very bad humor towards the whites, and for a long time refused to receive the goods taken to them by the agents." Neighboring members of the "upper Sioux tribes" disputed the terms of the treaty of 1858 and were "great displeased with the Yanktons for ceding their lands in Dakota Territory," arguing that "these lands were the common property of the several tribes of the Sioux, and that the Yanktons had no right to sell without the concurrences of all parties interested."

Despite their efforts to resist forced migration, the Yankton relocated to the hilly country along the Missouri River in southeastern South Dakota on July 10, 1859. The abundant grasslands that they left behind stretched to the horizon as far as the eye could see. But these plains would not remain empty for long.

ON A BRIGHT, pleasant morning in the autumn of 1879, a wagon carrying a woman, a man, four girls, and all their worldly possessions traversed the Big Sioux River in present-day South Dakota. The river's mouth marked one of the boundaries of the territory surrendered by the Yankton nation twenty years earlier, before the girls were born. As they squinted into the daylight, they surveyed what Laura Ingalls, then twelve years of age, described as "a big meadow as far as we could see in every direction." With the sun on their backs, a team of strong horses pulled the heavy load westward along a faintly visible path.

To the children in the family, what was then known as "Dakota Territory" appeared uninhabited by man or beast, save for the scores of birds migrating south, their loud calls drifting on the unceasing prairie winds. The girls spotted evidence of bison in the craterlike depressions where the massive mammals had rolled around in the dust to alleviate the discomfort from irritating insects. Now, these powerful creatures were rapidly disappearing from the Great Plains.

Curiosities like these caught the attention of the youngest members of the Ingalls family as they slowly made their way across the unfamiliar country. They knew little of the Yankton people who had once inhabited these grasslands but would soon notice the impressive mound that rose high above the banks of Spirit Lake nearby. Its earthen walls protected the bodily remains of those who had hunted, worshipped, and reared families for years on the sweeping prairies of Dakota Territory.

LAURA'S PARENTS, CAROLINE and Charles Ingalls, had led a peripatetic life. They moved their daughters from the forests of Wisconsin to Missouri, then Kansas, Wisconsin, Minnesota, Iowa, and then back to Minnesota between 1867 and 1879. During this time, Charles took advantage of the Homestead Act, a revolutionary piece of legislation enacted during the Civil War that granted public land to individuals. The war had been fueled by competing visions of the future of the West: slaveholders sought not only to preserve slavery in the US South and Chesapeake region but also to expand it westward. Before the outbreak of war, advocates of populating the western territory with farmers rather than slaveholders coalesced to form the Free-Soil Party, a political group most influential between 1848 and 1854. When the Republican Party was founded in 1854, its convention of delegates at the 1856 Republican National Convention, some of whom were influenced by the Free-Soil movement, produced a platform that staunchly resisted "the extension of Slavery into Free Territory."

The election of a Republican president, Abraham Lincoln, in 1860 and the secession of eleven states where slavery was legal paved the

way for Congress's passage of the Homestead Act in 1862. Its enactment struck a critical blow against those who envisioned slave plantations spreading from coast to coast. The Confederacy's defeat in
April 1865 resulted in the abolition of slavery and the discharge of
hundreds of thousands of Black and white soldiers, many of whom
looked westward in search of a new life as farmers. Wartime turmoil
also politically weakened Indigenous peoples living in the American
West. Their lives were upended by inflation, trade interruptions, and
political fraud on the part of Republican agents operating on the frontier. In the words of the historian Ryan Hall, they were "plunged into
an irreversible state of crisis and found themselves more vulnerable to
postwar American empire-building" as the nation sought to expand its
borders during Reconstruction.

The Homestead Act fueled westward expansion by offering, with
few restrictions, up to 160 acres of federal land to claimants: not only
male US citizens but also single women and immigrants intending
to become citizens could apply. After the passage of the Civil Rights
Act of 1866, which granted citizenship to African Americans born in
the United States, free and formerly enslaved Black men and women
were also able to participate. All applicants had to be at least twenty-
one years old and had to pay the fees associated with registering their
claim. Only those who had never borne arms against the United States
were eligible, meaning that Confederate veterans were excluded.

The law was intended to develop what some scholars have called a
middle class of farmers befitting a democratic republic. Indeed, 270
million acres of public land shifted into private hands between 1863
and 1986, fueling the dream of prosperity for four million claimants
from diverse backgrounds. Many had nothing to their names besides
the few possessions that fit into a horse-drawn wagon.

The prospect of a stable income and a desire to acquire arable land
prompted the Ingalls family to move once again. In 1879, Charles and
Caroline decided to leave their homestead in Redwood County, Minnesota, where Charles had filed a claim under the Homestead Act the

previous year. They would relocate with their daughters, Mary, Laura, Carrie, and Grace, to Dakota Territory. Charles had found work there through his sister as paymaster for the Chicago & North Western Railroad. During the decades that followed, the company would expand its lines across the western frontier. They advertised tickets to "homeseekers" interested in "excursions to the Northwest, West, and Southwest." Some 80 percent of the towns constructed in Dakota Territory during this period were built to support the railroad industry. Companies including the Omaha Railway and the Milwaukee Road sought to capitalize on the newly available land following the deadly Sioux Wars of the 1860s and 1870s.

Although the Yankton nation ceded the region that comprises eastern South Dakota in 1858, the territory of western South Dakota remained under the control of the Cheyenne and Lakota Sioux nations throughout the 1860s and 1870s. In 1874, however, American prospectors trespassing on the Great Sioux Reservation, land protected under the Treaty of Fort Laramie in 1868, found gold—the precious metal that had fueled global mass migration to California twenty-six years earlier. Within two years, approximately four thousand prospectors had flooded the Black Hills, angering the Sioux by squatting on their protected land.

It was plain to the Lakota people that greed triggered the settlers' repeated intrusions. One member of the group, Standing Bear, remembered how "we knew that there were forests, animals and gold [in the Black Hills] and the white people wanted these riches. They attained wealth and we were in great distress." The Sioux did not want to relinquish their land to the federal government, but the government implicitly admitted its disinclination to enforce the Treaty of Fort Laramie by acknowledging the "great temptation held out to emigrants and miners to occupy that country, and . . . the difficulties which have already surrounded the question of protecting the Sioux in their treaty-rights to that territory."

The conflict between the covetous settlers, the US military, and the Sioux came to a head in June 1876, when US troops, led by Lieutenant Colonel George Custer, attacked Lakota and Cheyenne Sioux warriors, led by Chief Sitting Bull, who refused to remain within the bounds of the Great Sioux Reservation. The Battle of Little Bighorn, fought on the grassy yellow hills of present-day Montana, was a rout of the Seventh Cavalry, but victory for the Sioux and Cheyenne was short-lived.

Within a year, US troops gained control of the region, and the federal government forced a small percentage of the Sioux to authorize a document giving up the Black Hills to the United States. Then, the floodgates opened. Thousands of Americans flocked to Dakota Territory, eventually claiming and settling 41 percent of South Dakota's land under the Homestead Act. Life would never be the same for Standing Bear and his people, who were removed from their homeland. "When I think back to the time when we were free and had stags, deer and buffalo, I feel very sad especially when I go to bed hungry," he later lamented.

THE INGALLS FAMILY set off for Dakota Territory on September 6, 1879, near the end of a booming decade for white settlement. Between 1870 and 1880, South Dakota's population grew sevenfold, reaching nearly a hundred thousand people. At the turn of the twentieth century, eleven years after South Dakota officially became a state, more than four hundred thousand people resided on its vast, treeless plains, where bison once roamed freely.

But these new homesteaders were surprised to discover that South Dakota's terrain was less than hospitable to farming. Although proponents of settlement, including geographers, journalists, railroad companies, and community organizations touted what Surveyor General George Hill called the region's "immense natural agricultural resources," in reality, South Dakota's land was arid and its weather extreme and unpredictable.

To the unwitting Ingalls family, Dakota Territory represented the chance for a new life. In her autobiography, Laura recorded her joy upon arriving in the "great, new country clean and fresh around us." Mere days had passed since she left home, but she "never gave a thought of regret to Walnut Grove, where the settlers, Pa said, were getting too thick." Like Laura, the other members of the family were pleased to have reached Dakota Territory at last; not one of them was "lonesome or homesick," she remembered.

A happy stranger in an unfamiliar land, Laura wondered at the barren landscape surrounding her, noticing only a single tree, a cottonwood, growing among the seemingly endless grasslands. She and her family had no farm to tend, no house of their own, no attic filled with preserved food to help them survive the coming winter.

In Dakota Territory, Laura Ingalls Wilder represented a new type of explorer. She came not to briefly survey the landscape or chart a route, but to reside—to build a home and cultivate the earth. Yet women like Laura have often been ignored in the history of western exploration and expansion. Far more attention has been paid to the frontiersmen who forged paths, trapped, and traded. History celebrates men like Daniel Boone and Davy Crockett, explorers whose stories of adventure and danger have been told, embellished, and told again over generations. Pioneer women have, by contrast, been overlooked as the "invisible helpmate" to male farmers and ranchers, according to the historian Margaret Walsh. But female homesteaders, most of whom never recorded their stories, nonetheless played a critically important role in exploring and settling the West.

Mothers, sisters, and daughters like Laura ran farms, managed homes under austere conditions, and created community organizations that enabled fragile towns to flourish. Unmarried or widowed women could own their own homesteads due to the language of the Homestead Act, which placed no sex-based barriers in the path of female claimants. While some farmed the land by themselves as they sought to prove their claims, others managed hired hands. They were single women and widows, immigrants and citizens, all looking for financial stability, independence, and a place to call their own.

When the Ingalls family arrived in Dakota Territory near what would become known in 1880 as the town of De Smet, there were no hotels or inns. And so the family of six slept in crude lodgings. They joined the Chicago & North Western Railroad Company's workers at their camp, which Laura described as a collection of "shanties" and a "bunk house" overlooking Silver Lake. Relatives of Caroline Ingalls already residing there welcomed Laura and her family, helping them navigate an unfamiliar world where the boisterous men who laid down tracks significantly outnumbered women and children. Although the Ingalls family's one-room house was set at a distance from the other structures, there was little privacy. After the sun set each autumn evening, Laura likely fell asleep not to the soft rustling of the prairie grass, but to the animated sounds of men exchanging stories, drinking, and gambling in the bunkhouse nearby.

As paymaster, Charles Ingalls managed the salaries of railroad workers laboring under exceedingly challenging conditions. John Grosvenor, an employee for the Central Missouri Pacific Railroad in 1880 and 1882, attested to the grueling nature of life on the Kansas prairie approximately four hundred miles south of where the Ingalls family resided. "We all worked 10 hours a day and they had to be 10 big ones," he remembered. Incessant rains, flooding, and hunger killed morale. "We had no raincoats only [boots]," Grosvenor recalled regretfully. "We were soaked through. Food was short . . . Many of the men were just dead on their feet and it was some job to wake them up." Earning about a dollar a day, Grosvenor used his extra income to support his wife and their homestead, since the couple was unable to survive on farming alone.

Dakota Territory was similarly undeveloped. Without stabilizing community structures, a sense of lawlessness like that of the California Gold Rush pervaded the Ingallses' railroad camp. Roaming bands of horse thieves presented a constant threat, and fights erupted among the men at the camp without warning. On one occasion, Laura saw a brawl develop in mere seconds. Outside her house, a group of railroad workers quickly formed a circle in the wake of a disagreement. Two opponents squared off and prepared to strike their first blows. In the nick of time, another man threw himself between the rivals and managed to talk them down.

Economic concerns also fueled discontent among the railroad workers. Laura recorded that, soon after her arrival in Dakota Territory, the "men became restless and unruly" when "some of them wanted to quit the job" and receive their salaries immediately. As paymaster, Charles Ingalls found himself in a dangerous position. When night fell, he headed to the camp store, fearing an outbreak of violence there. Before long, two hundred disgruntled workers surrounded the building, threatening to break down its door and ransack the place.

From inside, Charles listened as ear-splitting gunshots shattered the silence of the prairie night. He wondered what to do next. Appalled by the rowdy men and fearing for her father's life, Laura and her mother peeked through the door of their shanty and watched as Charles slowly opened the door of the store and confronted the men. When "two or three of the men started to talk ugly," she remembered, her father adeptly defused the situation. "Pa told them how foolish they were, for he couldn't pay them when the money was not there," Laura wrote, but "if they shot him and wrecked the store, still they wouldn't have their pay, but would be in trouble."

To her relief, Charles succeeded in convincing the men of their folly. At last, the crowd dispersed and her father returned to their shanty, a hero in the eyes of his family. To Laura, "the gun shots, the shouts and loud talking had not been the worst of the affair, but [it was] the low, ugly muttering [she] had heard as a sort of undertone." Her initial rage at how her father had been treated turned into cold fear after the conflict had been settled. Laura perched "on the edge of [her] bed shivering with fright long after [she] should have been asleep."

December 1879 brought the dispersal of the railroad camp, the departure of their relatives, and an improvement in their living conditions. Laura marveled at how peaceful Dakota Territory felt in the absence of "two hundred men with some women and children and teams and all the noise and confusion of the camp." Now all that remained were "abandoned shanties and the wind," which seemed to foretell the imminent onset of winter. But good luck befell Charles when the railroad company presented him with the opportunity to move out of the crude shanty and into the wooden house previously occupied by surveyors. With its spacious rooms and warm stove, the house felt "very comfortable and homelike" to Laura, offering her family a foothold as they prepared to search for a homestead site.

When railroad work ceased for the season, Charles Ingalls turned his attention to procuring additional provisions to support his family during the Dakota winter. That autumn, the skies were teeming with migrating birds that ranged from geese and ducks to pelicans and cranes. Charles went hunting frequently, bringing back a variety of game to feed the family. From the bird carcasses, Laura and her mother, Caroline, also "saved all the feathers," using them to make four pillows and a soft featherbed. Pelts were a valuable commodity, and Charles tracked the wolves and smaller game that roamed the banks of Silver Lake.

One day, while hunting on the prairie around Christmastime, Charles located an attractive plot of land. Weeks later, in February 1880, he journeyed to the town of Brookings, about forty miles east of the future town of De Smet. There, he filed a claim for 154 acres of land, paying the necessary fees to the Receiver's Office. It was a gamble, as the terms of the application made clear: "It is required of the homestead settler that he shall reside upon and cultivate the land embraced in his homestead entry for a period of five years from the time of filing the affidavit, being also the date of entry." The Ingalls family had never lived in one place for more than three years at a time. Would they be able to successfully grow crops on the arid grasslands of Dakota Territory, improving the land to the satisfaction of the US government in such a short period of time?

DAKOTA TERRITORY FELT especially wild to Laura and her family during their first winter, when they anticipated striking out on their own as homesteaders. Isolated in the surveyors' house, amid what

Laura described as "the vast stillness and quiet of the empty prairie, with the cold and whiteness of winter around us," they spent months without glimpsing their closest neighbor, an unmarried man who lived six miles away. The brutal weather made journeying to other even more distant homes impractical. Laura recalled how "the cold shut down, the snow fell and blew and drifted into huge drifts . . . the lake froze over and was a smooth sheet of ice."

The only signs of life during the darkest months of the year were the coyotes and skulking gray wolves whose nighttime howls surely sent shivers up Laura's spine. Laura recalled a conversation with her father about the advent of the railroad and its ominous implications: that "the crowds on the railroad had frightened the buffalo herds away." As predators of the bison, the wolves may have found themselves desperate for food in the winter of 1879. Forty miles away, near the town of Farmington, wolves were reported to have "made an attack upon a flock of 600 sheep . . . killing thirteen and injuring many others."

One evening, Laura and her younger sister Carrie were playing on the frozen lake when they found themselves in a dangerous situation: "When we reached the end of a particularly long slide and looked up, we were at the very edge of a shadow cast by the southeast bank and on top of the bank sat a wolf." Immediately, the two girls fled for the safety of the surveyors' house. Behind them, the gray wolf's lonely cry pierced the icy night. When Laura dared to cast her eyes in its direction, she "saw the wolf still sitting on the bank, with his nose pointed at the moon." Although the event impressed upon Laura the dangers of the Dakota prairie, it was among the few times she glimpsed a gray wolf near De Smet. The arrival of settlers and the expansion of the railroads would so severely disrupt the ecosystem that the gray wolves were, in Laura's words, "never seen there again."

When spring finally came, the town of De Smet sprang forth like a sturdy pasqueflower. It was a time of rapid population growth in Kingsbury County. One newspaper reported, "Every day sees long trains of wagons stretching across our prairies bringing their living freight of muscle, brains, and enterprise to gather the treasures of the fertile soil and make homes for this and coming generations." Laura's future sister-in-law, Eliza Jane Wilder, also homesteaded near the future town of De Smet and recorded that, in the spring of 1880, "Hundreds of men from the east were coming every day on foot or by private conveyance, as workmen on the RR in every way, all seeking land."

Anticipating the demands of these energetic men, one entrepreneur erected the first structure in De Smet: a saloon to replace the campsite's bunkhouse as the central gathering place. Other buildings appeared, constructed by new settlers arriving in droves. The Ingalls family soon joined the action. When the railroad company's surveyors reclaimed the house in which they had been living, Charles built two simple structures and moved the family into town. Another milestone occurred in late April 1880, Laura recalled, when "the first train came . . . and then De Smet was headquarters for the work farther west." By late spring, the town had "a bank, lumber yard, a livery stable, drug store, dry goods store, another grocery and hotel." The population would reach 116 people that year. Dakota Territory's days as one of the last American frontiers were numbered.

That spring, news of a claim jumper who shot a family friend after attempting to steal his land raised fears among the members of the Ingalls family that they might lose their homestead if they did not quickly relocate there. The theft of legally held land was not uncommon in Dakota Territory in 1880, when an ever-growing number of settlers arrived. Rather than risk legal or physical confrontation with a potential claim jumper, the Ingalls family acted decisively, going out from town "a mile out onto the sweet prairie," Laura remembered fondly.

The claim was situated on a sunbaked, treeless hill overlooking miles and miles of grasslands. Their lodging was modest: a shanty comprising a single room, with a curtain that served as a makeshift wall to create privacy at night. The family wasted no time in starting to cultivate the land per the terms of the claim. Charles immediately dug a well, planted trees, and began growing turnips. When the native grasses matured, he sliced them down to make hay. Laura and her mother aided in this crucial work; they "helped load it on the wagon and unload and build it into large stacks to feed [their] horses and two cows through the winter that was coming." Like Sacagawea, Beckwourth, and the other explorers who preceded them, the Ingalls family prepared far in advance for surviving periods of scarce resources and unforgiving weather.

As homesteaders, their primary goal was to endure: to withstand the backbreaking labor that cultivating the land required; to persevere through severe weather, limited funds, and social isolation; and to build a stable life in which they might flourish. The Ingalls family shared these aspirations with thousands of other settlers who would move across the country, from hardship to hardship, in search of opportunity over the next few decades.

Homesteaders came from the upper and lower South, leaving behind sun-parched plantations, dilapidated cabins, and memories of the overseer's whip. Some were veterans, carrying trunks that may have contained tattered blue uniforms stained with sweat and blood from days spent on Civil War battlefields. Others traveled to the Midwest from around the world, packed together on ships with men, women, and children speaking a multitude of languages. Female homesteaders were single and married, sometimes becoming widowed during the period required for proving their claims. They faced the prospect of failure, starvation, debilitating accidents or disease, and death.

Disaster almost immediately befell the Ingallses on their homestead. In early October 1880, Laura opened her eyes one morning to discover a blizzard assailing the thin walls of their shanty on the claim. The weather was "frightfully cold," she recalled, and temperatures may have descended to fifteen degrees Fahrenheit. The family members huddled under their bedcovers or crouched around the stove in a futile attempt to keep warm. The blizzard lasted three seemingly interminable days, its icy blast killing livestock in the hundreds and several people. One local paper called it a "storm without a parallel in the history of the territory," producing snow that "was whirled by the wind into drifts in many places ten and twelve feet deep."

The so-called October Blizzard caught homesteaders off guard; it appeared to come out of the blue. The newspaper sought to reassure its readers, many of whom were new homesteaders or recent migrants, that "while this storm is unseasonable and to those unfamiliar with Dakota weather may presage an early winter, its discomforts will be more than compensated for by the benefit which the soil will derive from the moisture it has furnished." It seemed impossible for such unseasonable weather to persist, the newspaper concluded, heartening readers that "we feel assured that we have yet in store for us many weeks of warm and delightful weather."

Laura and her family were shaken by the blizzard, however, and worried that even worse weather lay ahead. Their fears appeared to be confirmed when they learned that an elderly Indigenous man "passing through town warned the people that a terrible winter was coming." The Ingalls family moved to the comparative safety of town shortly thereafter, not wanting to spend the winter isolated in the uninsulated claim shanty. November brought another three-day blizzard that "filled the cuts on the railroad," Laura remembered. The storm caused such trouble that it took "snow plows and men with shovels [days to] . . . clear the track so trains could run again."

During the wintry weeks that followed, all of De Smet's able-bodied men tried to clear the tracks, but their efforts to keep the trains moving on the Dakota Central line from Tracy, Minnesota, to the vicinity of De Smet were no match for the endless succession of violent storms. When snowfall buried the railroad tracks, it prevented the arrival of badly needed supplies. "The officials of the R.R. and the citizens along the route united in their efforts to keep it open but 'twas useless. In December they gave up the battle and from that time until May no trains reached De Smet," Eliza Jane Wilder recalled.

The residents of De Smet were woefully unprepared for a winter of hardship. Many new homesteaders had not yet built up a store of food, and they lived in hastily built lodgings that could not withstand the extreme conditions. When blizzards assaulted the town during that cruel winter, Laura recollected, "We would lie in our beds those nights, listening to the wind howl and shriek while the house rocked with the force of it and snow sifted in around the windows and through the nail holes where nails had been withdrawn." It became clear to Laura that she and the members of her community were in the midst of "a hard, long battle," a "war with the elements."

In the months ahead, when temperatures dropped to thirty-two degrees below zero Fahrenheit, the townspeople came close to being totally deprived of essential supplies. The men of De Smet could not safely hunt on the prairie in near constant wintry conditions, and provisions of fruits and vegetables quickly ran out. The Ingalls family's starved dairy cow produced almost no milk, and the town stores sold all their sugar and flour. In town, Laura and her family used up the coal to heat their small stove. Without its warmth, they would surely perish.

Wood was not available, since trees were few and far between on the prairie. So the Ingalls family began making twists out of hay, one of their last remaining resources. "If well done," Laura wrote, "it made a hard twist from 18 inches to a foot long. . . . We called them sticks of hay and they could be handled like sticks of wood. They made a surprisingly good fuel, a quick, hot fire, but burned so quickly that someone must be twisting hay all the time to keep the fire going."

That burden fell on Laura and her parents, who already had their hands full with other tasks. Unfortunately, they were unable to enjoy the fruits of this labor because they had begrudgingly agreed to accommodate houseguests who were down on their luck: an old acquaintance, George Masters; his wife, Maggie; and their infant, Arthur. George and Maggie did not try to earn their keep. Instead, they positioned themselves in front of the stove, occupying the warmest spot in the house. Day after day, they sat there, refusing to help with chores and consuming outsized portions of food at mealtime.

If these houseguests epitomized selfishness during a time of crisis, Almanzo Wilder, Laura's future husband, and his friend Cap Garland embodied selflessness. By January 1881, the town's situation was dire. The general store had no flour and private caches were also running low. Eliza Jane recorded that "one [local] family . . . lived on boiled turnips for months" while another "had nothing but wheat cracked in a coffee mill and made into mush with water for food." Starvation loomed and the mood grew desperate.

Word began circulating about a local farmer who possessed a good amount of grain, but his house was twelve miles away. Anyone who attempted the journey there risked getting caught in one of the ferocious blizzards that had isolated De Smet. "It was dangerous to go after it and no one wanted to go," Laura remembered, "but finally the youngest Wilder boy and Cap Garland each with one horse, on a sled used to haul hay from the slough, started."

Eight years apart in age, the two young men had moved to Dakota Territory as homesteaders in 1879: Garland with his family and Wilder as an independent homesteader. Together they successfully made the trek across "sloughs where horses would break through and have to be dug out," Laura recorded. Hours before another storm swept across the prairie, the boys returned with enough wheat to feed the hungry townspeople. With extra nourishment, Laura, her family, and the other residents of De Smet felt they could endure a few more weeks of winter until the snow melted enough for the trains to begin running again.

But time seemed to drag on. Day after day, Laura twisted hay into sticks to feed the stove in the kitchen, the single room occupied by the entire family. To feed the large household, she spent hours grinding seed wheat by hand in a coffee mill. It was "slow work grinding enough wheat to make flour to make bread to feed eight people," Laura recalled. Afterward, her mother used the flour to concoct "mush or biscuits raised with soda and souring," a rather unappetizing dish but one that provided the hungry family with nourishment. Her birthday passed in February without fanfare. By March, they had nearly given up hope that winter would come to an end. "We were getting short tempered," Laura admitted.

At last, the first signs of spring arrived in April. "Bare prairie showed in spots and farmers went back to their claim shanties and began their spring work," Laura wrote. On May 9, the townspeople heard the sharp whistle of a long-awaited train. A "mob of men" surrounded the train, ransacking the cars, which were largely filled with agricultural equipment, in search of food. The last car held a veritable bounty: "Seed wheat and potatoes" and enough provisions for each man to go "home carrying his share, a little sugar, some flour, a bit of salt pork, some dried fruit, and a little tea." Laura and the other members of the family eagerly anticipated Charles's return from the depot, and when he reached the house, the mood was joyful. Laura, however, was surprised by what struck her most in that moment: the recognition of the mental toll that the endless winter had taken on everyone. "I think none of us had realized the strain we had been under until it broke," she later reflected.

If the winter of 1880–81 had tested the endurance of the settlers throughout Dakota Territory, their livestock had suffered even more. Once the snow melted, the casualties became increasingly evident. A correspondent in Cheyenne reported in June 1881 that he had "never before seen so many wretched, hide-bound cows and steers, thousands of them almost too weak to crop the grass within their reach." They were the lucky ones; "decaying carcasses of animals lay within sight of the railroad track," the journalist continued, "giving a sad intimation of the mortality that must have overtaken the unhappy brutes during the depth of snow of a [savage] winter." Homesteaders like Laura recognized anew how lucky they were to be alive. Recovery from that winter would be slow, but she remained optimistic about her future in Dakota Territory. Like Rip van Winkle, the prairie shook off the long winter slumber; the spring grasses burst into "a beautiful green,"

Laura remembered. It appeared as though the family's hardest days as homesteaders were behind them, once and for all.

DURING THE YEARS that followed the memorable winter of 1880–81, Laura witnessed the transformation of Dakota Territory. The population exploded, more settlers established homesteads, and the community of newcomers became firmly established. During her childhood years in Dakota Territory, Laura alternately welcomed development and celebrated the wilderness, expressing her "great relief" at the construction of the first hotel in the town of De Smet while also admiring the natural vegetation of the prairie. Wolves were a danger to families living on isolated farms, but when these animals eventually disappeared, it saddened her.

As to the increasing number of migrants, Laura had distinctly regretful feelings, noting in her autobiography, "I didn't much care for all these people. I loved the prairie and the wild things that lived on it, much better." Indeed, the native fauna of the grasslands remained forever impressed upon her memory. She wrote admiringly of the yellow-breasted meadowlarks, the patterned squirrels, the orange-striped garter snakes, and the long-legged jackrabbits that enlivened the landscape of her youth.

Laura grew from a teenager to a woman in Dakota Territory. In a modest one-room schoolhouse, she learned to eloquently express her thoughts in writing. Community gatherings introduced her to a widening array of men and women who, like the Ingallses, had moved to the region in search of economic stability. She became better acquainted with Almanzo Wilder, the homesteader who risked his life to bring wheat to the hungry townspeople during the winter of 1880–81.

After earning her teaching certificate in December 1883, she taught several local children in what she described as "an abandoned claim chanty, one thickness of boards with cracks between, through which the snow blew." She and Wilder began a courtship in 1884 and

married in August 1885, when Laura was eighteen years of age. Together they moved onto Almanzo's tree claim, where he intended to
plant trees to fulfill the terms of the Timber Culture Act (1873), rather
than onto his homestead land. The setting filled Laura with happiness. Now she had "a house and a home of her own."

Yet, like Caroline and Charles Ingalls, Laura and Almanzo faced
a hard road as homesteaders. They would endure brutally cold winters
and arid, scorching summers. In fact, when she became Almanzo's
fiancée, Laura harbored serious reservations about marrying a farmer.
Early on, she told him that "a farm is such a hard place for a woman.
There are so many chores for her to do, and harvest help and threshers
to cook for." Acknowledging her concerns about committing to a life
as a female homesteader, he proposed a compromise: "If you'll try it
for three years and I haven't made a success in farming by that time,
I'll quit and do anything you want me to do. I promise that at the end
of the three years we will quit farming if I have not made such a success that you are willing to keep on."

Laura carefully considered his words. On the one hand, she
dreaded the manual labor that homesteading would entail. On the
other, she was drawn to the wildness of the prairie, with its "freedom
and spaciousness . . . with the wind forever waving the tall wild grass
in the sloughs and rustling through the short curly buffalo grass." It
was so different from town, where they would have had "neighbors so
close on each side." In the end, she agreed to Almanzo's proposition
and prepared to begin married life as a homesteader.

During the years ahead, Laura spent her days "cooking, baking,
churning, sweeping, washing, ironing, and mending" while Almanzo
worked in the fields. But, like her parents, Charles and Caroline, they
were unable to make a success of farming. Severe weather, poor har-
vests, and falling grain prices prevented them from enjoying financial
stability. These hardships were exacerbated as their debt grew because
they had to borrow money from banks to purchase farm tools and other
supplies. In the summer of 1886, a hailstorm destroyed the Wilders'
magnificent crop of wheat, obliterating thousands of dollars' worth

of crops. They were subsequently forced to move from their house on Almanzo's tree claim to a humbler shanty on his homestead. There, Laura gave birth to their daughter, Rose, a joy but also another mouth to feed.

Diphtheria struck the family in March 1888, weakening Laura and ultimately contributing to what Almanzo's doctor described as his "slight stroke of paralysis." The health-related expenses proved unmanageable, and the couple was forced to sell the labor-intensive homestead claim, moving back to their smaller tree claim in the spring of that year.

Drought conditions depleted their wheat crop on their homestead in the summer of 1888, but Laura agreed to remain on the farm for a fourth year. It was nearly impossible to keep up with her chores, however, as Rose was becoming a curious toddler requiring constant attention. Overwhelmed, Laura began to "hat[e] the farm and the stock and the smelly lambs, the cooking of food and the dirty dishes . . . . She hated it all, and especially the debts that must be paid whether she could work or not."

Their fourth year of farming ended in tragedy. In 1889, only about eight inches of rain fell during the prime growing period. "Hot winds" scorched the Wilders' oats, wheat, and trees that summer, ruining their harvest. Almanzo filed a pre-emption claim on their property since he was unable to show that he had improved the land, an act that necessitated their purchase of the homestead for $200 within six months. Hopeful that they could stay on their land with their growing family, Laura gave birth to a baby boy in July. Less than a month later, "the baby was taken with spasms" and stopped breathing. His death shattered the Wilders. "The days that followed were mercifully blurred," Laura recalled. "[I] only wanted to rest—to rest and not to think."

Laura's overwhelming grief and the strain of her postpartum recovery may have distracted her on the afternoon of August 23. She began warming a tea kettle over the kitchen fire, then stepped into another part of the house. She saw and heard nothing out of the ordinary.

When she returned to the kitchen, however, the room "was ablaze: the ceiling, the hay, and the floor underneath and wall behind." Within minutes, the entire house was engulfed in flames, and nearly all the Wilders' possessions were gone.

In the aftermath of the fire, Laura and Almanzo took stock of their situation. It was clear that farming had not been a success. As homesteaders, they had faced drought, disease, debt, and death with fortitude, but in the end, they lost their battle in Dakota Territory. In November 1889, Almanzo and Laura announced to the Ingalls family their plan to sell their land and look elsewhere for opportunity the following spring.

They mirrored other settlers in this decision to flee Dakota Territory. The extreme droughts that drove them away during the late nineteenth century were in fact due to the homesteaders' destruction of native grasses. The historian Caroline Fraser writes that, ironically, the "drought was in large part created by the settlers themselves." By seeking to improve the land per the terms of the Homestead Act, migrants like the Ingallses and the Wilders "tore away . . . protective [native] grasses and their roots, exposing bare soil to intense heat, evaporation, and drying winds." Many thousands of impoverished homesteaders had failed at farming in Dakota Territory. They had faced a nearly impossible task.

LAURA, ALMANZO, AND Rose departed De Smet for good in 1894, having spent the previous four years with relatives in Spring Valley, Minnesota, and Westville, Florida. In 1892, they had returned to De Smet to live near Caroline and Charles Ingalls, who had recently exchanged life on their burdensome homestead for a place in town. Saving up their money by doing odd jobs during the economic crisis that followed the Panic of 1893, the Wilders prepared for what they hoped would be their final move: to Mansfield, Missouri.

Mansfield's promoters in the railroad and farming industries touted its advantageous climate and abundant crops in advertisements and pamphlets. Rose remembered the family's arrangements for the 670-mile trek, when Laura "baked two dozen hardtacks for the journey" and Almanzo "painted [their covered wagon] . . . . shiny black." Bidding farewell to the Ingalls family early on the morning of July 17, Laura, Almanzo, and Rose climbed into their wagon and called out final, bittersweet words as they departed, intending to "make haste, driving every day to reach The Land of the Big Red Apple."

As they made their way south, Laura wrote in a journal, commenting on the "poor wheat" and the "grain . . . burned brown and dead" that littered the landscape. Later pausing to take in the receding view of Dakota Territory after crossing the James River, Laura wondered at the beauty of "the river winding down the valley, the water gleaming through the trees that grow on the bank. . . . [and] the bluffs [that] rose high and bare, browned and burned, above the lovely green of trees and grass and shining water." In that moment, she thought of the Native Americans who had long inhabited South Dakota prior to the arrival of white settlers. "If I had been the Indians I would have scalped more white folks before I ever would have left it."

Why had so many settlers destroyed the native grasses, killed the bison, and driven away the wildlife, only to forsake the land and community where they had so desperately tried to build a life? Prosperity remained out of reach for homesteaders like Laura, but their efforts to establish farms, build towns, and expand railroads permanently altered the landscape, with far-reaching consequences for dispossessed Native Americans, wildlife, and natural vegetation.

When Laura and Almanzo Wilder returned to South Dakota later in life, they found the terrain almost beyond recognition. In 1946, in a letter to a friend, Laura lamented that "on our several visits back . . . we came away still unsatisfied; the country and the town are so changed from the old, free days, that we seem not able to find there what we were looking for. Perhaps it is our lost youth we were seeking in the place where it used to be. But there is something about the boundless sweep of the prairie that makes it unforgettable."

In the end, Laura found success not as a homesteader, as she and Almanzo had so desperately wanted, but as a chronicler of exploration and pioneer life. At the age of sixty-five, from her home in Mansfield, Missouri, Laura published her first book, *Little House in the Big Woods* (1932), a dramatization of her childhood as the daughter of pioneers. She went on to write eight additional works that would comprise the Little House series of children's literature. These stories captured the imagination of twentieth-century Americans and transported them from their comfortable, modern homes to the rustic log cabins of the past. Selling more than sixty million copies to date, the Little House books were translated into forty-five languages. Perhaps, when Laura thought back to all those years of deprivation and hardship in Dakota Territory, she realized a surprising truth. The land had not failed her after all.

4

John Muir

THE PRESERVATIONIST

In June 1889, California's Yosemite Valley awakened to the thrill of the summer sun, which penetrated its dense coniferous forests and danced on the surfaces of its cold, clear streams. Wildflowers opened their petals, their vibrant colors contrasting sharply against the valley's gray granite walls. When the soft wind gently shook the branches of the native sequoia trees, the rustling of their needles intermingled with the sweet sounds of birdsong. Yosemite Valley was, in the words of John Muir, a "temple of Nature in the heart of the mountains." But during the summer of 1889, the destructive tendencies of loggers, shepherds, and ranchers threatened one of the nation's most remarkably beautiful regions.

At first glance, visitors would have considered the majestic valley, carved millions of years earlier by glaciers and the powerful Merced River, unsullied. But Muir, who had spent the prior two decades exploring Yosemite's groves of sequoias, soaring waterfalls, and towering cliffs, noticed signs of damage to the valley's fragile ecosystem. He warned the urban readers of the *San Francisco Daily Evening Bulletin* of the "trampling down of the rich beds of flowers and grasses" and the construction of "miles of fences . . . around hay-fields and patches of kitchen vegetable." Horses and sheep were devouring the native vegetation and "run[ning] loose over the unfenced portion of the valley year after year until in many places it look[ed] like a dusty, exhausted wayside pasture." Worst of all were the loggers, who had decimated the invaluable "grand old oaks and pines that required centuries of growth." Without immediate intervention to protect unique Yosemite Valley, Muir feared, its days as one of the country's last wild places were numbered.

A SCOTTISH IMMIGRANT who came to the United States at the age of eleven, John Muir harbored a different perspective of the American wilderness than most. Born in 1838 in Dunbar, a small coastal town in southeastern Scotland, Muir wrote in his memoir that he "was fond of everything that was wild" in his native country. His hometown overlooked red sandstone cliffs, sandy beaches, and the dark gray waters of the North Sea. Called Johnnie by his boyhood friends, Muir "loved to wander in the fields to hear the birds sing, and along the seashore to gaze and wonder at the shells and seaweeds, eels and crabs in the pools among the rocks when the tide was low; and best of all to watch the waves in awful storms thundering on the black headlands and craggy ruins of the old Dunbar Castle when the sea and the sky, the waves and the clouds, were mingled together as one."

An athletic child, he honed his climbing skills by ascending trees, or the slippery, moss-covered walls of neighboring gardens, or even the roof of his family's three-story house on the town's main thoroughfare, High Street. Johnnie and his playmates passed the time by staging races, "running on and on along a public road over the breezy hills like hounds. . . . [thinking] nothing of running right ahead ten or a dozen miles before turning back." Scotland's brown skylarks, with their delicate voices, also fascinated Muir. "Oftentimes on a broad meadow near Dunbar," he remembered, "we stood for hours enjoying their marvelous singing and soaring." A budding naturalist, he delighted in capturing chicks, observing them, and releasing them. "Wildness was ever sounding in our ears," Muir recalled, and these "first excursions [were] the beginnings of lifelong wanderings."

One dark evening in February 1849, when Muir was almost eleven years old, his father made a surprising announcement. When the sun rose the next morning, the family would set off on a one-way journey to North America. This decision came near the beginning of the Gold Rush, just one year after John Marshall found gold in Coloma, California. News of this event rapidly traveled around the world, attracting hundreds of thousands of immigrants to the United States in the months that followed. Some men and women, like the mountain man James Beckwourth, would try their luck in the gold fields of California's Mother Lode country, while others would seek prosperity elsewhere as farmers, ranchers, or entrepreneurs.

Muir's father was driven to immigrate not by a desire for earthly riches but for religious freedom. A strict Protestant whose views clashed with those of the Presbyterian Church in Scotland, he hoped to join a settlement of the Campbellite Disciples of Christ somewhere in North America. For young John, North America's reputed natural marvels made the journey seem worthwhile. He recalled his desire to taste the sweet syrup that came from the maple trees growing in New England and the northern Midwest, to see the "hawks, eagles, [and] pigeons, filling the sky," and to explore the continent's "boundless woods full of mysterious good things."

The next morning, John, his father, and two of his six siblings boarded a train bound for Glasgow, from which they would embark upon their transatlantic journey. The rest of the family planned to follow once a homesite was established. As Muir looked back at the shores of his native country, he felt only joy and anticipation for experiencing what he had only read about in his schoolbook: the "American wilderness." In the naivete of his youth, it was impossible for him to "know what we were leaving, what we were to encounter in the New World, nor what our gains were likely to be." Like so many other nineteenth-century immigrants, Muir made a bittersweet bargain. He exchanged the familiarity of a home he might never see again for the chance of a better life in an unknown land.

The sea crossing lasted more than six weeks. John kept busy by befriending the captain and crew, peppering them with questions about the marine life teeming in the cold Atlantic waters off the coast of North America. Meanwhile, his father, Daniel, holed up below deck. There, he debated with the other immigrant families as to which region of this new land offered the best prospects. Although Daniel had intended to settle in present-day Ontario, they convinced him that "the States offered superior advantages, especially Wisconsin and Michigan," Muir recalled.

When the Muirs disembarked in New York City, they immediately headed for the Midwest via the recently completed Erie Canal. Upon reaching Wisconsin, a state admitted to the Union just one year earlier in 1848, Daniel Muir worked with a land agent to quickly obtain 160 acres of land in Marquette County for $200. There, he constructed a sturdy little house made of oak while the children spent their time "wandering in the fields and meadows, looking at the trees and flowers, snakes and birds and squirrels," Muir remembered.

About half a year after his arrival, Daniel's wife and the rest of his children joined them, a sign of the family's commitment to their new life in the United States. It was a heady time for young John, who rejoiced in the new and undeveloped environment so different from that of Dunbar, a bustling port town of thirty-five hundred residents with a new railroad station. While much of America still contained pristine woodlands, Scottish forests had been decimated over the centuries and would cover only 5 percent of its territory by the end of the nineteenth century. For a budding naturalist like Muir, Wisconsin was paradise. "This sudden splash into pure wildness—baptism in Nature's warm heart—how utterly happy it made us! Nature streaming into us, wooingly teaching her wonderful glowing lessons. . . . Oh, that glorious Wisconsin wilderness!" he remembered.

The Muirs reached Wisconsin during a decade of explosive growth. Between 1840 and 1850, the state's population increased by 900 percent to nearly 300,000 people. Immigrants came from the British Isles, Germany, Norway, and Sweden between 1830 and 1860, drawn by the availability of agricultural land and, like the Muirs, the freedom to worship as they pleased. Their arrival contributed to the displacement of Wisconsin's Meskwaki, Ho-Chunk, Sauk, and Menominee tribes, a pattern that had continued since European missionaries and fur traders first arrived in the seventeenth century. Between 1804 and 1854, the US government pressured Wisconsin's tribes to cede millions of acres of land through seventy signed treaties, acquiring most of the state's territory for settlement and agricultural development. By the 1840s, when the Muirs established their homestead, Wisconsin's Native population had been decimated, with only tens of thousands of people remaining.

Muir came of age in central Wisconsin without a serious awareness of the consequences of settlement for the local Indigenous population. As a young adult, he helped his father on their farm, which sat one hundred miles northeast of Galena River, where Beckwourth had mined for lead among the Sauk and Meskwaki people twenty-seven years earlier. The Muirs grew corn and raised livestock on land that was sacred to members of the Ho-Chunk nation who had been driven from their homeland by the federal government in 1840; Muir described riding a pony purchased from the Ho-Chunk or Menominee people atop an Indian mound containing the remains of deceased Native Americans. It rose from the earth just feet from their shanty. His frequent childhood encounters with Indigenous people living in Wisconsin nonetheless remained firmly imprinted on his memory, and he associated the concepts of wilderness and wildness with them.

Muir marveled at the Native Americans' foraging and hunting abilities; they clearly possessed knowledge about the natural world that far surpassed his own. On one occasion in his youth, he recalled how men from the Menominee or Ho-Chunk nations went "direct to trees on our farm, chop[ped] holes in them with their tomahawks and [took] out coons, of the existence of which we had never noticed the slightest trace." When the weather turned cold, Muir continued, "we frequently saw three or four Indians hunting deer in company, running like hounds on the fresh, exciting tracks. The escape of the deer from these noiseless, tireless hunters was said to be well-nigh impossible; they were followed to the death."

Life on the farm, however, did not appeal to Muir in his adolescence. Year in and year out, Muir's father made his children contribute to the homestead by performing hard labor: hoeing corn, planting and harvesting crops, and chopping wood. For about a decade Muir spent his days working from dawn until dusk for a parent who rarely expressed appreciation. In the end, according to his biographer Donald Worster, Muir "repudiated, on the whole, the tradition of agrarian patriarchy from which he had suffered so grievously." At the age of twenty-two, he decided to leave the family farm and try his luck as an inventor. Carrying fifteen dollars and a primitive time-keeping invention comprising "a package made up of the two clocks and a small thermometer made of a piece of old washboard," he headed to the State Fair in the capital city, Madison. There, he discovered the University of Wisconsin, enrolled as a freshman, and dabbled in a range of fields, including physics, math, geology, and botany during six terms.

In 1863, Muir embarked upon a new adventure, leaving the University of Wisconsin for what he called "the University of the Wilderness." Reflecting on this, with the benefit of a lifetime's hindsight, he remembered simply "wander[ing] away on a glorious botanical and geological excursion, which has lasted nearly fifty years and is not yet completed, always happy and free, poor and rich, without thought of a diploma or of making a name, urged on and on through endless, inspiring, Godful beauty." Then and in his later life, Muir found himself magnetically drawn to the wilderness. His goal was to observe, not conquer, the natural world. But little did Muir realize that, at this exact moment, many of the varied landscapes he longed to see were on the verge of destruction.

TO THE YOUNG Muir, central California represented one of America's most magnificent areas of wilderness. He desperately wanted to see the Yosemite Valley, located more than 175 miles inland, which had long been the home of the Ahwahneechee people. A tribe indigenous to the Sierra Nevada, its members called Yosemite "Ahwahnee," or "gaping, mouth-like place." During the warm months of summer and autumn, they hunted for game and harvested acorns from black oak trees, which they later stockpiled in granaries. Brush houses sheltered the Ahwahneechee from the sun during the day and let the cool, evening breezes pass between the branches. They traded goods with other Native Americans for pale clam shells from the Pacific Ocean and fragments of glassy black obsidian that were used to fashion arrowheads. The Ahwahneechee people had lived this way for thousands of years. But then came the Gold Rush. The Mother Lode country did not stretch eastward into the Yosemite Valley, but as settlements expanded, drawing ever closer to Ahwahneechee land, the new arrivals began pressuring the state government to remove the Indigenous population from central California entirely.

The year 1851 brought violence to the Ahwahneechee people and news of the Yosemite Valley to the wider world. One year earlier, the newly formed California State Legislature passed the Act for the Government and Protection of Indians, a deceptively named law that accelerated the loss of Native land and authorized the forced labor of indentured Indigenous children. Emboldened settlers saw the law's enactment as signal of government support for Indian removal. Not long after its passage, a group of volunteer soldiers called the Mariposa Battalion arrived in Yosemite Valley with one goal: to subdue and forcibly remove the Indigenous people who lived in the region. The militia set fire to the Ahwahneechee's lodgings and food stores, driving them out of the valley in a cruel show of military power. In the months that followed, some of the Ahwahneechee people returned to live in the valley, but a rival tribe, the Monos, killed many Ahwahneechees during an 1853 attack that further weakened them.

One member of the Mariposa Battalion, a soldier named Lafayette Bunnell, published an article in *Hutchings' Illustrated California Magazine* in 1859, touting the Mariposa Battalion's "discovery" of the Yosemite Valley eight years earlier. Word of the valley began to circulate during the five years that followed. As the American public became increasingly aware of its exceptional beauty, tourists began trickling into the remote region to see its gigantic sequoias and breathtaking views. While the Civil War raged in 1864, President Abraham Lincoln took time away from military planning to sign an act that granted the state of California control over the valley "upon the express conditions that the premises shall be held for public use, resort, and recreation."

The war's end one year later heralded the return of veterans to their homes and a period of national rebuilding that saw the long-awaited completion of the transcontinental railroad. Stretching from coast to coast, the route made possible the movement of people and goods across vast distances, at record speeds. Many travelers from the eastern United States who visited California hoped to see its natural wonders with their own eyes.

John Muir would first reach Yosemite Valley in 1868, after an accident had given him a new appreciation for the gift of sight. After leaving the University of Wisconsin, Muir found himself torn between two desires: to work with the new automated machines that were converting natural resources into mass-produced materials and powering a second industrial revolution in the United States, or to further explore America's remaining areas of wilderness. Between 1863 and 1867, he took on a variety of odd jobs that provided him with the income to fund a few small exploratory expeditions.

While working in a carriage factory in Indianapolis in March 1867, Muir noticed one day that a machine's belt needed an adjustment that required the use of a dangerously sharp file. As Muir handled this tool, it suddenly "slipped and pierced [his] right eye on the edge of the cornea." Immediately he felt intense pain. Muir blinked as fluid seeped from the wound. Then came "perfect darkness."

In that instant, Muir recognized the accident's potential consequences. "My right eye is gone," he whispered to himself, "closed forever on all God's beauty." Within hours, Muir lost sight in his other eye, a reaction called sympathetic ophthalmia, which likely resulted from a mistargeted immune response to the injury. He may have feared that he would never again admire the dazzling wildflowers whose scent filled his nose, glimpse the sun as it warmed his skin, or spot the songbirds whose melodies flooded his ears as they darted from tree to tree.

But with the passage of time, Muir's eyes miraculously healed. He credited one of his mentors, the educator Catharine Merrill, for nursing him back to health. While Muir recuperated in Indianapolis, this friend "came to my darkened room like an angel of light, with hope and cheer and sympathy purely divine, procured the services of the best oculist and the children she knew I loved. And when at last after long months of kindness and skill [Merrill] saw me out in Heaven's sunshine again . . . her joy was as great as my own," Muir later wrote of her tender care. His brush with blindness changed him. No longer would he tinker with inventions or spend his days in a factory. From now on, Muir promised himself, he would "devote the rest of [his] life to the study of the inventions of God."

Muir would subsequently embark upon what he called a "long excursion, making haste with all [his] heart to store [his] mind with the Lord's beauty and thus be ready for any fate, light or dark." Departing from Louisville, Kentucky, in September 1867, Muir began a thousand-mile journey on foot that took him through the cool pine forests of Kentucky and North Carolina to the humid swamps of Georgia and Florida. He passed through the Cumberland Mountains where the frontiersman Daniel Boone had blazed a trail nearly a century earlier; Muir in turn followed in the footsteps of the Shawnee, the Cherokee, and herds of bison migrating west. In Georgia, a state devastated by fighting and the Union general William Sherman's March to the Sea during the Civil War, Muir noted with sorrow that "traces of war [were] not only apparent on the broken fields, burnt fences, mills, and woods ruthlessly slaughtered, but also on the countenances of the people."

Pressing on through the watery marshlands of Florida, Muir finally approached the Gulf of Mexico. Having been landlocked for so many years since he first emigrated from Scotland, the scent of the salty ocean immediately reminded Muir of home. "[The smell] suddenly conjured up Dunbar, its rocky coast, winds and waves; and my whole childhood," Muir later remembered, "[places] that seemed to have utterly vanished in the New World, now restored amid the Florida woods by that one breath from the sea." In the midst of the damp, flower-filled woods, Muir "could see only dulse and tangle, long winged gulls, the Bass Rock in the Firth of Forth, and the old castle, schools, churches, and long country rambles in search of birds' nests."

Reaching the Gulf Coast did not quench Muir's thirst for travel; rather, he craved new adventures. After a short stint in Cuba, Muir abandoned his original plan to explore South America and decided instead to travel to California by way of New York. In what he described as the noisy, overwhelming "big ship metropolis," Muir purchased tickets on a vessel bound for the Isthmus of Panama, a route taken by thousands of gold-seekers and immigrants before him. After crossing the Isthmus, he would resume his maritime journey to San Francisco by boat before heading up into the Sierra Nevada.

MUIR REACHED CALIFORNIA in March 1868. It was a marvel to him. With his companion Chilwell, a fellow immigrant from Great Britain, he departed from bustling San Francisco in search of the famed Yosemite region. His memories of the trek stayed with him throughout his life; Muir later described how he "set out afoot from Oakland, on the bay of San Francisco, in April. It was the bloom-time of the year over all the lowlands and ranges of the coast; the landscape was fairly drenched with sunshine, the larks were singing, and the hills were so covered with flowers that they seemed to be painted." En route, he admired immense sequoias and "mountain streets full of life and light, graded and sculptured by the ancient glaciers, and presenting throughout all their courses a rich variety of novel and attractive scenery."

Once Muir arrived in Yosemite Valley, he looked for work. After trying his hand at a few odd jobs, he accepted a position as a shepherd, caring for 1,800 sheep owned by an Irishman named John Connel at the rate of one dollar per day on the plains of Twenty Hill Hollow. By summer, he took responsibility for a new flock of 2,050 sheep and departed for the High Sierra on June 3, 1869. Muir was accompanied by the owner of the sheep, a former gold-seeker named Pat Delaney, who found husbandry more lucrative than mining. Also, a shepherd named Billy, a Chinese immigrant, and an Indigenous man, with whom Muir shared responsibility for the animals, came along.

At first, Muir spent his days climbing higher and higher through the unfamiliar territory, in search of meadows where his flock might graze on a sweltering summer day. But he often found that his surroundings distracted him from the task at hand. Standing at a lookout point above California's Merced Valley one day, Muir reached down and untied the notebook that was attached to his belt to record his impressions. Below him stretched "a glorious wilderness," he wrote, where the "heaving, swelling sea of green [was] as regular and continuous as that produced by the heaths of Scotland." He "gazed and gazed and longed and admired until the dusty sheep and packs were far out of sight, made hurried notes and a sketch, though there was no need of either, for the colors and lines and expression of this divine landscape-countenance [were] so burned into mind and heart they surely [could] never grow dim."

During the next few months in the high mountains, Muir fully immersed himself in his new surroundings. He documented his observations of the "airy and strangely palm-like" gray pine trees, orange lilies, and hardy, prickly shrubs, taking notes that would later appear in his book *My First Summer in the Sierra* (1911). The world around him brimmed with life and Muir found himself overcome with the desire to study its every aspect. "How interesting everything is," he enthused. "Every rock, mountain, stream, plant, lake, lawn, forest, garden, bird, beast, insect seems to call and invite us to come and learn something of its history and relationship." His life's vocation had been revealed; Muir wanted nothing more than to spend his days as "a servant of servants in so holy a wilderness." Never before had he felt so magnetically drawn to a physical landscape, and he vowed to return to it.

But even in 1869, Muir noticed that human activity had begun to endanger the picturesque mountains. He realized that the domesticated sheep he was charged with protecting were an environmental nuisance. A "dusty, noisy flock," they appeared to be "outrageously foreign and out of place in these nature gardens" where they "trampl[ed] leaves and flowers and imperiled fragile seedlings," Muir remembered. "Should the woolly locusts be greatly multiplied," he warned, "as on account of dollar value they are likely to be, then the forests, too, may in time be destroyed." He longed to relinquish his task of caring for the animals, which provided him with the requisite income for food and supplies but furthered the destruction of the place he had grown to love. "The harm they do goes to the heart," Muir lamented, "but glorious hope lifts above all the dust and din and bids me look forward to a good time coming, when money enough will be earned to enable me to go walking where I like in pure wildness, with what I can carry on my back."

Muir's sentiments were radical for the day. The first national park, Yellowstone, would not be established for three more years. Even then, the scholar Roderick Nash has argued, Yellowstone's preservation at the federal level "was almost accidental and certainly not the result of a national movement." What Yosemite and other wilderness areas required, Nash contended, was an advocate who could explain why wilderness was inherently worth protecting. Writers and artists of the mid-nineteenth century had begun to articulate such ideas in literature and paintings, but Muir would express them persuasively enough to launch America's environmental preservation movement.

Although Muir harbored an instinctive desire to protect the mountains' natural character and prevent the disruption of its ecosystems, fighting the forces of development that increasingly menaced the Yosemite region during the decades ahead would be a daunting task. During numerous excursions to Yosemite Valley between 1869 and 1890, Muir became enamored of its soaring domed rocks, deep glacial basins, and fragrant meadows. He journeyed there regularly, continuing to make the 150-mile trip after settling in Martinez, California, with his bride, Louisa Strentzel, in 1880. Packing little more than bread and tea, he traveled across the landscape lightly and often camped without so much as a blanket to keep warm at night. Muir learned the "glad cascade songs" of Yosemite's waterfalls, the calls of its birds, and the "stone sermons" of its peaks.

Putting pen to paper, Muir sought to capture Yosemite's beauty and, for the first time, share it with the public. In 1872, he began writing a handful of short pieces about different locations, including the Tuolumne Canyon and the Hetch Hetchy Valley. These essays were published in the *Overland Monthly*, a literary magazine launched in San Francisco four years earlier. Then, in 1874, Muir wrote a collection of seven essays titled "Studies in the Sierra." A unique blend of science and poetry, the articles contained Muir's hypotheses about Yosemite Valley's glacial origins. In the years ahead, Muir would hone his skills as an author until he mastered the art of nature writing.

But it was with apprehension that Muir also witnessed the relentless building of houses and businesses, which he feared would "bedraggle the valley from end to end, making it appear like the raw pine towns of a new railroad." Settlers began placing fences around the sweeping meadows and cutting down trees with impunity, actions that Muir argued would lead to the annihilation of natural beauty in the name of "vulgar mercenary 'improvement.'" Year after year, Muir watched what appeared to be the unstoppable development of the Yosemite Valley. In 1889, he decided that the moment had come for action.

THE ENVIRONMENTAL CATASTROPHE facing Yosemite Valley at the turn of the twentieth century, when Muir penned editorials alerting the public to disaster, mirrored that of other regions across the country. Human settlement and the ecological consequences of farming and industrialization had taken a heavy toll on the American landscape during the twenty years since Muir first walked the foothills of the Sierra Nevada. In 1890, continued westward migration prompted the superintendent of the census, Robert P. Porter, to make a significant announcement. Considering recent census data on the American West, he reported that "up to and including 1880 the country had a frontier of settlement, but at present the unsettled area has been so broken into by isolated bodies of settlement that there can hardly be said to be a frontier line."

Porter's statement formally marked the closing of the American frontier, a significant event that the historian Frederick Jackson Turner argued would permanently reshape American life. Westward expansion had for three centuries been "the dominant fact in American life," he mused in his famous 1893 essay "The Frontier in American History." It had influenced the nation's institutions and its peoples' collective identity. Turner knew not, however, how Americans' relationships with the land and the goals of exploration might change in the years ahead.

The ecological consequences of settlement on the western frontier were indisputable. The bison population, a critical source of food for the Arapaho, Cheyenne, Lakota, and other Plains Indians tribes, shrank from approximately ten million in 1850 to several hundred by century's end. Their decline was caused by indiscriminate hunting for sport and trade in bison hides, the establishment of cattle ranching, and the construction of railroad tracks that crisscrossed the land. In 1862, the US Army lieutenant George Brewerton described the Great

Plains prairies as populated by thousands of buffalo, "found at times in such immense herds that their huge forms darken the plain as far as the eye can reach, while the very earth seems trembling beneath the shock of their trampling hooves, as they rend the air with deep-mouthed bellowings."

But by 1889, their numbers had declined due to what the naturalist George Bird Grinnell called "scenes of butchery" that saw "buffalo . . . shot down by tens of thousands, their hides stripped off, and the meat left to the wolves." After a century of unrelenting westward expansion, he lamented, "of the millions of buffalo which even in our own time ranged the plains in freedom, none now remain." Starvation loomed for Great Plains Indians for whom the bison was, in the words of Grinnell, "the staff of life. . . . their food, clothing, dwellings, tools." Their hunger was compounded by the loss of their territory, which reduced their ability to hunt and gather other sources of sustenance.

Zitkala-Ša, a Yankton Dakota Sioux woman living on the Yankton Reservation in South Dakota in the 1880s, remembered her family's efforts to survive. Her mother "dried many wild fruits—cherries, berries, and plums" and hung rings of sliced pumpkin "on a pole that stretched between two forked posts" on a grassy area close to their family's tipi. "From a field in the fertile river bottom," Zitkala-Ša recalled, her "mother and aunt gathered an abundant supply of corn. . . . spread a large canvas upon the grass, and dried their sweet corn in it." For dispossessed Sioux families like Zitkala-Ša's, the unceasing arrival of settlers in the Great Plains made for an increasingly precarious way of life.

And then there were the missionaries. In South Dakota, the arrival of Quaker evangelists prompted noticeable alterations to the Sioux's traditional ways of life. "First," Zitkala-Ša observed, "it was a change from the buffalo skin to the white man's canvas that covered our wigwam. Now [my mother] had given up her wigwam of slender poles, to live, a foreigner, in a home of clumsy logs." Like other children in her community, Zitkala-Ša left her family to attend an out-of-state boarding school called the Indiana Manual Labor Institute, where instructors taught her musicianship and the English language. They also sought to erase aspects of her cultural inheritance, calling her Gertrude Simmons and lopping off her long, dark hair. When she returned to the Yankton Reservation three years later, she found herself unsure of her new identity and her place among the Sioux people. "During this time I seemed to hang in the heart of chaos, beyond the touch or voice of human aid," Zitkala-Ša remembered. "I was neither a wee girl nor a tall one; neither a wild Indian nor a tame one. This deplorable situation was the effect of my brief course in the East, and the unsatisfactory 'teenth' in a girl's years."

The same year that Zitkala-Ša graduated from the Indiana Manual Labor Institute, Congress passed the Dawes Act (1887), which granted Native Americans citizenship and urged Indigenous people to adopt the non-Native practice of private landownership. The act divided reservation land into allotments, which accelerated cultural loss and weakened community ties. On the Great Sioux Reservation, tensions grew between Native Americans, settlers, and federal troops during the late 1880s, after the Dawes Act led to the loss of 80 percent of the land promised to Native Americans following the Treaty of Fort Laramie of 1868. Members of the Lakota Sioux began practicing the Ghost Dance religion, whose prophets foretold the imminent death of white settlers and the return of Native lands. Following a struggle to disarm and subdue the Lakota people in late December 1890, US soldiers shot and killed approximately three hundred men, women, and children in the tragedy known as the Wounded Knee Massacre. A devastating episode in the long history of violent Native American land dispossession, Wounded Knee would be one of the last coordinated efforts of Indigenous people to resist the US military in the nineteenth century.

When Muir reached his moment of crisis in 1889, industrialization had also dramatically changed the western landscape. Fueled by new production technologies, foreign investment, and a steady supply of laborers, industrialization led to explosive economic growth, both out west and back east. As railroads expanded across the western frontier, Americans more easily accessed the nation's natural resources. The giant fir and pine trees of the Pacific Northwest provided lumber for the construction of houses, business, and railroad tracks. Miners extracted zinc and copper, which were used to further electrification and make brass products for American and international consumers. Entrepreneurs invested in steam engines to mine coal, an energy source for trains, steamships, and factories and an essential product for making steel. In what became known as the Gilded Age, businessmen, whose revenues contributed to a rise in GDP of 2.5 percent per year, held little regard for the environmental consequences of indiscriminate logging and mining.

Most American gold was found in the Sierra Nevada, Montana, Colorado, Oregon, and Idaho, supplying world markets during the mid-nineteenth century. Placer mining was the predominant technique employed by prospectors to separate gold from earth using water power. Miners could pan for gold or construct sluice boxes to increase their efficiency. When prospectors engaged in machine-power hydraulic mining and dredging, they significantly eroded the soil and further reduced the natural vegetation. Miners also chopped down trees for their houses and prospecting operations, leaving bare mountains dotted with tree stumps. Frequently they soon abandoned these hastily constructed camps and moved on.

Loggers similarly decimated American forests to supply eager buyers with timber during the second half of the nineteenth century. Forests across the country were being destroyed at a shocking rate without any sign of abatement. In 1865, a federal report estimated that a hundred million acres of forest, an area greater than the state of Montana, would likely be cut down by settlers during the next decade for use as farmland. By the 1870s, even trade journals that generally supported the timber industry, such as the *Lumberman's Gazette*, began sounding the alarm. "The extent of the demand for lumber, created by the extension of railroads in the west," two residents of Saint Louis testified, "is now a matter of much importance." The consequences of unmitigated demolition were fast becoming clear to eyewitnesses. In Ohio, it was reported that the destruction of regional forests was causing climate change by increasing wind and the subsequent evaporation of water. Across the country, national deforestation led to what one man called the "drying up of the streams" and an increase in the rate of evaporation, two consequences of "the ignorance, cupidity, or recklessness of men who 'lifted up axes on thick trees' far up the mountains."

Back east, industrialization was transforming the way Americans lived—and where they lived. Entrepreneurs like Andrew Carnegie, John David Rockefeller, and Henry Clay Frick invested in heavy industries such as coal, steel, and oil, creating jobs in mining, drilling, and refining. Urban centers rapidly expanded, both vertically and horizontally, to accommodate millions of rural migrants and immigrants from around the world. Between 1870 and 1890, when the US population grew by 63 percent, the total urban population grew from 9.9 million to 22.1 million, and the number of cities with more than 100,000 residents doubled. People who chose to live in cities rather than on farms adopted new ways of living that no longer centered on planting, harvesting, and the turn of the seasons. But these changes also led to a sense of disconnection from the land, priming city dwellers for a new kind of relationship with the countryside: an escape from the hustle and bustle.

AT THIS CRITICAL moment, John Muir faced a seemingly unsurmountable challenge. He needed to convince both the public and the federal government to shift their view of the wilderness from something to be developed to something that must be preserved. Muir railed against the prevailing mindset: "No dogma taught by the present civilization seems to form so insuperable an obstacle in the way of a right understanding of the relations which culture sustains to wildness as that which regards the world as made especially for the uses of man." To Muir, who possessed an instinctive understanding of ecosystems long before the term was coined in 1935, nature was worthy of preservation in part because of the interconnectedness between all living things. "Plants, animals, and stars are all kept in place, bridled along appointed ways, *with* one another, and *through the midst* of one another—killing and being killed, eating and being eaten, in harmonious proportions and quantities," he mused.

The natural world, left uncultivated and wild, also benefited humanity in other ways. While working among machines and contemplating a career as an inventor, Muir felt a constant urge to escape into nature. In the end he found it impossible to resist. For Muir, the great outdoors was a place of reinvigoration, introspection, and closeness to God, whom he increasingly viewed during his lifetime as the divine Creator of nature's beauty. Having found personal fulfillment in nature and devoted his life to its study, Muir believed that the wilderness offered redemptive qualities to all people. "In God's wildness lies the hope of the world—the great fresh unblighted, unredeemed wilderness," he wrote in his journal. There, he concluded, "the galling harness of civilization drops off, and wounds heal ere we are aware."

To persuade as many people as possible of the glories of the so-called wilderness, Muir put pen to paper. Writing primarily about California's natural wonders for a mass readership, he published article after article in popular illustrated journals and newspapers between the 1870s and 1890s. "As a publicizer of the American wilderness," Roderick Nash writes, Muir "had no equal" during the late nineteenth century. Muir brought readers along on his exploratory adventures as he scaled cliffs, searched for places to camp among the tall, fragrant pine and fir trees of the Sierra Nevada, and tracked wildlife. Hiking, camping, and botanizing were activities that all Americans could enjoy, Muir suggested, inviting his readers to experience for themselves the majesties of central California's wilderness.

Muir sought to democratize the act of exploration. For those tired of life in cities, where smog filled the air and waterways brimmed with grime, Muir's wholesome depictions of the wilderness seemed appealing. For instance, one late-nineteenth century observer described the

growing pollution in Cincinnati, Ohio, a problem that plagued industrializing cities across the East and Midwest. "Cleanliness in either
person or in dress is almost an impossibility," wrote Willard Glazier
in *Peculiarities of American Cities* (1883). "The smoke of hundreds of
factories, locomotives and steamboats arises and unites to form this
dismal pall, which obscures the sunlight, and gives a sickly cast to
the moonbeams." To Muir, nature offered the antidote to these aspects of modern living. "Pollution, defilement, squalor are words that
never would have been created had man lived conformably to Nature,"
he once speculated while contemplating the life cycles of the natural
world.

Largely missing, however, from Muir's descriptions of the California wilderness were discussions of the Indigenous people who had
long made such regions their home. In his essays, Muir spilled far
more ink on the destruction of the environment than on Native Americans' forced dislocation from places like Yosemite Valley and other
regions of California. When he did record his encounters with people
indigenous to the Sierra Nevada, he sometimes described them using
insensitive language that reflected his biases and a lack of compassion.
In moments like these, the preservationist failed to make the key connection: American settlement had not only ravaged the environment
but also violently upended the lives of Native Californians.

In 1889, Muir focused his public commentaries on the environmental crisis facing Yosemite Valley: its material destruction at the
hands of loggers, ranchers, and shepherds. After witnessing twenty
years of continuous damage, Muir came to a radical conclusion. The
most effective way to protect the magnificent region, he argued, was
to harness the power of an unlikely partner: the federal government.
Although the United States had historically encouraged the settlement
and development of the American West, Muir was determined to
make the case for Yosemite's preservation. But he had to find a collab-
orator who shared his love of the California wilderness and possessed
the political savvy to muscle essential legislation through Congress.

In June 1889, an associate editor of *Century Magazine,* Robert Un-

JOHN MUIR [107]

derwood Johnson, made his way to California. His assignment was
to locate and interview individuals who had participated in the Gold
Rush forty years earlier for a series called "Gold-Hunters." While in
San Francisco, Johnson arranged an introductory meeting with Muir.
The two men decided to go camping for a few days in Yosemite Valley. Johnson found himself overcome by the staggering beauty of the
mountains and impressed with his "enthusiastic and ideal guide," who
"loved this region as a mother loves a child . . . [knowing] every foot of
it, having made the Valley his headquarters for many years." Johnson
described Muir as a veritable "John the Baptist. . . . spare of frame,
full-bearded, hardy, keen of eye and visage, and on the march of eager
movement."

A prophet of the conservationist movement, Muir had found an
eager convert in Johnson. Horrified by the destruction of the exquisite
wildflowers about which Muir had written years earlier, Johnson proposed that the two men team up to help enact federal legislation protecting Yosemite as a national park. They would model the proposal
on the one drawn up for Yellowstone, founded in 1872 under President
Ulysses S. Grant and the only national park in existence in 1889. Muir
agreed to promote the creation of a national park in *Century Magazine,*
while Johnson would "go to Congress . . . and advocate its establishment before the committees on Public Lands," where he had political
connections. Upon returning from their camping trip, the two men
set to work, executing "the programme [Johnson] outlined at the
campfire . . . to the letter," Johnson later remembered.

In *Century Magazine,* Muir rhapsodized about Yosemite's natural
wonders and explained how readers could experience them for themselves. In text that accompanied illustrations of Yosemite and a map
of the new park's potential boundaries, Muir described "camping-
grounds all along the meadows" and the "capital excursions to be
made" to the region's mountains and canyons. "All of these are glorious," Muir promised his readers, "and sure to be crowded with joyful
and exciting experiences."

But American tourists would never be able to enjoy Yosemite's bounties if the government failed to act quickly. He passionately warned readers of the "ravages of man" in Yosemite and advocated for the immediate passage of a protective bill. "Unless reserved or protected the whole region will soon or late be devastated by lumbermen and sheepmen," he cautioned, "and so of course be made unfit for use as a pleasure ground." Even now, Muir explained, "the ground is already being gnawed and trampled into a desert condition, and when the region shall be stripped of its forests the ruin will be complete." To preserve Yosemite Valley from further devastation, Muir argued for extending the park's existing boundaries and harnessing the power of the federal government to protect the land in a way that the state of California had not.

At first, legislative progress was slow. But the following year, Johnson made a powerful presentation to the House Committee on Public Lands about the necessity of preserving Yosemite. Armed with Muir's articles and magnificent pictures, Johnson sought to persuade the members of Yosemite's value as a region of unparalleled natural beauty. He also had an in: Johnson's father was a friend and former colleague of committee member Representative William S. Holman. Years later, Johnson confessed that he "felt sure that Holman would be predisposed to the scheme." Behind closed doors, select business interests also made the case for the valley's protection. Southern Pacific Railroad hoped to generate additional revenue through tourism to the new park, and businessmen providing water to farmers in the San Joaquin Valley recognized the benefits of safeguarding the Merced River, a tributary of the San Joaquin River.

Their collective efforts finally led to the realization of Muir's dream. On October 1, 1890, Congress passed an act creating Yosemite National Park, and granted the secretary of the interior the responsibility of "provid[ing] for the preservation from injury of all timber, mineral deposits, natural curiosities, or wonders within said reservation, and their retention in their natural condition." It was a significant moment in the history of western expansion: the federal government recognized the wilderness as something worth preserving for its own sake.

JOHN MUIR'S FIGHT to protect American landscapes was only beginning. Two years after Congress announced its designation of Yosemite Valley as a national park, Muir cofounded the Sierra Club, an organization that promoted the preservation of the Sierra Nevada during an era of increasing development. With Muir serving as its first president, the Sierra Club spearheaded efforts to further safeguard vulnerable forests in California, to share scientific knowledge with the public, and to broadly promote the idea of preservation rather than conservation, which permitted controlled logging and drilling. Although the members of the Sierra Club failed to care for the concerns of the Indigenous people who had lived on the lands that its leaders aimed to preserve, Muir did gain a growing understanding of and appreciation for Native Americans through his subsequent travels to places like Alaska Territory and the Bering Strait, where he spent time among the Tlingit, Haida, and Inupiat people.

In the end, John Muir's fight to preserve Yosemite heralded a shift in the way Americans perceived the wilderness: from a place to be conquered or settled to one worth preserving in its original state. His victory in 1890 encouraged members of the nascent conservation movement, who subsequently fought to establish national parks, including Mount Rainier National Park (1899) in Washington and Grand Canyon National Monument (1908) in Arizona.

Throughout his efforts as a champion of environmental preservation, Muir succeeded in reaching the "thousands of tired, nerve-shaken, over-civilized people" whom he hoped would "find out that going to the mountains is going home; that wildness is a necessity; and that mountain parks and reservations are useful not only as fountains of timber and irrigating rivers, but as fountains of life." An explorer of the American wilderness who wielded a pen rather than a gun, he created images in writing that depicted the nation's natural wonders, to "show forth the beauty, grandeur, and all-embracing usefulness of our wild mountain forest reservations and parks, with a view to inciting the people to come and enjoy them, and get them into their hearts, that so at length their preservation and right use might be made sure."

America's places of beauty were saved. But what would happen to the country's wildlife, increasingly endangered by Gilded Age elites in the name of fashion?

5

Florence Merriam Bailey

THE CONSERVATIONIST

A national crisis loomed in 1885, when a young woman named Florence Merriam began her final year at Smith College. It was the heart of the Gilded Age, a time when urban populations swelled and many Americans felt more distant from the natural world. The wealthiest residents of growing cities like New York and Boston spent much of their time navigating a complex metropolitan world with distinct yet evolving customs. For the wives and daughters of prosperous businessmen who hoped to maintain their elite position in society, it was essential to keep up with the latest sartorial trends. In the 1880s, that meant hats—big, ostentatious hats, adorned with impressive plumes or even entire preserved birds. But women's insatiable demand for feathers was causing an ecological disaster that Merriam, a budding conservationist, was determined to end.

The *New York Sun* described this catastrophe as the nation's descent into a "desert condition," which would leave its woodlands devoid of birdsong. The American bird population was rapidly dwindling; one late nineteenth-century report estimated that 50 percent had vanished over the past fifteen years. This decline mirrored that of the American bison, the *Sun* lamented. "Lewis and Clark had to set a guard at night at every camp they made in the valley of the upper Missouri, to avoid being trampled to death by the thickly crowded animals," the *Sun* remarked, but less than a century later, "these royal beasts are gone." Like the slaughtered bison, American birds faced destruction from Gilded Age human activities, including "fashion . . . and murderous sport." Using language that echoed that of the era's nascent conservationist movement, the *Sun* warned that "to avert this disaster, legislation and social morality ought to be promptly invoked." Only a significant change in the way Americans interacted with wildlife could prevent the extinction of countless species of American birds.

Florence Merriam, then twenty-two years old, seemed an unlikely candidate to lead such a transformation. In the 1880s, less than 1 percent of Americans were college-educated, the great majority of them men. Merriam, however, was an exception. At Smith College, a new school for women in Massachusetts, she would advance her knowledge of ornithology, found an Audubon Society chapter, and publish essays about birds that denounced what one contemporary described as the "wholesale bird slaughter [which] was then at its height due to the enormous demand for plumage by the millinery trade and the traffic in birds' eggs and skins which was fostered by dealers and taxidermists." But merely exposing her peers to the world of birds was not enough to effect widespread change. After leaving Smith College, Merriam realized she needed to convince Americans of all backgrounds, not only educated young women, that avifauna were worthy of preservation. To take on this challenge, she would have to leave behind all that she had known in the East. She would have to become an explorer.

BORN IN LOCUST Grove, New York, in 1863, Merriam came of age during an era of national rebuilding, industrialization, and westward expansion. The Civil War's end in 1865 heralded a tumultuous period known as Reconstruction. The ratification of the Thirteenth Amendment that year abolished slavery, liberating four million African Americans from bondage. During the decade that followed, African Americans would gain citizenship and, for men, the franchise. In the US South, however, former Confederates sought to regain political power and white supremacists terrorized freedpeople who strove to exercise their right to vote, to learn to read, to establish schools, and to forge strong communities. The US economy, shattered by four years of civil war, lay in shambles in 1865. But between 1870 and 1900, westward expansion and industrialization fueled significant economic growth, leading to urbanization and the creation of a new class of elite Americans concentrated in East Coast metropolises.

Amid this backdrop, Florence Merriam spent her childhood in rural upstate New York, where she was largely sheltered from the industrial upheaval that was transforming cities, the violence against Indigenous people that accompanied the settlement of the West, and the political and racial conflicts in the US South. Her hometown, Locust Grove, was nestled between great green forests and the winding Black River, an idyllic setting for youthful days spent exploring the countryside near the Adirondack Mountains. Florence's parents, Clinton Levi Merriam and Caroline Hart Merriam, supported their daughter's interest in the outdoors. Caroline, an avid astronomer who owned a telescope, introduced her curious daughter to the constellations and the moon that shown brightly above the peaceful hills. And Florence's father, Clinton, was a friend of John Muir.

The two men first met in 1871 in California, where Muir spent his days exploring the Yosemite Valley, and they found that they had much in common. Clinton, who had reached California by way of the recently completed transcontinental railroad, initiated a friendship with Muir that would last through the next few decades. Together they hiked up the mountain "Cloud's Rest," where they gazed upon the valley's spectacular dome rock formations and deep blue lakes. At the summit, Muir eagerly expounded upon his largely accurate theories

about the glacial movements that shaped the Yosemite's unique landscape, ideas that Muir would later share directly with Florence and one
of her older brothers, Clinton Hart, known as Hart.

Florence and her father regularly strolled through the forest surrounding Homewood, their three-story white clapboard house situated on a hundred acres of land. In her childhood diary, Merriam
recorded her pleasure in excursions on foot or horseback. On one
occasion, she set off with her father in search of a yellow canary to
keep as a pet. Although they failed to find one, Florence remembered,
her disappointment was tempered by their enjoyment of the great outdoors. When she later acquired a canary, she expressed an early love of
avifauna, describing the bird as "the dearest little thing [that] ever lived
I think." On other occasions, she marveled at the landscape around
her, admiring a "lovely evening" when "the Sun set beautifully" and a
splendid morning when the sun was "shining very bright indeed."

Her brother Hart cheered on Florence's interest in the environment.
Eight years her senior, he was a budding naturalist who would later
become the head of the US Biological Survey. Both Florence and Hart
wondered at the colorful birds that filled the trees of the Adirondack
foothills. While they ate in their dining room, they gazed out the large
window at the blue jays and woodpeckers that flew from branch to
branch. Yet their approaches to the study of birds were quite different.
Florence, who felt a tender sensitivity to the creatures, taught herself to
observe birds in their natural setting. Her teenage brother instead used
his gun, adding specimens to his extensive private collection.

MOST NATURALISTS OF the nineteenth century followed Hart's
credo: kill first, then observe. When Meriwether Lewis and William
Clark led the Corps of Discovery across the continent in 1805, they
shot birds and mammals to study the characteristics of what were, to
them, new species. On the Oregon coast at Fort Clatsop, Lewis described how one member of the group brought down a "bird of the
Corvus genus, which was feeding on some fragments of meat near

the camp." After retrieving the body, Lewis was able to inspect its beak, talons, coloring, and the unique features of its jaw, which Lewis noted were "perceptible only by close examination." During the early nineteenth century, the naturalist John James Audubon explored the North American continent while sketching its avifauna, a practice that required first shooting and killing his specimens to create life-size portraits showing their distinctive features. "I shot, I drew, I looked on nature only; my days were happy beyond human conception, and beyond this I really cared not," he wrote of his artistic process.

In Florence Merriam's day, decades later, the practice of shooting birds and preserving them for observation and identification continued. Popular journals like New York's *Turf, Field, and Farm* advocated taxidermy, which its editors noted was a practice "growing in favor . . . as a valuable means of acquiring a correct knowledge of natural history" in 1874. Some taxidermists set up shop in New York City, where business was brisk. One man, Edwin Harris, created a memorable window display for his store at 177 Broadway. One journalist described the realistic setting he created for dead English woodcock and European quail "so skillfully prepared and set up as to be actually alive" against a backdrop of scenes depicting their habitats. The observer wondered at how he had "never seen anywhere, at home or abroad, anything to compare with these as specimens of taxidermy, or as pictures illustrative of game birds and their habits." Naturalists could learn more from this type of display "than by a perusal of all the ornithological works and essays ever published," he argued, predicting that orders from gun enthusiasts would soar.

Sportsmen were even worse than taxidermists where birds were concerned. Hunting, historically a pastime of the European aristocracy, was attracting Americans flush with new Gilded Age wealth. They hoped to signal their rising social status to others by pursuing the hunt not as a means of feeding their families but as a leisure sport. In New York, men established private clubs where the pursuit of waterfowl, foxes, and other types of game was the featured activity. Many were located on Long Island, where the city's wealthiest residents spent the summer. By 1900, sporting enthusiasts across the country could select from thirty-nine journals catering to their interests; some are still published today.

As hunting took off, bird populations declined. Passenger pigeons had filled the skies of North America by the hundreds of millions in the early nineteenth century, when Audubon recalled how, in Kentucky, "the air was literally filled with Pigeons; the light of noonday was obscured as by an eclipse." But by the 1870s, when the *Cultivator and Country Gentleman* described their "annual flights in long, wide flocks" as commonplace, the population was dropping precipitously.

Yet many voices were raised in support of this form of recreation. A proponent of pigeon hunting described its popularity in 1878, when twenty-thousand birds were killed at a single sporting convention. But this number paled in comparison to the number of birds slaughtered to meet consumer demand. In New York State, one could "find these birds hung up in bunches of a dozen, for the convenient carrying of purchasers," the man told hunting enthusiasts who wondered whether their actions might be contributing to the growing crisis. "Our friends of the sporting clubs . . . need never fear that the birds which they shoot can affect the supply," he concluded, "so go on and enjoy yourselves . . . [because] all the legitimate shooting in the world will not decrease the numbers of *Ectopistes migratorius*." But this confidence was misplaced. Indiscriminate killing and the destruction of natural habitats due to industrialization and urbanization were decimating passenger pigeons. By 1914, only a single survivor, Martha, then in the care of the Cincinnati Zoo, remained in the entire United States. In September, Martha died at the age of twenty-nine, and the passenger pigeon was gone.

As Florence Merriam was coming of age, other bird species were threatened with annihilation. Long-legged wading birds with delicate plumage, including herons and egrets, were increasingly targeted by hunters at the turn of the century. This time fashion, not sport, created the problem. During the 1870s and 1880s, milliners created hats for women that contained exquisite feathers plucked from magnificent birds. This choice of ornamentation stemmed not only from urban Americans' growing interest in recreational activities like hiking, hunting, and fishing but also from the cultural influence of realism and naturalism in art and literature.

Popular illustrated periodicals like *Godey's Lady's Book*, led by the formidable editor Sarah Josepha Hale, presented extravagant hats and bonnets adorned with feathers to capture women's fancy. Female readers seeking to purchase what the journal called "walking dresses," conspicuous attire in which to be seen by others, could accessorize a delicate black dress with a "white chip bonnet, trimmed with black velvet and feathers" or a ruffled gray costume with a "gray chip hat, trimmed with gray ribbon and feather." Formal evening attire could include decorative plumage, such as a dress accessorized with a hairstyle of "puffs and curls, with gold beads and pink feather in it." Descriptions for fashion plates rarely named the species, a striking omission that left readers unconnected to actual avifauna, with no sense that their purchasing decisions might have serious consequences for the birds whose feathers they loved.

In the name of fashion, demand for dead birds of all kinds rose dramatically throughout the 1870s and 1880s. By the mid-1880s, five million birds were being killed annually to decorate hats and dresses. Popular awareness of the growing ecological crisis, however, was limited. One of the first to sound the alarm was Mary Thatcher, whose *Harper's Bazaar* article "The Slaughter of the Innocents" decried the use of plumage for fashion. "Tall women and short women, richly dressed women and shabbily dressed women, little girls and big girls, have decorated themselves with these spoils of the forest," she observed in 1875. "Not only in the street, but in the ball-room, on head-dresses and in the hair, these feathered ornaments have been worn," she informed her readers. Thatcher may have been the first American to publicly condemn the decline of grouse in Kentucky, passenger pigeons in Massachusetts, and ducks on the East Coast. She connected the massacre of bird populations to "the wholesale slaughter of buffaloes on the Western plains . . . another instance of [Americans'] folly and reckless waste of life" at a time when most people cared little about the environmental consequences of their fashion choices and leisure activities.

The ubiquity of feathered hats among Gilded Age women became increasingly apparent to the public following a revelatory report in 1886. Frank Chapman, an ornithologist born just one year after Merriam, decided to survey the fashion scene near Union Square. "At that time," Chapman later remembered, "Fourteenth Street was the most frequented shopping thoroughfare" where, with "notebook in hand, I recorded . . . the names of birds which, usually entire, were seen on the hats of passing women." Over the course of two afternoons, Chapman played the role of observer, writing down no fewer than forty different species, ranging from robins and bluebirds to grouse and woodpeckers. The most common bird was the cedar waxwing, whose red-and-yellow-tipped feathers adorned the hats of twenty-three women. More rarely spotted were the elegant green heron or the northern saw-whet owl, a miniature night bird. A stunning 77 percent of the seven hundred hats Chapman saw contained feathers. After he published his data, the ecological crisis fueled by hunting and women's fashion choices was harder to ignore.

FLORENCE MERRIAM WOULD play a critical role in bringing national attention to the decline of the bird population during the late 1880s. But her transformation from adolescent naturalist to adult conservationist was gradual. When Merriam matriculated at Smith College in Northampton, Massachusetts, in 1882, she had not yet decided what to do following the completion of her studies. At the age of eighteen, Merriam knew that she loved the great outdoors. But beyond the forests of the Adirondacks stretched the arid deserts of the Southwest and the towering mountains of the West, distinctive regions that sheltered diverse populations of birds. Merriam had not yet traveled to places like these, and Northampton, a small town in rural Massachusetts, seemed far from the nation's remaining areas of wilderness. But her worldview, and her sense of future possibilities, would expand in college.

During her first two years at Smith, Merriam dabbled in a range of subjects. By her third year, however, she increasingly focused on ornithology. Outside the classroom, Merriam observed local species and corresponded with her brother, Hart, whose career as a naturalist was rapidly advancing. Like his father, Hart would also befriend John Muir, whom he described as "a famous wanderer" who "left a trail that [was] well worth following." In 1883, Hart helped establish the American Ornithologists' Union (AOU) and, two years later, nominated his sister, Florence, as its first female associate member. Back on campus, Merriam noticed her peers wearing the feathered hats that graced the pages of women's fashion magazines and began reading widely about the scale of the crisis. In autumn of 1885, she prepared an article about the demise of avifauna and sent it to a family acquaintance, Beman Brockway, then an editor of New York's *Watertown Daily Times*. Then, with nervous anticipation, she waited.

Brockway accepted the piece and ran her private correspondence with him as an introduction to her article. In doing so, Brockway revealed to readers Merriam's bold objective in writing the essay: to "put a stop to the outrageous work that is being done . . . [by creating] public sentiment on the subject." Merriam appealed to both logic and emotion. She made an ironclad empirical case, quoting statistics that explored the dynamics of supply and demand in national and international feather markets—but also sought to instill a sense of shame in those who killed birds for the sake of fashion.

As she put it, "Every time that we buy a wing or head, and part of what once held a happy bird life, in order to add to our own attractiveness, we are not only committing a crime against the bird world, not only violating our best natures, but we are retarding the progress of civilization by an act of barbarity." Although women lacked the right to vote in 1885, Merriam argued that her female readers were nonetheless politically empowered. She urged women to "awake to a realization of the gravity of the case" to make possible "game laws [that would] . . . protect the birds and reduce this wholesale slaughter." With hard work, Florence believed that she and other young women could change the minds of their peers and successfully launch a movement to conserve American birds.

Although Merriam realized that saving America's birds required behavioral changes on a national scale, she started small, with the three hundred undergraduates at Smith College. In March 1886, she established a formal club to bring together women interested in learning about and preserving birds. She planned to follow the example set by Dr. George Bird Grinnell, an explorer, ornithologist, and AOU member who had founded a national organization to save avifauna just one month earlier. Grinnell's group was called the Audubon Society for the Protection of Birds, named after the early nineteenth-century artist and naturalist. In an article for *Audubon Magazine*, Merriam described her strategy for attracting new members from a group of elite women at Smith College who, for the most part, had not "acknowledged to themselves any especial interest in birds." Rather than bore them by "reading prosy descriptions from ornithological tomes," she decided to take her peers "into the fields" to "let them see how the birds look, what they have to say, how they spend their time, what sort of houses they build, and what are their family secrets."

Merriam asked the famous naturalist John Burroughs, a friend of her brother, Hart, to lead a bird-watching excursion composed of forty Smith College women. Like John Muir, whom Burroughs would later call "the truest lover of Nature," Burroughs adored the outdoors. Keenly aware of the crisis facing American birds, he wanted to inform women of the ecological consequences of wearing "head gear adorned with the scalps of our songsters." Visiting the women at Smith College provided the perfect opportunity. Merriam later remembered the magic of that rainy morning. "With gossamers and raised umbrellas we would gather about him under the trees, while he stood leaning against a stump . . . interpreting the chippering of the swift as it darted about . . . with the simplicity and kindliness of a beneficent sage," she recalled. Burroughs inspired them with his sincere passion for birds, and by the time the day had ended, the girls had "all caught the contagion of the woods." It was a winning approach, and one that would inform the rest of Merriam's career.

The Smith College Audubon Society's first meeting was also a success. With a friend, the future ornithologist Fannie Hardy, Merriam had prepared by spreading the word around campus and putting together an engaging program. After the gathering ended, Merriam remembered proudly, "a city milliner inquired anxiously if the college authorities had forbidden the use of birds, [since] so many hats had been brought to her to be retrimmed." Membership grew substantially afterward. Notebooks in hand, the Smith College women headed into the nearby woods, where they observed different species of birds. A guest speaker helped improve their bird-watching and classification skills. By the end of the semester, nearly one-third of the entire student body had joined the Smith College Audubon Society. Merriam's two-pronged strategy of "proselytizing" and encouraging "field work" had generated among her peers a love of birds and an awareness of their plight. Buoyed by her accomplishments, Merriam began to look beyond the brick walls of Smith College as she considered her next move.

AFTER MERRIAM DEPARTED Smith with a certificate of completion, she went to New York City during the winter of 1887. It was a metropolis of extreme wealth and dire poverty, a place where industrial titans constructed sprawling mansions along Fifth Avenue while immigrants crammed together in downtown tenements. When the reformer Jacob Riis investigated the living conditions of the city's poorest residents in the late 1880s, he discovered appalling levels of overcrowding in unsanitary housing, findings he would expose in his 1890 book, *How the Other Half Lives*.

Riis focused one chapter on the tens of thousands of women who toiled endlessly in factories and stores throughout the city. In the attic of one East Side building, Riis met a widowed woman who made more than three thousand paper bags each day for just twenty-five cents. Her situation was not unique. "There is scarce a branch of woman's work outside of the home in which wages, long since at low-water mark, have not fallen to the point of actual starvation," Riis lamented, although he expressed his hope that "a better day is dawning" due to the efforts of female reformers to uplift impoverished women.

In New York City, Florence Merriam found herself drawn to the cause of improving the lives of those less fortunate. During the winter of 1887, Merriam began volunteering at the newly established Association of Working Girls' Societies. Organized by the reformer Grace Dodge, these clubs sought to bridge the socioeconomic divisions that existed between working-class women and those whom Dodge described as "wealthy young ladies . . . willing to give not only their means but their time, [and] their talents." As a recent college student acutely aware of both the environmental and social problems of the Gilded Age, Merriam devoted many hours over the next two years to supporting New York City's impoverished wage-earning women while maintaining her passion for birding.

Using notes from days spent in the woods of Locust Grove or Northampton, Merriam published a few articles in her spare time in *Audubon Magazine* on common birds of the region. She may have hoped that these two seemingly very different paths—ornithology and social work—would one day collide. Merriam recognized that, as individuals dependent on miniscule wages for survival, New York women faced a bleak future trapped in factories, attics, and industrial shops. If Merriam could share her knowledge about birds through accessible, appealing writing, she could perhaps instill a love of the natural world to uplift those working in a stifling urban environment.

When Merriam was not volunteering for the Association of Working Girls' Societies, she spent time among the members of New York City's upper class, a world of rigid rules, particularly for young unmarried women like her. Although Merriam was reared in Locust Grove, her father had made his fortune in New York City during the 1850s and 1860s as a merchant and banker. His network of acquaintances included not only the financiers of the Empire State but also the powerful politicians of Washington, DC, where he served in the House of Representatives between 1871 and 1875. As the daughter of a prominent political figure, Merriam would have been keenly aware of societal expectations. Gilded Age mores required women to dress in specific ways for a range of occasions, including carriage rides, receiving guests, visits to the opera or theater, and evening balls.

Leisure activities could be just as onerous. Horseback riding in Central Park was not just a form of exercise; it was a performative act that required a lady to exhibit perfect posture while on public display. Even walking down the street could be a burden. In his humorous etiquette book, *Manners for the Metropolis: An Entrance Key to the Fantastic Life of the 400,* Frank Crowninshield warned female pedestrians that it could be "difficult to determine whether the fast-approaching queen of fashion is going to bow or not," offering them a clever way to avoid social humiliation should the reader experience a "snub."

This was the society to which Florence Merriam belonged. During the daytime, she went to the Metropolitan Museum of Art or attended lectures on topics ranging from the political causes of the English Reformation to the ancient Greek poet Sappho. She was naturally intrigued by the concept of transcendentalism and once enjoyed an Emersonian sermon delivered by the Boston clergyman Phillips Brooks. Merriam also attended events alongside the city's brightest thinkers. At one reception, she marveled at the celebrities in attendance, including Frederick Law Olmsted, whom she described as "the man who planned Central Park—quite noted in his way." On another evening, Merriam attended a lively dinner party, with food that she claimed rivaled the cuisine at Delmonico's, one of the city's finest restaurants, and "fun & rollicking merriment by music" among the young guests. Much of Merriam's time in New York City was filled with intellectual stimulation and invigorating conversation, luncheons with friends, and thrilling nights at the opera or theater.

Though she was in many ways a model young woman of the Gilded Age, there was one social convention she could not abide. As Merriam walked down the crowded avenues of Manhattan, she felt deep displeasure at the ornate feathered hats worn by fashionable women. At a time when New Yorkers were amassing wealth to an unprecedented degree, many ladies thought it essential to spend the family fortune on expensive clothing that signaled their social status. The Gilded Age journalist James McCabe even opined that "to be the best dressed woman at a ball, the opera, a dinner, or on the street, [was] the height of . . . ambition" for a "New York woman of fashion." Her primary goal was "to outshine all other women in the splendor of her attire," an accomplishment that would "render her supremely happy." Not Florence Merriam. As she matured in her early twenties, she began to search for sincere happiness elsewhere—in the natural world of forests and fields, deserts and mountains—where her treasured birds dwelled.

CAROLINE MERRIAM, FLORENCE'S mother, suffered from tuberculosis, a bacterial infection of the lungs, prompting her to recuperate at Homewood in the late 1880s. On at least one occasion in 1887, Florence left New York City during the social season in order to help out at home. She was glad to be by her mother's side but felt burdened by caregiving and tending to the family's affairs. In May of that year, she found herself solely responsible for running the household while her mother was out of town. It was an overwhelming experience, one that prompted her to record upon her mother's return her "great relief after [a] week of keeping house, house-cleaning, looking after serving girl & gardener, & two children: getting up to observe [birds] at early dawn." Although she disdained particular fashion trends, she still cared about her appearance and noticed that her "hair [was] turning gray" as "life [was] getting more intense."

Domestic tasks took up most of her time, but she still found moments to hone her observational skills and take notes. From the bay windows of the house, she watched gray squirrels munching on corn, robins and blue jays darting through the trees, and ruffed grouse covering themselves in the fragrant conifers. Florence went on walks in the forest with her aunt, spotting a thrush with a melodious song, a bright red-winged blackbird, and a scarlet tanager. As the concerns of adulthood increasingly weighed upon her, Florence saw with growing clarity her true passion: birds. But to become a full-fledged ornithologist, she needed to escape the confines of Homewood and see the broader world.

In February 1889, she and her parents departed on a grand adventure, one that would unexpectedly launch her career as an explorer. They decided to leave the cold, damp climate of upstate New York for the arid mountains of southern California, where Clinton Merriam's brother homesteaded in Twin Oaks. The move was intended to benefit not only Caroline but also Florence, who had likely contracted tuberculosis from her mother, a subject she avoided discussing publicly. On the West Coast, Florence saw for the first time the grassy hills north of San Diego, where her uncle's house was nestled in a sunny valley. She wondered at the rugged peaks of the mountains that cut into the blue horizon, a varied terrain that supported abundant birdlife. The place made an indelible impression on Merriam during the spring of 1889; she would return to it in the years ahead. But by summer, she and her family were on the move again, heading north to the Territory of Washington before returning to New York. There, she threw herself into writing; later that year she published her first book, *Birds Through an Opera Glass*.

A groundbreaking tome, *Birds Through an Opera Glass* spoke to a popular audience of young people, particularly women, not ornithologists, naturalists, or hunters. Its focus was the pleasures of bird-watching. Based on the articles she had previously published for *Audubon Magazine*, the book described common birds of the Northeast. In it, Merriam urged her readers to "go to look for your friends" in fields and meadows, a pastime that would yield many rewards. But at the forefront of her mind were the female laborers she had come to know through her charitable work in New York City. "It is not merely those who can go to see for themselves I would tell of my walks," she wrote; "it is above all the careworn indoor workers to whom I would bring a breath of the woods, pictures of sunlit fields, and a hint of the simple, childlike gladness, the peace and comfort that is offered us every day by these blessed winged messengers of nature."

At that time, few Americans cared to distinguish between bird species. Fewer still considered the complex, rich lives of the creatures that, at best, flitted about in the background of their busy days, or at worst, adorned their elaborate hats. Merriam knew that to change people's attitudes about the value of birds, she needed to stress their anthropomorphic characteristics: their habits, preferences, and relationships. In *Birds Through an Opera Glass*, she described the humble robin as "a domestic bird, with a marked bias for society . . . [like a] self-respecting American citizen." Of the unappreciated crow, Merriam wrote that he was "one of our most interesting birds. . . . Though the crow has no song, what a variety of notes and tones he can boast! In vocabulary, he is a very Shakespeare among birds." Waxwings, one of the birds most commonly slaughtered in the name of fashion, were "elegant, delicately-tinted birds . . . silent and retiring," which "practice[d] among themselves amazing courtesy and gentleness." Implicit in this description was the idea that destroying such gentle, feeling creatures would be an act of cruelty.

Merriam's most important piece of advice was perhaps the most radical. She encouraged her readers to "carry a pocket note-book, and above all, take an opera or field glass with you." Here, she diverged from Audubon, Lewis, Clark, and even her own brother, Hart, men who employed firearms in their study of birdlife. By contrast, Merriam advocated a pacifistic approach that reappropriated one of the Gilded Age's most distinctive objects—the opera glass—in the service of exploring the natural world.

Reviews of the book were largely positive. The *Atlantic Monthly* praised Merriam for "bringing out freshly the more salient features" of birds and categorizing them by "trait, habit, or haunt." Merriam was in good company; the journal's review of a book by her mentor John Burroughs appeared just below her own. With her first successful book behind her, Merriam pressed on to expand her ornithological knowledge. She continued to divide her time between New York City, where she worked on her literary endeavors, and Locust Grove, where birdlife abounded. Then, her mother's health worsened in the winter of 1893. Florence and her father accompanied Caroline to Florida, where they hoped the warm air would restore her lungs. But it was too late. Tuberculosis overpowered her, and Caroline Hart Merriam passed away on March 28, 1893. Her death was a devastating blow to the family. They brought her body back to Locust Grove, where she was buried on the peaceful grounds of Homewood.

DURING THE WEEKS that followed, Merriam struggled not only to come to terms with the loss of her mother but also to deal with her own suspected tuberculosis. In New York City, where spring had recently arrived, she felt trapped. Possessing a "feverish longing" to flee from the bustling metropolis to the countryside, Merriam remembered how the "brick horizons and squares of sky had irritated [her] tired spirit," while the "lonely groups of trees, turning green on the outskirts, made [her] only more restless." When her friend and fellow ornithologist, Harriet Mann Miller, proposed a trip to find "new worlds to conquer," Merriam leaped at the opportunity. They chose Utah Territory as their destination, a region with a dry climate salubrious to Merriam and an abundance of unique birdlife. Although New York society tongues may have wagged at the prospect of two women going on an unchaperoned journey, Merriam and Miller set off by train for the west, reaching a small village at the foot of the Wasatch Range without incident. There, they made their base in a boardinghouse set among fragrant lilacs and rosebushes.

Liberated from the confines of New York society, Merriam rejoiced in her new surroundings. "Now at last I [am] free," she wrote, celebrating her newfound independence and immersion in the Utah wilderness. Far from the noisy streets of Manhattan, where the shouts of vendors selling their wares intermingled with the sounds of carriages rattling over cobblestones, Merriam felt renewed. Sitting among pale-green sage bushes near a cool stream one day, she remembered how her "spirit rose exultant, catching inspiration from nature in its purity, strength, and radiant joy." Like her father's friend, John Muir, Merriam found tranquility far from the trappings of civilization. Utah Territory provided an opportunity for both introspection and external observation; the more time she spent outdoors among wild birds, the more she believed that ornithology was her calling.

Nearly twelve thousand feet in height, the Wasatch Range gave Merriam ample opportunity for exploration and research. From the Mormon residents of the village where she was staying, she purchased a horse named Jumbo. She and Jumbo spent "long happy mornings together . . . wading through the pebbly brooks . . . and rambling along, enjoying the fresh air and sunshine, the mountains and the meadows," she later recalled. Merriam reveled in the desolate, expansive landscape, where the mountainside's "restful silence [was] jarred by no footfall; its only suggestion of the figure of a man a lone tree against the sky." The sole evidence of past human activity was an abandoned mine, a relic of Utah's rush for gold during the 1860s, and the corroded wheel of a wagon. Although there were no other people, Merriam was surrounded by birds. Tiny towhees emitted long, loud calls from the shrubbery, and an opportunistic bird of prey circled overhead. She spotted elegant herons, a scuttering sage-grouse, and, near a shimmering lake, a massive flock of blackbirds.

As time passed, however, signs of western settlement became more apparent. Immigration to Utah had significantly increased during the second half of the nineteenth century. Between 1870 and 1890, the population more than doubled, a phenomenon driven both by high birthrates and the arrival of people looking for religious freedom or work as miners. Back in the village, Merriam marveled at the constant drama of wagon trains; she wrote of carts carrying impoverished immigrants who had been traveling for weeks on end. Most were men, and some were prospectors. "The silver question was the excitement of the summer," she remembered, and eager men told tall tales about the abundance of minerals supposedly waiting to be discovered in the region. Shattering the peace of the local village, the express train's piercing whistle was a reminder of the rapidly industrializing world from which she had fled.

An excursion to Alta, located about eight thousand feet above sea level at the highest reaches of the Wasatch Range and once a silver-mining boomtown, confirmed the worst of Merriam's fears about the future of Utah Territory's wilderness. While exploring the area, Merriam met a group of male tourists looking for adventure. Like Merriam, they were attuned to the activities of local birdlife; they had noticed grouse nearby. Their interest in avifauna, however, was rather more sinister: the men were "calling for a gun" and bragged to Merriam about their "recent achievements—they had killed a badger, a deer, and an eagle within a few days."

"It was a rude shock to me," Merriam recalled, "and I thought bitterly that even these wild grand mountains would soon be 'civilized' by the pleasure-seekers who destroy all they can of the nature they come to enjoy; leaving the country lifeless and bare, after having had the refined satisfaction of taking pleasure in giving pain, of taking life to evade the tedium of an idle hour." Like Muir, she recognized that human settlement in areas of wilderness often meant wanton destruction and the depletion of natural resources. Merriam, however, was more conscious than Muir of the ways in which environmental destruction also devastated wildlife—especially birds—and was determined to take action.

When Merriam's summer in Utah Territory ended, she decided against returning to the East Coast. Instead, she headed to California. After a short stint at Stanford University, she made her way back to Twin Oaks, her uncle's ranch in the San Marcos hills, which served as the base for her bird-watching excursions. As in the past, Merriam refused to carry a gun; instead, she was "armed with opera-glass and note-book."

On horseback, she roamed the hills and dales that teemed with birdlife from sunrise to sunset. Warblers and towhees, thrashers and hummingbirds sang their distinctive music from the thick brush that covered the valley. She encountered grim turkey vultures, fierce great-horned owls with piercing eyes, and red-headed woodpeckers. Her travels took her to the top of mountain peaks, into treacherous quick-sand, and through turgid waters filled with snakes. As in Utah, she enjoyed her unfettered life as an explorer and observer of birds. "The world was mine. I never spent a happier spring," Merriam later told her readers. "The freedom and novelty of ranch life and the exhilara-tion of days spent in the saddle gave added zest to the delights of a new fauna." In her writing, she surely intended to inspire others to enjoy the natural world.

Merriam recounted her experiences in California for East Coast readers in *A-Birding on a Bronco*, published in 1896. In it, she described the birds of the West Coast and made the case for their preservation. One of her chief purposes in penning *A-Birding on a Bronco*, however, was to convince naturalists that it was not necessary to hunt birds in order to identify them. "In my small valley circuit of a mile and a half," she wrote of her home base, "I made the acquaintance of about seventy-five birds, and without resort to the gun was able to name fifty-six of them."

Her task was not without its difficulties. She acknowledged to her readers that "you can identify perhaps ninety percent of the birds you see, with an opera-glass and—patience; but when it comes to the other ten percent, including small vireos and flycatchers, and some others that might be mentioned, you are involved in perplexities that torment your mind and make you meditate murder; for it is impossible to name *all* the birds without a gun." Faced with the choice of shooting a bird to determine its species or letting it live and remain a mystery, Merriam chose mercy. "I pondered long and weighed the matter well, trying to harden my heart; but the image of the winning trustful birds always rose before me and made it impossible," she explained.

The book's pacifism was radical for the time without being didac-tic. Florence led her readers by example, showing them, through descriptions of her own ornithological adventures, that they could learn about birds without wielding a firearm. Just as important, by sharing the details of birds' habits and proclivities in anthropomorphic language, Merriam made the case for their protection. One reviewer of *A-Birding on a Bronco* marveled at the way in which she "seem[ed] to have been an almost welcome guest at birds' nests" who achieved an "intimacy of feathered folk" that was rare at the time. By entering the birds' private world, Merriam acquired new knowledge about avifauna to disclose as "revelations" to readers in the service of "break[ing] down the barriers of ignorance that divide her bird friends from the human." To another reviewer, *A-Birding on a Bronco* was nothing less than groundbreaking; it was in fact "the first book of real popular interest and value which . . . [the world had] on the subject." In the end, Merriam's literature not only pleased and informed readers. It also helped change popular perceptions of birds as awareness of the bird crisis continued to grow.

After Merriam returned to the East Coast from excursions in Utah and California, she moved to Washington, DC, where she wrote and published. But many of the fashionable and politically connected women in her orbit continued to wear feathers. The bird population may have declined by as much as 50 percent during the last decades of the nineteenth century, yet plume hunters continued to target domestic birds, ranging from gulls and terns to blackbirds and crows. Living in the nation's capital provided Merriam with a unique opportunity for political involvement. She joined the American Ornithologists' Union's Committee on the Protection of North American Birds. The scholar Harriet Kofalk calls the founding of that committee "the first direct action taken by ornithologists to combat the millinery trade in birds."

The group took a two-pronged approach. First, it worked its way into the halls of federal power by helping create the Department of Agriculture's Division of Biology Survey, where members of the AOU tracked animal populations. According to the historian Theodore Cart, their findings served as "the most influential source of advice in the drafting" of a groundbreaking bill meant to preserve American birds. The legislation was named after the Iowa conservationist and representative John Fletcher Lacey. Gathering public support for the bill was also essential. The AOU sought to inform Americans about the destruction of the bird population and to counter the work of interest groups like milliners and purveyors of animal skins and feathers, which were then lobbying the government. Together, AOU members reported "exert[ing] all their influence in behalf of" this new bill.

In 1900, Congress passed the Lacey Act, which safeguarded birds through legal text that penalized those who sought to illegally acquire, transport, or sell protected wildlife. It was a seminal moment in the history of the US conservation movement. It recognized birds, and not only landscapes like those championed by John Muir, as worthy of federal protection. The effects were immediate: the AOU reported on the governmental seizure of thousands of gulls and terns, as well as the criminal charges filed against millinery firms that year. In 1901, just months after the law went into effect, South Dakota's *Pioneer Press* spread word of what was "believed to be the greatest seizure of game birds on record. . . . more than 22,000 quail, grouse, and ducks" from forty-eight male dealers. The federal appetite for protecting American avifauna by curbing supply grew in the years ahead: plume hunters were dealt another blow in 1918 with the passage of the Migratory Bird Treaty Act, a law that would keep millions of birds from slaughter by shielding certain species from capture or killing.

One year before the passage of the Lacey Act, Florence married Vernon Bailey, a friend of her brother, Clinton Hart. He worked for the US Biological Survey and the two shared a love of wildlife and the natural world. With Bailey, she returned to California and explored new parts of the Southwest. There, she expanded her ornithological knowledge and shared it with her readers through books like the *Handbook of Birds of the Western United States*, *Birds of New Mexico*, and *Among the Birds in the Grand Canyon Country*. By the end of her life, Florence Merriam Bailey was widely recognized as a leading conservationist and an explorer driven by a desire to protect American avifauna, rather than profit from its destruction. Through her ornithological discoveries and activism, she changed the way Americans viewed the creatures of the natural world. Support for the preservation of endangered wildlife increased.

In the years that followed the closing of the frontier in 1890, the conservation movement grew. And then, a new generation of adventurers began to look beyond the boundaries of the United States in search of unfamiliar regions to explore. What human stories and natural wonders might they find in the jungles of Africa and South America, the Arctic tundra, or beyond the limits of the earth's atmosphere?

PART II

6

William Sheppard

THE MISSIONARY

On the morning of February 26, 1890, as the pale winter sun slowly climbed above the New York City skyline, the crew of the *Adriatic* prepared to embark upon a rocky transatlantic crossing. Aboard the single-funneled iron steamship was a twenty-four-year-old African American missionary named William Sheppard, a young man who had never been abroad. The port city of Liverpool, however, would not be his final destination. From England, he would make his way to the West African seaport of Banana, from which he would travel up the Congo River into the heart of the continent on a journey to share the gospel with the Congolese people.

As a small tugboat dragged the *Adriatic* from a pier in Lower Manhattan, Sheppard may have wondered whether he would ever see America again. The well-wishers on shore slowly faded from view as the ship passed Governors Island and the recently erected Statue of Liberty, its copper robes gleaming in the sunlight. After passing through The

Narrows, the deep-blue tidal strait where the Hudson River meets the sea, Sheppard saw the Atlantic Ocean stretching endlessly before him. While salty waves splashed against the sides of the *Adriatic*, Sheppard considered the distance separating the United States from Africa, a watery expanse that would prevent him from seeing his family and friends for many years. "The *Adriatic* soon put a vast space between us and our home people like a blank page in existence," he later remembered. Privation, disease, and danger awaited him in what was then known as the Congo Free State. The first white Protestant missionaries had arrived twelve years earlier, but no African American missionary had set foot there. Yet Sheppard was undaunted. "It is sad to leave home, friends and native land and seek a home among strangers, yet it [is for Jesus's] sake." Unbeknownst to Sheppard, his excursion would lead not only to the establishment of a new Presbyterian mission but also to the discovery of a terrible secret.

BORN IN VIRGINIA to Fannie and William Sheppard on March 8, 1865, about one month before the Civil War ended, William grew up in a household steeped in faith. His father, who served as a sexton of the local Presbyterian church, regularly led the family in prayer. When night fell, his mother knelt by his bedside, helping him say the words "'Now I lay me down to sleep" before tucking him in. Sheppard felt secure and loved, despite the political and social upheaval that surrounded him in the postbellum South.

Sheppard entered a world shattered by the Civil War and grew up in a time of uneasy peace, punctuated by white supremacist violence against members of the Black community. His hometown, Waynesboro, was located in the Shenandoah Valley, just sixty miles north of Appomattox, the site of the Confederate general Robert E. Lee's surrender to the Union general Ulysses S. Grant. One hundred miles to the southeast lay Richmond, the capital of the Confederacy during the Civil War. Slavery drove the economy of Augusta County, where Waynesboro was located; when the war began 20 percent of the population was enslaved, growing wheat, corn, rye, and oats. Although it is unknown whether Sheppard's father was born into slavery, his mother, Fannie, was a member of Augusta County's free Black community, which numbered 586 people in 1860.

The ratification of the Thirteenth Amendment legally liberated Virginia's enslaved people, some of whom had fled to territory controlled by the Union army prior to the war's end. After 1865, freedpeople gained literacy by establishing schools, and they founded businesses in pursuit of financial independence. During his childhood, William Sheppard Sr. had worked as a barber while Fannie was a bath maid at a local spa. Although they were members of Waynesboro's Black middle class, the Sheppards' economic circumstances remained tenuous, and their children had to work. Young William remembered "carrying water from a pump a block away morning and evening" for a local woman in the thick heat of summer. When he became old enough to take on more responsibility, he was sent to the nearby town of Staunton with his older sister, Eva. There, he cared for the livestock owned by a dentist. William's faith was always at the forefront of his mind; it was the lens through which he viewed the world.

One day, while William was running around barefoot, a kind woman from the community approached him. "William, I pray for you, and hope some day you may go to Africa as a missionary," she said. He paused to consider the idea. He did not know Africa existed, but wondered what the place was like and began to consider a future there. "Those words made a lasting impression," he remembered, and sparked his interest in the distant continent.

Africa seemed a world away from rural Virginia, but Sheppard was determined to obtain an education that expanded the possibilities for his future. When he heard about the Hampton Normal and Agricultural Institute, a vocational school for African Americans in Hampton, Virginia, he pored over its brochures and began putting away money from odd jobs so he could enroll. Founded in 1868, during Reconstruction, Hampton represented a unique opportunity for fifteen-year-old Sheppard. The school uplifted African Americans, many of them newly emancipated, by teaching them trade skills. One student whose life was transformed by his experience there was Booker T. Washington.

Born into slavery, Washington entered the school in 1872 with "only a small, cheap satchel that contained what few articles of clothing I could get," he later wrote. At the end of his schooling, he had "learned to love labour, not alone for its financial value, but for labour's own sake and for the independence and self-reliance which the ability to do something which the world wants done brings." Perhaps more fulfilling than work itself was the good that Washington recognized he could do for others. "At that institution I got my first taste of what it meant to live a life of unselfishness, my first knowledge of the fact that the happiest individuals are those who do the most to make others useful and happy," he wrote in his autobiography, *Up from Slavery*. That realization would influence the course of Washington's life as an educator not only at Hampton, where he returned to teach after graduation, but also as the founder of the Tuskegee Normal and Industrial Institute in Alabama.

Like Washington, William Sheppard left Hampton believing it was his life's purpose to help others. Yet he would ultimately pursue a different, more grueling path: that of a missionary working in a foreign land. In 1880, Sheppard enrolled in the program's night school, then run by Washington himself. He worked on the school's farm and in its bakery during the daytime before attending classes after the sun set. In his free time, he visited the "Curiosity Room," which housed objects from other continents in a museum-like setting. But it was the institution's religious work that most attracted him. Sheppard experienced a revelation of sorts following an invitation from the school's chaplain to establish a Sunday school for impoverished Black children in a nearby town. He took on only the role of transporting hymnals and Bibles, but after witnessing the chaplain's interactions with community members, he "felt from that afternoon that [his] future work was to carry the gospel to the poor, destitute and forgotten people."

Sheppard graduated in 1883 and returned to Waynesboro, where the pastor of the First Presbyterian Church urged him to attend the Tuscaloosa Theological Institute, a new seminary for Black ministers where he could prepare for service in Africa. Soon after, he moved to Alabama to begin pastoral training with fifteen other men in a humble structure. Outside the classroom, Sheppard gained practical skills doing what he called "missionary work around the town, visiting and praying with the sick—a work which [he] enjoyed so much." But the possibility of traveling to Africa, an enormous continent filled with diverse peoples about whom he knew so little, loomed larger than ever as graduation approached. During his examination at the institute, Sheppard's interlocutors tested his willingness to go abroad. "If you are called upon to go to Africa as a missionary, would you be willing to go?" they asked. Without hesitating, Sheppard replied, "I would go, and with pleasure."

But Sheppard would not be called for four years. After becoming an ordained minister, he first served in Southern Presbyterian churches in Montgomery and Atlanta. William still felt a keen desire to obtain a position abroad, but his letters to the Presbyterian Foreign Missions Board, requesting a post in Africa, were turned down. It was the era of Jim Crow, when laws enforcing racial segregation proliferated across the US South and white supremacists terrorized Black communities. Some white Southerners, including the Alabama senator John Morgan, even advocated the establishment of a colony in the Congo Free State that would receive African American immigrants, a plan intended to consolidate white power in the South. Men like Senator Morgan believed that missionaries from Southern churches might lay the groundwork for Black emigration.

By contrast, William's interest in traveling to Africa remained firmly rooted in his desire to share the gospel with Africans. Yet despite his sincere intentions, Sheppard faced significant obstacles. At the time of his ordination, the Southern Presbyterian Church was largely composed of white congregants and headed by white men who refused to permit Sheppard to lead a mission to Africa. With little clout, Sheppard was forced to wait until another Presbyterian mission-ary, a white man, emerged as a potential partner. In the meantime, Sheppard directed his attention to the predominantly Black members of his congregation at the Zion Presbyterian Church of Atlanta, tend-ing to their needs while setting aside his own desires.

Unexpected news arrived in 1890. "Joyful tidings came that Rev. S. N. Lapsley, a young white man of Anniston, Ala., and I had been appointed as missionaries of the Southern Presbyterian Church to Af-rica," he recalled. Lapsley was the son of a former colleague of Senator Morgan, who may have recommended him for the role. The two men, just one year apart in age, had been directed to make the arduous voy-age to the Congo Free State in Central Africa, a region then controlled by Leopold II, the king of Belgium.

Sheppard had hoped for this moment since his youth. But now he would have to leave behind all that he knew. By steamship, he would journey to the western shores of Africa. From the port town of Ba-nana, where the Congo River poured its reddish waters into the Atlan-tic Ocean, Sheppard would travel upstream by boat. Once he reached the Congo Free State, a region largely unknown to Americans at that time, he would work to establish Southern Presbyterian missions. But how would the Congolese people perceive him—a Black man born and reared in the United States? Could he rely on Lapsley, a Southern white missionary whose father had been a slaveholder, as a spiritual partner in a foreign land? Although these questions may have troubled Sheppard, he nonetheless answered the call.

IN THE YEARS before the turn of the twentieth century, Africa was a continent of immense demographic diversity, cultural wealth, economic possibility, and political strife. The brutal transatlantic slave trade, fueled by demand from New World landowners and facilitated by European, American, and African traders, decimated Africa's pop-ulation and exacerbated divisions among its peoples. Approximately

12.5 million Africans were captured and forced onto ships heading to North America, the Caribbean, and South America between 1501 and 1866. Ten million people survived the horrific journey, known as the Middle Passage, during which they were shackled below the decks of slaving vessels for weeks on end. The United States and Great Britain banned participation in the international slave trade at the turn of the nineteenth century, but the Spanish Empire continued to capture Africans and transport them to its colonies until 1867. Europe's withdrawal from this trade resulted in gradual economic shifts in regions like West Africa, where farmers and merchants began exporting resources, including peanuts and palm oil. But the practice of slavery within Africa itself, which had roots in premodern civilizations, persisted in some regions.

Near the turn of the twentieth century, avaricious European powers increasingly viewed Africa as a place for conquest. During the 1880s and 1890s, nations including France, Great Britain, and Belgium competed to snatch up and control vast swaths of land in a frenzy that became known as the "Scramble for Africa." One particularly contested region was Central Africa, where the Congo River wound through the fertile interior of the continent. Populated by millions of Bantu-speaking peoples from about two hundred distinct groups, the Congo River basin had remained largely isolated from the outside world due to its distance from the coasts.

After Henry Morton Stanley traversed Central Africa during the mid-1870s and wrote about the journey in his book *Through the Dark Continent*, European leaders began paying closer attention. Stanley, a Welsh-born American immigrant who had traveled throughout the US western frontier as a journalist, wrote of Central Africa's varied resources: robust tributaries that enriched the land, abundant wildlife and livestock, and mineral deposits. "The productions of the land are of great variety, and, if brought within reach of Europeans, would find a ready market—ivory, coffee, gums, resins, myrrh, lion, leopard, otter, and goat skins, ox-hides, snow-white monkey skins, and bark cloth, besides fine cattle, sheep, and goats," he assured his readers.

On November 12, 1877, Stanley penned a letter for London's *Daily Telegraph*, commenting on the fact that "no European Power" yet controlled the Congo River basin, which he believed was a missed opportunity. "I could prove to you that the Power possessing the Congo, despite the cataracts, would absorb to itself the trade of the whole of the enormous basin behind," he argued, because the Congo River "is and will be the grand highway of commerce to West Central Africa."

Leopold II, then the ruler of Belgium, saw potential in exploiting the region. He founded the Comité d'Études du Haut Congo that month and paid Stanley handsomely to start building infrastructure along the Congo River. Seven years later, Leopold prepared to show the world that he was ready to stake his claim to one million square miles of territory where twenty million people lived. In the winter of 1884–85, representatives of several European powers, including Great Britain, France, and Germany, gathered in Berlin to discuss partitioning the African continent. No Africans were invited to the meeting. By the conclusion of what became known as the Berlin Conference, King Leopold, who followed proceedings from afar but exerted his will through proxies like Stanley and Belgian officials, had achieved his goal: to eliminate opposition to his plan to treat the Congo River basin as his private colony. On February 5, 1885, Leopold formed the Congo Free State, a region twenty-six times the size of Belgium containing nearly 10 percent of the African continent's territory. Now he could accelerate the extraction of its natural resources, which he would do without regard for the sanctity of human life.

ON THE EVE of his long-anticipated departure to Africa, William Sheppard knew little about the people of the Congo Free State and the environment in which they lived. During a meeting with the Southern Presbyterian Church's Executive Committee in Nashville shortly before the *Adriatic* set sail, Sheppard struggled to answer the questions posed to him about "the new country to which we were going," he wrote. His central source of knowledge was the stories told by those who had traveled to Africa before him, which focused on the dangers faced by missionaries. "In Great Britain, Sweden, and America [missionaries] were told that the climate was deadly; that they would be pelted by the rains, scorched by the sun, and murdered by the natives," he later recorded.

Sheppard sought to walk in the footsteps of those who had preceded him: "in full knowledge of these conditions and with hearts imbued with the Spirit of God they went forth on their mission of love." Sheppard felt he had been called by God to embark upon this mission. Fellow Presbyterians shared his excitement and enthusiasm as the date of his departure approached. Shortly before setting sail to England, Sheppard delivered a lecture to a large crowd in Nashville, where, a reporter observed, his words "brought [tears] to many eyes." As the first African American missionary to travel to the Congo Free State, Sheppard's journey would be "watched with no ordinary interest by the Presbyterian world."

At long last, the *Adriatic* pushed off on the eleven-day journey to Liverpool. Aboard the ship, Lapsley and Sheppard slowly became better acquainted. Born in Selma, Lapsley was twenty-three years of age when he agreed to join Sheppard. Pastoral work was a family tradition; both of his grandparents had been Presbyterian ministers. A man of deep faith, Lapsley strove to be righteous, to quell what his brother called the "doubts and fears and evil thoughts" that had plagued him since childhood. After graduating from Chicago's McCormick Theological Seminary in 1889, Lapsley began to consider mission work in Africa, though he knew it was a dangerous choice due to the harsh climate and threat of disease. Nonetheless, he volunteered to go to the Congo because, his brother remembered, "he had reason to believe he was fitted for the work, and he knew that he was willing to give himself up to it."

Against the odds, Sheppard and Lapsley became fast friends. They shared not only meals and evening prayers but also bouts of seasickness, as the *Adriatic* bounced and rocked its way across the Atlantic Ocean. By the time they reached Liverpool, Lapsley had concluded that Sheppard was "very modest, and easy to get along with," making him an ideal partner for shared work. It was the beginning of a relationship that would flourish far from the poisoned earth of the Jim Crow South, where interracial mixing in public spaces was banned and African Americans were treated as second-class citizens. Soon the two men would rely on each other to survive entirely new conditions in a region where people viewed the color of Sheppard's skin as a shared trait representing kinship, not a mark of racial inferiority.

Sheppard and Lapsley traveled by train from Liverpool to London, where they "exchang[ed] most of [their] American money for cowrie shells, beads, salt, and brass wire" to use as currency, as well as "flour, sugar, butter, and lard; also thin linen clothing and helmets for the sun." Lapsley got to know well-connected men in London who could advise him on how to ensure their mission's success. He even traveled to Brussels, where Henry Shelton Sanford, the former US minister to Belgium, introduced him to King Leopold.

The Belgian king much surprised Lapsley by expressing interest in the Southern Presbyterian Church's mission. "I wonder at his kindness and freeness in talking with me, and questioning me," he wrote to his mother later that day, "for I was not well-dressed nor courtly in manner, and worst, I had no special business with him." Young Lapsley suspected nothing of Leopold's grand vision to dominate the ironically named Congo Free State. He failed to realize that he was merely "a pawn in the king's own scheme," as the historian Pagan Kennedy argues, a missionary whose foray into the Congo would pave the way for Belgian merchants. As Lapsley left the meeting with King Leopold, he felt buoyed by the powerful man's enthusiasm and apparent support; Leopold had praised the younger man's "courage, enterprise, and Christian pluck." With renewed determination, Lapsley reunited with Sheppard and prepared to depart.

On April 16, 1890, the Dutch merchant vessel *Afrikaan* set sail from Rotterdam for the Congolese seaport of Banana with Sheppard and Lapsley aboard. After three weeks of violent weather, which tied Sheppard's stomach in knots, the *Afrikaan* reached its destination.

Sheppard recorded his first impressions of Africa in his autobiography, thrilling at the sight of the "dark, tea-colored water" of the Congo River, which was "swarm[ing] with man-eating sharks." Along its banks stood hundreds of mangrove trees, their roots reaching deep into the muddy waters and their branches creating a thick, shady covering. Living up to its name, the city of Banana was dotted with thousands of banana trees.

Numerous people from the kingdom of Loango stood on the shore. Lapsley admired the women's "wide, gay colored bandanas, secured below the shoulders, and falling in drapery very picturesque and graceful," while the men sported "a simple breech-cloth." European traders and missionaries also populated the towns along the Congo River, which flowed westward from the continent's interior. In Boma, then the Congo Free State's capital, there were as many as seventy white residents by Sheppard's count.

Days later, on May 20, Sheppard and Lapsley began their expedition from Banana to the interior. They planned to travel by steamboat up the Congo River, into a region where no evangelist had trod, to establish their first mission. It was an arduous journey. One hundred and eighty miles from the coast, malaria cut down the two missionaries. Weakened by high fevers, Sheppard and Lapsley put themselves under the care of Baptist evangelists whom they had just met. They were given "five grains of calomel and five grains of jalap each" before being sent to bed. "We were covered with six blankets, hot tea, cup after cup, followed in quick succession, and soon we were like two ducks in a puddle of water," Sheppard wrote of the trial. After they had endured almost unbearable heat for days on end, their fevers broke. Malaria would claim the lives of many missionaries in the Congo Free State, but this time, it did not claim theirs.

Having regained their strength, Sheppard and Lapsley recommenced their journey along the Congo River. Prevented from heading farther upstream by the massive cataracts of Livingstone Falls, they arranged to march deeper inland on foot, a task made more difficult by a lack of pack animals to carry their supplies. On June 27, they set off with a group of twenty-five Congolese men, who carried their "loads of tent, beds, bedding, trunks, chairs, guns, corned beef, hard tack, lard, salt, tea, coffee and sugar." In the days that followed, they woke at 5 a.m. to cover as much ground as possible before the blazing sun made marching unbearable. Then they rested during the hottest part of the day before resuming their trek until the golden sun dipped below the thick canopy of trees.

During their two-week journey, Sheppard and Lapsley encountered evidence of the passage of explorers who had taken the same treacherous route before them. "We found a road had been made by Mr. [Henry Morton] Stanley and saw some of Stanley's heavy iron wagon wheels lying by the roadside," Sheppard wrote. Ominously, alongside the debris lay "sun-bleached skeletons of native carriers here and there who by sickness, hunger or fatigue, had laid themselves down to die, without fellow or friend." Not long after they made this discovery, strange sounds echoed through the jungle. The group stopped to investigate the dense bushes that lined the road for the source of the sound. "On making our way into the thicket found a man dying of smallpox," Sheppard recalled. "We longed to help him," he continued, "but there was nothing we could do, and our head man hurried us away, fearing we ourselves might catch the disease."

On July 18, Sheppard, Lapsley, and the other members of the expedition reached Leopoldville, a town on the western banks of present-day Pool Malebo, then known as Stanley Pool. There, European merchants had begun setting up substantial trading posts to transport ivory by steamship along the Congo River and its tributaries. Foreshadowing economic shifts to come, Lapsley noticed that "the supply of ready ivory seems to be running short . . . rubber is already taking its place." As Europe and America continued to industrialize, rubber became essential. It was used to make waterproof garments, tires, and casing for the electrical wires in lighting, telephones, and other equipment. European rulers' pursuit of rubber, however, would come at a terrifying human cost.

Sheppard and Lapsley stayed at Pool Malebo for several days, becoming acquainted with local missionaries as well as the local Bantu-speaking Teke people, whose main occupation appeared to be trading goods like ivory and wood. The Teke acquired these materials from other Congolese groups residing upstream and exchanged them for goods from European merchants at Pool Malebo. The Teke people bestowed upon Sheppard a unique name, Mundele Ndom, or "black white man," Lapsley wrote to his aunt. The moniker reflected Sheppard's unusual position in a land populated by Black Africans and a handful of white Europeans. Sheppard shared the Indigenous Africans' skin color but spoke a different language and had western habits that seemed strange to the local people. "The Bateke think there is nobody like [him]," Lapsley mused.

Before long, Lapsley left Sheppard on his own at Pool Malebo to visit with Baptist missionaries in a town farther up the river. During his absence, Sheppard learned that the local villagers were facing severe hunger. To help them, he set off in search of a hippopotamus with a Congolese guide. Ninety minutes later, the men spotted twelve of the formidable gray-backed beasts in the distance. "Some were frightened, ducked their heads and made off; others showed signs of fight and defiance," Sheppard remembered. He lifted his gun, took careful aim at one of the swimming creatures, and fired, killing it. Seconds later, he shot a second hippopotamus, much to the excitement of his guide. Their carcasses sank into the water, but the men patiently waited. Before long, the massive carcasses resurfaced. Fifty Teke men worked together to wind a rope around the animals' feet and pull them to shore. The successful hunting trip attained about five tons of meat for the famished villagers. Soon enough, they'd prepared a meal and, Sheppard happily recalled, "the missionaries enjoyed a hippopotamus steak that day also."

By late autumn, the time had come for Sheppard and Lapsley to push onward to establish a new mission for the Southern Presbyterian Church. Months of subsequent travel through the Kwango River region convinced them that the Kasai River offered better prospects for a mission site. "The country [near the Kwango River] was too low and swampy," Sheppard recorded, "the villages small and far apart." On March 17, 1891, Sheppard and Lapsley climbed aboard the *Florida*, a steamship bound for the Kasai River. Violent storms threatened to capsize the vessel, but at last, Sheppard saw "the red waters of the Kasai running into the Congo like a mill race."

As they proceeded up the Congo's tributary, they encountered few animals. Forays into the jungle for buffalo or other substantial game were fruitless. Starvation began to set in. "Our men were so weak from hunger they could gather but little wood last night," Sheppard lamented. The *Florida's* captain warned Sheppard that "if he did not soon get food for his men he would go crazy." Finally, on the morning of March 26, the outlines of a village appeared up ahead on the riverbank. The *Florida* pulled up and the men quickly disembarked to search for food. Miraculously, the villagers were well-stocked. "Chickens, ducks, hogs and a number of dogs were bought; also bunches of plantains and bananas," Sheppard wrote. Reinvigorated, the men continued their search for a suitable mission site.

On April 17, the *Florida* reached a junction where the Kasai River met one of its tributaries, the Lulua River. It was a breathtaking region, and Sheppard would soon call it home. "The whole country was filled with palm trees; the hills and valleys and everywhere beautiful palms," Sheppard remembered. "There were numerous fishing traps along the bank, canoes skimming over the water, paddled by excited natives, getting out of the way of the big steamer, and plenty of natives and small towns on the right bank." Sheppard and Lapsley unloaded their supplies near the Luebo Rapids, where the *Florida* could pass no farther. With trepidation, they bid farewell to the vessel's captain and crew. "At this point we are 1,200 miles from the coast and 800 miles from the nearest doctor or drug store," Sheppard wrote, "but we were comforted by these words, 'Lo, I am with you alway.'"

THE LONG JOURNEY from the United States to the interior of the Congo Free State was over. But Sheppard's role as a missionary was just beginning. The years ahead would transform his life. He would begin to speak the language of the local people, learn their cultural practices, and form relationships with them while working "as best we could in promoting the Lord's kingdom," he later wrote. Converting men, women, and children to Christianity remained Sheppard's primary goal, but he did not seek to do so by force. In fact, he harbored deep respect for the people among whom he lived—and their material culture. During Sheppard's time in the Congo, he collected over four hundred works of art, including knives, cups, fabrics, rugs, mats, and masks. He would later sell them to his alma mater, the Hampton Institute.

At first, all seemed well as Sheppard and Lapsley set up what would become known as the American Presbyterian Congo Mission, in a region where more than a thousand Africans lived. In the village of Luebo, the missionaries acquired houses made of palm leaves and bamboo. Eager to become acquainted with the local people, Sheppard remembered how they "made daily visits to the village, mingling with the people, learning their language and curious customs." He observed how "many of them cultivated the ground, raising manioc, peas, beans and tobacco, and others spent their time hunting and fishing" by day, then gathered as a community to dance at the center of the village when night fell.

Sheppard wrote positively of their reception as foreigners, recording that "the people had given Mr. Lapsley the name 'Nto-manjela,' meaning a path-finder, for he had found his way into their country, their homes, their language, and into their hearts." Lapsley's pale skin was a source of great interest to them. The Congolese insisted he remove his shoes so they could examine his bare white feet. "Mr. Lapsley was ticklish under the bottoms of his feet and this caused him to join in with the admirers in a hearty laugh," Sheppard explained. So great was their amusement that "this exhibition had to be repeated for the newcomers a number of times daily." Sheppard was also well liked by the Congolese. In a letter Lapsley sent home to Presbyterian Church leaders, he described Sheppard's popularity among the local people and affirmed that he possessed the "constitution needed, and the gift of getting on in Africa."

Before long, the two missionaries began formulating a written version of the Bantu language. "We went into the town and with pencil and book in hand pointed at objects," Sheppard recalled, writing down words like *mbuxi* for "goat" and *muntu* for "person," with the help of the Congolese people. From their stock of supplies, Sheppard and Lapsley offered medicines to villagers experiencing ailments ranging from indigestion to malaria. And they shared the gospel with villagers who had never read or heard the stories of the Bible. "We sat under a large palm tree and began in the most simple way possible to teach them about God and His great love for everybody in sending His son Jesus to save them," Sheppard recalled. "What a strange story! How they looked at each other, touched each other, and laughed."

Yet episodes of danger and violence punctuated their days. Not long after his arrival, Sheppard awoke in the dark of night to a terrifying sensation: thousands of insect legs were crawling all over his body. "They were in my head, my eyes, my nose, and pulling at my toes," Sheppard remembered of his encounter with a colony of vicious driver ants. "In an incredibly short space of time they can kill any goat, chicken, duck, hog or dog on the place," he wrote. But the villagers quickly came to Sheppard's aid, saving him from the ants.

The next day, Sheppard learned of a terrible custom among the local people. They explained that "when there are triplets born in a family it is considered very bad luck, so one of the babies is taken by the witch doctor and put into a deep hole where these ants live and the child is soon scented by them and eaten." The two missionaries also discovered that cannibalism was sometimes practiced nearby. They themselves met a group of native people who had killed and eaten the flesh of an enslaved African woman because she was unable to complete a forced march. Such encounters shocked Sheppard, and he became alert to acts of violence against vulnerable people in the Congo Free State.

As time passed, Sheppard and Lapsley grew closer. But the physical strain of life in the Congo began to take its toll on Lapsley. After returning from a solo expedition to meet the residents of a region tens of miles south of Luebo, Lapsley made his way back to the mission. "With the big ivory horn blowing and the drums beating," Sheppard recalled, "we ran down the banana walk to greet and welcome him home." But Lapsley looked utterly depleted. "He was tired, worn, and weary, and walked with a limp." Sheppard feared the worst. Over-powered by emotion, he retreated to a private spot amid the greenery. There, Sheppard "broke down in spirit and wept," thinking of Lapsley's sacrifices and of his family in Alabama who loved him.

There was little time for Lapsley to recuperate before he was called away to the coast by Belgium's governor general. "The beach was crowded with natives to wave him good-bye," Sheppard remembered. He was touched by how "the stranger who had come to their land on a strange errand was now known and loved." During what Sheppard thought would be a quick period of separation, Lapsley rapidly fell ill from the cumulative effects of exposure to malaria. Near Matadi, a town upriver from Boma, Lapsley developed a low-grade temperature. While under the care of local Baptist missionaries, he fell gravely ill with malarial hemoglobinuria, known as blackwater fever for the dark urine it produces. Within a week, Lapsley was dead, and Sheppard was on his own in Luebo.

When Sheppard learned the terrible news, he was overcome. "My knees smote together, I staggered from the deck, threw up my right hand to the hundreds of assembled natives and called out, 'Ntomanjela wa kafua'" [Mr. Lapsley is dead], he remembered. The villagers shared Sheppard's grief, collectively "weeping and wailing" to express their sadness. Meanwhile, Sheppard crept into the jungle to come to terms with the devastating news. "I had nursed Mr. Lapsley in all his fevers, and he in turn had nursed me; and now the Master had separated us more than a thousand miles apart and had called him, the better pre-pared of the two, to himself," he thought. In that moment, it seemed an insurmountable challenge to continue the mission work on his own after losing the partner who had almost become a brother.

In the days after Lapsley's passing, Sheppard pondered how he would complete their projects alone. One of their central goals had been to undertake an expedition to Mushenge, the capital city of the Kuba people and home to their notorious sequestered king. Could Sheppard complete such a trip? It would be dangerous; no westerner had yet made the journey, as the king was famously hostile toward outsiders and had promised to lop off the head of anyone who led strangers into his domain. But Sheppard decided to seek Mushenge.

With a group of native Africans, Sheppard trekked across miles of countryside and through neighboring towns. But soon word reached the king that a foreigner was approaching his city. When he sent his warriors to meet them, Sheppard knew that his life was in danger. Speaking in the language of the Kuba people, he sent the warriors back to the king with a message intended to establish trust, along with a gift: an enormous cowrie shell. Before long, the king sent back additional men to speak with Sheppard.

Because Sheppard spoke their language and had dark skin, the men believed that he was "a Makuba, one of the early settlers who died, and whose spirit went to a foreign country" and subsequently come back. The king wanted to meet him. After days of travel, Sheppard finally entered Mushenge, a city of five thousand people, with wide streets filled with curious onlookers. Wearing a stained white-linen outfit and a pith helmet, Sheppard went to greet the king, Kot aMweeky, who sat on an impressive throne amid his advisers and three hundred wives. "I judged him to have been a little more than six feet high and with his crown, which was made of eagle feathers, he towered over all," he remembered. The king stood up and said simply, "Wyni" [You have come]. Sheppard replied, "Ndini, Nyimi" [I have come, king].

During the next four months, Sheppard spent time with the king and explored Mushenge. Before returning to Luebo, he acquired what the scholar William Phipps described as "the only information about the Kuba people before the twentieth century," forming a "highly significant [record] because it dates to an era when the delicate web of culture had not yet been torn by foreign imperialism." He also brought back to Luebo "Bakuba curios, cloth, rugs, masks, mats, hats, [and] cups," evidence of a magnificent society.

Sheppard would later share with the world his ethnological descriptions of the Kuba people's beliefs and rituals, as well as his geographic findings in the Congo Free State. These discoveries quickly made him an international celebrity. Sheppard gained recognition from Britain's Royal Geographical Society, which elected him as its first African American fellow. In London, Queen Victoria presented him with a medal at Buckingham Palace. When Sheppard left Africa in 1893 to spend a year back home in the United States, he was invited to meet President Grover Cleveland at the White House. During his time in America, he married his fiancée, Lucy Gantt, and spoke widely to the American public about his experiences abroad. But before long, he and his courageous wife returned to the Congo Free State, where they continued mission work together.

LIFE FOR THE people of the Congo Free State changed rapidly during the 1890s, as Leopold II consolidated his power. His agents ruthlessly pursued the collection of rubber, which merchants could sell in European markets and make a return on their investment of 1,000 percent. Although Leopold II never went to the Congo Free State, Belgian officials acted in his stead. Members of the *Force Publique*, a military group comprising white officers and conscripted Africans, were tasked with carrying out the state's commands. They employed especially cruel methods, even in the context of a continent and a century that had seen an extraordinary amount of brutality. Leopold's agents used torture and other forms of violence to compel people to harvest rubber. Between five and ten million Congolese people died during King Leopold's brutal campaign, while millions more were physically mutilated.

Near the end of the nineteenth century, stories of this state-sanctioned violence began to leak out of the Congo Free State. One of the first people to report on Leopold's reign of terror was George

Washington Williams, an African American journalist who published an open letter in 1890 after a visit to the region. He accused
King Leopold's government of "deceit, fraud, robberies, arson, murder, slave-raiding, and general policy of cruelty . . . to the natives."
His report briefly gained traction, but most readers did not believe
his claims or considered them exaggerations. Remembered as the first
whistleblower in the Congo Free State, Williams tragically died one
year later, leaving his campaign unfinished.

By 1899, Leopold's reign of violence had reached the Kasai region,
where Sheppard was then working at a new mission site in the village of
Ibanche. Nearby, a group known as the Zappo Zaps, who were Songye
people, inflicted terror on any Congolese people who dared to resist
Belgian officials' orders. The Zappo Zaps had historically supported
the Arab slave trade but were now agents of the Congo Free State and
employed to carry out their cruel work. Sheppard described them as
cannibalistic, with teeth "all filed to a sharp point and their faces tat-
tooed." Although he had sought to avoid dangerous encounters with
them, in 1899 Sheppard received orders from the concerned leaders
of the American Presbyterian Congo Mission (APCM) to investigate
the shocking reports of state-sanctioned violence near Ibanche. It was
a terrifying assignment, one that he knew might cost him his life.

With a group of courageous Congolese men and women, Sheppard
embarked upon the investigation on September 14, 1899. During days
of marching toward the camp of the Zappo Zaps, they saw smolder-
ing villages and mutilated people. Suddenly, "at a curve in the forest,"
Sheppard recorded in his report for the APCM, "we met face to face
with sixteen Zappo Zaps, who, with lightning speed, cocked their
guns and took aim." Recognizing Chembamba, one of the members of
the group, Sheppard stepped forward with characteristic bravery and
shouted, "Don't shoot, I am Sheppard!" In a stroke of luck, Chembamba remembered him. Sheppard had given him first aid two years
earlier at the mission. The Zappo Zaps lowered their guns and agreed
to bring Sheppard to their camp. Its periphery reeked of the rotting
corpses.

Quickly winning the trust of their chief, M'lumba N'kusa, he began questioning him. M'lumba N'kusa immediately acknowledged the Zappo Zaps' connection to the Belgian state, whose agents provided them with weapons and ammunition. Most recently, M'lumba N'kusa's men had murdered members of the region's Pyaang people, putting eighty or ninety of them inside a fenced area and slaughtering them en masse.

The chief proceeded to show Sheppard around the campsite, where gruesome evidence of cannibalism abounded. Forty-one bodies lay on the ground, many of them sliced into pieces. The Zappo Zaps had already consumed the other corpses. A skull rested in the dirt, evidently repurposed as a bowl for tobacco. Many bodies were missing hands. When Sheppard inquired why, N'kusa told him that "they always cut off the right hand to give to the State on their return." These body parts, the scholar James Campbell argues, "offered authorities proof that the mercenaries in their employ were using [Belgian] ammunition wisely, not simply squandering it on hunting or horseplay." Sheppard also discovered more than sixty women still alive, who were being held as prisoners. The evidence was clear: King Leopold was perpetrating genocide. But what could Sheppard do to stop it?

Sheppard went to work on a report, which the APCM forwarded to the Southern Presbyterian Board of Missions in the United States. To affirm the veracity of his disturbing findings, the APCM also mailed corroborating testimony from additional Presbyterian Church missionaries working in Luebo. In January 1900, American newspapers began alerting their readers to the atrocities. "Fourteen Villages Destroyed and Some of the Prisoners Roasted and Devoured— Missionaries' Stories," read one headline in the *Indianapolis Journal*. "A Slaughter by Cannibals: Congo Free State Natives Murdered for Not Paying Taxes," blared the *Evening Times* of Washington, DC. The newspapers credited Sheppard for discovering the atrocities and placed blame squarely on the Belgian state.

European newspapers also picked up the story. The following month, the *Times* of London described how "Mr. Sheppard, a member of the Royal Geographic Society and of the Southern Presbyterian Mission, was sent out to investigate the affair." Detailing his astonishing findings, the journalist wrote that Sheppard had seen "14 villages burned and plundered and discovered that these atrocities had been committed almost under the eye of State officials at Luluaburg, who did their best to shield their agents." Belgium's *Le Peuple*, a socialist newspaper, printed a translation of the *Times*'s article days later. The politician Léon Meysmans called the accusations "excessively serious" in *Le Peuple*, criticizing the Belgian state for engaging in colonial enterprises and competing with other European countries such as England, which Meysmans condemned as having "scattered capitalist development with pools of blood."

While European audiences grappled with the truth about the brutal consequences of colonization in Africa, American readers received news of the massacre in the Congo Free State against the backdrop of the Jim Crow era. In the case of the *Indianapolis Journal*, a newspaper with a predominantly white readership, Sheppard's story ran alongside a column describing a brutal lynching in Virginia, the missionary's home state. "Negro Shot to Death," the headline read, "Taken from Jail at Newport News, Fastened to a Tree, and at Least Forty Bullets Fired into His Body—Accused of Assaulting a White Woman."

EXPLORER, MISSIONARY, HIPPOPOTAMUS hunter—Sheppard's life in the Congo Free State had been filled with adventure. Now he had also become a writer and activist who used his pen and his voice to raise awareness of King Leopold's genocide. As Leopold's agents continued to terrorize the people of the Kasai, APCM missionaries broadened their campaign to galvanize the international public. Sheppard courageously participated, speaking to multiple audiences in the United States in 1904 while on leave and visiting Theodore Roosevelt's White House in January 1905 to share his knowledge of the Congolese people and their trials.

In 1906, Sheppard went back to Luebo, where he was devastated to see that conditions had further deteriorated. He wrote a condemnatory essay, published in the *Kasai Herald* in January 1908, arguing that the Compagnie du Kasai, a rubber-harvesting company in which the Belgian state owned a majority share, abused Congolese laborers. Infuriated, the Compagnie du Kasai publicly rejected Sheppard's claims and tried to intimidate their accuser. The company requested $20,000 in payment for defamation and took Sheppard and William Morrison, another courageous Presbyterian missionary who had been working to expose Leopold's atrocities, to trial in September 1909. It was a dramatic occasion. Sheppard brought many Congolese witnesses to speak and he, too, stood up to the powerful company. In the end, he was acquitted. Public outrage against King Leopold finally reached a boiling point, leading to the Belgian parliament's seizure of control over the Congo Free State in 1908. But little changed in the years ahead for the Congolese people, who lacked political power and whose forced labor on farms and in mines enriched Belgium.

The year 1909 was ultimately a time of great transition for Sheppard. His affection for the Congolese people and his love of Africa had transformed him from a young man with no knowledge of the great continent to a confident missionary fluent in the Kuba language. A hardened adventurer, a practiced hunter, and now a successful international human rights advocate, Sheppard represented a new kind of American explorer at the turn of the twentieth century: one who sought not to conquer Indigenous people but rather to live among them, learn from them, and protect them from colonization's devastating effects.

But the moment had come for Sheppard to return to the United States for good. He asked the APCM for permission to resign due to poor physical health and a desire to live closer to his daughter, who was residing in Virginia. News was also circulating within the APCM about his infidelity; he had fathered a son with a local Congolese woman. Nonetheless, in response to Sheppard's letter of resignation, members of the APCM wrote of their "sincerest and deepest regret." They felt that their Congo mission would be "much crippled; for

[Sheppard] had been an apostle among [the Congolese] for so long." In 1910, he and his wife, Lucy, departed Africa, never to return.

In the years ahead, Sheppard continued to live out his Christian faith in the United States as a minister at Grace Presbyterian Church in Louisville, Kentucky. But his memories of Africa would remain with him for the rest of his life. He would never forget striving to minister to the Congolese people and learning their languages, customs, and culture. It had been a time of tragedy and joy, hardship and overcoming, a baptism by fire for an inexperienced, idealistic young missionary and explorer who ultimately laid bare Leopold II's outrages.

Like Sheppard, other American explorers of the early twentieth century began expressing concern about the effects of colonization, imperialism, and globalization on Indigenous peoples. Some adventurers would travel to distant regions to document their traditions and ways of life before they were lost to a modernizing world. To reach such far-flung places, they would endure extreme climates, navigate uncharted terrain, and build trust across cultural divides.

7

Harriet Chalmers Adams

THE GEOGRAPHER

A revolt in Panama at the turn of the twentieth century caught the world by surprise. On November 4, 1903, American newspapers broke the news of a rebellion in the former Spanish colony, then under the control of Colombia. "The independence of the Isthmus was proclaimed at 6 p.m., yesterday in Panama," blared the *Alexandria Gazette*. The US government was decidedly on the side of the revolutionary Panamanians. Flexing its growing muscle, the industrializing nation had recently shed its isolationist traditions during the Spanish-American War of 1898. Swift victory over Spain in the weeks-long conflict resulted in the US annexation of the Spanish colonies of Guam, Puerto Rico, and the Philippines. Now the United States, led by President Theodore Roosevelt, saw itself as a global power with the potential to expand its profits from international trade.

While European nations enriched themselves through the colonization of parts of Africa and Asia in this era of imperialism, Roosevelt thought that the United States could gain a geopolitical and trading advantage by building and controlling a canal in Panama, where a French company had begun but failed to finish work on such a project twenty years earlier. During the California Gold Rush, no waterway crossed the fifty-mile-wide isthmus, so people had to travel by foot or on mule. The land route required a dangerous, weeks-long trek through the jungle, which exposed travelers to what one migrant described as Panama's "oppressive" heat as well as the threat of death from malaria or cholera. The construction of the Panama Railroad in 1855 significantly reduced the time it took to traverse the isthmus, but shipping corporations longed for a more efficient way to transport goods.

The US government supported the rebellion in Panama by ordering "American naval commanders in isthmian waters to prevent the transportation of Colombian troops across the isthmus," reported the *Alexandria Gazette*. At least one newspaper claimed that the revolt was prompted by the Colombian Senate's recent rejection of the Hay-Herrán Canal Treaty, a document that would have leased a canal zone in Panama to the United States. The following day, Panamanian revolutionaries sent cables to the secretary of state, John Hay, reported Washington's *Evening Star*, "making formal announcement of the establishment of a new government at Panama and giving assurances of its ability to maintain order." Seeking to justify US intervention, the *Evening Star* concluded that the United States was acting "in the interest of the general good . . . to maintain open transit across the isthmus" and to "prevent bloodshed at any cost."

By the time a young woman named Harriet Chalmers Adams made land in Panama City in February 1904, the US-backed revolution was complete and the United States had already begun plans for the construction of a canal. On February 26, the United States and Panama announced the Hay–Bunau-Varilla Treaty, a document that gave the United States "in perpetuity the use, occupation and control of a zone of land and land under water for the construction, maintenance, operation, sanitation and protection of said canal." In return the United

States promised to preserve the republic's independence and pay Panama $10 million with future annuities of $250,000. It was a pivotal moment in American history, one in which the United States justified military intervention in order to secure trade benefits.

At this historic juncture Adams arrived in Panama from San Francisco, having traveled across more than three thousand miles of Pacific Ocean, on a path well-traveled a half-century before, during the Gold Rush. Thousands of gold-seekers had made a similarly arduous journey from New York City to San Francisco by way of Panama.

In the aftermath of the revolution, Adams found herself in a changing region. But her activities diverged sharply from those of the US government. Driven by an interest in geography and human culture, she was on a mission to observe and record her impressions of the Indigenous people and the natural landscapes of Central and South America. "There is a great work to be done in comparative studies of words and peoples," she later said. "Travel and exploration open up for me an immense field of inquiry along paths ethnological and etymological." It was Adams's chief hope that, by exploring new regions of the world and sharing her findings with others, she would expand geographic knowledge.

NEW FRONTIERS WERE always on the horizon for the Chalmers family. In 1864, Harriet's father, a Scottish immigrant named Alexander Chalmers, boarded a ship in New York City bound for San Francisco by way of Panama. Sixteen years had passed since James Marshall discovered gold at Sutter's Mill, but Alexander was determined to try his luck out west. He made his way to Coloma, where he befriended Marshall and searched for gold in the American River, but ultimately decided to establish himself in Stockton as a merchant, a career that provided a steadier income, as earlier settlers like James Beckwourth had found. Alexander married Frances Wilkins, the daughter of settlers who traveled across the continent from New Hampshire to California during the Gold Rush. She gave birth to their daughter,

Harriet, on October 22, 1875, when Reconstruction's collapse was nearly complete and tensions between settlers clamoring for land, forceful federal soldiers, and frustrated Native Americans simmered in the Great Plains.

Frances and Alexander chose to remain in Stockton, California, to rear Harriet and her sister, Anna. Originally established as a supply town during the Gold Rush, Stockton served prospectors working on the rim of the San Joaquin Valley. During the second half of the nineteenth century, the town flourished due to its advantageous location as a port of trade located on a channel that fed the San Joaquin River, which connected towns across the region. Recognizing the potential to profit by selling goods to local residents and immigrants, Chalmers opened a shop near the town's Main Street, then a dusty dirt road. The Chalmers Brothers Dry Goods and Carpets store was located on the ground floor of a two-story Italianate building, with tall glass windows giving passersby a glimpse of the variety of wares within. The family home, a modest house surrounded by a white picket fence, was situated five streets away in a quiet neighborhood lined with trees.

Although Frances and Alexander had decided to settle down and join Stockton's growing upper class, life on the frontier was anything but boring for young Harriet. With its rugged terrain, California provided her with a dynamic environment to explore. Much of her childhood was spent outside, where she honed her athletic abilities with the support of both parents. When she was only eight years old, Harriet and her father, who must have held liberal views for the day about rearing girls, made a weeks-long horseback excursion together throughout the state. It was a transformative experience that "made me over," she later said, "from a domestic little girl fond of knitting and skipping rope to one who wished to go to the ends of the earth and to see and study the people of all lands."

Her aquatic skills were as impressive as her equestrian prowess. At the age of ten, Harriet performed a public feat that revealed her strength and endurance, swimming five hundred yards in the open sea to the astonishment of onlookers. Two weeks later, Harriet impressed a journalist from the *Santa Cruz Surf* with "daring diving from the rafts and the spring-board" that, he reported, "excel[ed] all the young misses." Harriet's abilities in the water distinguished her from other children her age, particularly in Stockton. There, one journalist reported in 1890, "drownings [were] terribly frequent," especially during the summertime among boys who engaged in risky behavior.

The daughter of a successful entrepreneur, Harriet grew up among Stockton's emerging elite at the onset of the Gilded Age. She attended a local school until she turned eleven and studied with private instructors thereafter. As she reached womanhood, her name began to appear more frequently in the society pages of local newspapers. With chocolate-colored hair, gleaming brown eyes, and an even complexion, Harriet's charm was undeniable. Highly sociable and well-connected, she enjoyed games of whist and parties at private homes as a teenager. An accomplished musician, she also played the piano in concerts and at social events. Young Harriet adeptly straddled two worlds: the rugged natural environment and the upper-class domestic interiors where much was expected of her by way of manners and social graces.

Despite Stockton's attractions, Harriet's eyes were always turned toward the outside world. From her parents, she had gained what the explorer John Oliver La Gorce later described as "the urge to seek and find the mysterious something which lies beyond the seven seas, the snow-capped mountains, and the arid deserts of the world." Men like Henry the Navigator, the fifteenth-century Portuguese prince who studied geography to launch exploratory trips abroad, loomed large in her imagination. California's diversity and history of colonization by Spain between the sixteenth and nineteenth centuries also interested her greatly. "We lived in California," she once said, "where I acquired my love of all things Spanish." Hispanic settlers, Indigenous peoples, and immigrants from all over the world populated her home state in the late nineteenth century. Harriet was worldly, especially for her era, studying both Spanish and French and later vowing that she would travel to every country colonized by the Spanish.

At the tender age of thirteen, when other girls began turning their attention to matters of the heart, Harriet prepared for her greatest adventure yet. With her father, Harriet, barely five feet tall, finalized plans for an excursion to the mountains that stretched from Oregon to Mexico. The pair set out on horseback, with one pack animal to support them. It was a time when, Harriet later remembered, "the high places, now well known, were virtually unexplored." She and her father traversed the Cascade Range of northern California that spring up before crossing the Sierra Nevada, a region under threat from loggers, miners, and shepherds, in summer and autumn.

These mountains were particularly magnificent. The Scottish immigrant John Muir, then seeking to generate support for California's preservationist movement, described them in rapturous terms. Muir called the Sierra Nevada a "Range of Light," with "white beams of the morning streaming through the passes, the noonday radiance on the crystal rocks, the flush of the alpenglow, and the irised spray of countless waterfalls." Harriet's time in California's mountains instilled in her a passion for frontier living, or what one interviewer later described as a propensity for "the big, broad, unconventional West, the tang of which is in her blood, making her love the saddle, the outdoor life and the excitement of finding new, hard experiences in a strange land better than the stay-at-home comfort which most women find necessary to their happiness." Harriet, who had acquired her father's adventurous spirit, was hooked. It was from Alexander Chalmers, La Gorce would later observe, that she "learn[ed] the secrets of Nature at first hand."

THE HAPPY YEARS of Harriet's childhood flew by and before long, the time had come for her to consider marriage. A young woman with a sparkling intellect, she also possessed what one reporter described as "a sweet, musical voice, a dash of originality and an appreciation of the things that lie beneath the surface," qualities that were further enhanced by her natural beauty. Although Harriet was comfortable among Stockton's elite, able to slip out of dust-covered riding boots into delicate shoes for an evening party, there was still something unconventional about her. She would not be content to live the life of a society wife in her hometown. She needed a partner whose ambitions matched her own.

When Harriet was twenty-four, a local man seven years her senior caught her eye. His name was Franklin Pierce Adams and, like many of their generation in Stockton, he was the child of California settlers. They were connected through Harriet's cousin, Etta: Frank was the brother of Etta's husband. With a broad, straight nose and smiling eyes that expressed his amiable personality, Frank would be a sturdy companion to Harriet. He worked for the Stockton Gas and Electric Company, which was housed in an impressive neoclassical brick building near the town's center. He also possessed a keen wit and observant eye, writing about social events for Stockton's *Evening Mail* in his late twenties and early thirties. One reporter even joked that Frank was the "Ward McAllister" of Stockton, a man whose wedding to Harriet Chalmers was "the marriage of the season among [Stockton's] four hundred," a clever reference to the individuals governing Gilded Age society in New York City. Their engagement was announced just three weeks before the wedding, which was held at Harriet's childhood home on October 5, 1899.

It was a joyful occasion, one lovingly prepared for by her parents, who bedecked their house in greenery and white flowers. After the ceremony, the newlyweds celebrated their marriage with a festive meal and dancing among their eighty guests. Harriet wore an elegant white satin-and-lace dress with a delicate veil, a sharp contrast to the rough-and-tumble clothes of her youth. But even on the eve of her wedding, new frontiers were never far from her mind. Rather than establish a home of their own and start a family as newlyweds, she and Frank planned to rent a room in a local house after returning from a brief vacation because, in the words of Harriet's biographer Durlynn Anema, they were "thinking first of travel."

The chance for a grand adventure was not long in presenting itself.

Frank's familiarity with California's mining industry positioned him to capitalize on US companies' growing interest in Central and South America. The end of the Spanish-American War, just months before their wedding, marked a crucial shift in US international relations. The United States' quick defeat of Spain and acquisition of new territories inspired the federal government to turn its attention farther south. It was a moment when, the historian Robert Hannigan argues, "Washington's aspirations were to dominate the development of [South America] commercially, to shape the political future of the region, and, ideally, to organize South America as a bloc behind the US in world affairs." The United States did not want to directly establish colonies, as European nations had done centuries earlier; rather, it sought influence and a regional trading advantage.

During this time of rising interest in Latin America, Harriet and Frank made their first trip to Mexico. In 1900, Frank completed a survey of regional mines while Harriet studied local historic sites connected to Mexico's ancient Mayan kingdoms. The trip whetted Harriet's appetite for travel abroad, and she encouraged Frank to search for additional opportunities that might send them to far-flung regions. US companies would continue to seek access to South American mines in the years ahead, a search largely driven by the demands of industrialization in the United States. Frank's skills as a surveyor and his growing cultural knowledge made him an attractive prospective employee. It was only a matter of time before Harriet and Frank returned south for further exploration.

In 1904, Frank embarked upon a new business venture that would change Harriet's life. The Inca Mining and Rubber Company, based in Pittsburgh, wanted to send Frank to inspect its holdings in South America. Competition for trade advantage was fierce, particularly in Bolivia and Peru, where mineral deposits lay deep within the Andes. "Not only is American capital being invested in Peru," the *New York Tribune* reported in 1906, "but American engineers, prospectors, and specialists have invaded the country, and are busy developing new schemes of industry." Approximately $25 million in American investments had been made primarily in mining and rubber companies, the *Tribune* estimated, an amount that positioned the United States to effectively compete against German and English investors. US trade with Peru had grown by 166 percent since 1899, a staggering increase that even surpassed the growth rates of England and Germany. The *Tribune* emphasized that businesses, including the Inca Mining and Rubber Company, were eagerly searching for new contractors, men like Frank who spoke the language and knew mountainous terrain.

Harriet was thrilled. She imagined an adventure far beyond the scope of Frank's work for his new employer. It would involve exploring tens of thousands of miles of uncharted territory and meeting members of the Indigenous communities whose lives had been shaped by Spanish rule. It would be an expedition unlike any other, one that would test her stamina, language training, navigational skills, and ability to survive the elements.

ON JANUARY 9, 1904, Harriet and Frank Adams set sail from San Francisco aboard the steamship *Peru* on what would be a two-year journey. With twenty other passengers, Harriet bid farewell to the crowd of people who had gathered at Pier 38. Holding bouquets of carnations and violets presented before her departure, Harriet felt deeply the pain of her impending separation. "I wish all the ones I love were with us," she reflected later. But adventure beckoned and Harriet knew she was ready to go. It was a "perfect day," with clear skies, cool weather, and a view of the sparkling Pacific Ocean from the ship's deck.

Although Frank was "a fine sailor," seasickness incapacitated Harriet by mid-afternoon on their first day. "Very rolling sea," she jotted in her journal. "Capt. says worst pitch in a year." For much of the second day, she lay in her bedroom below deck with a hot-water bottle pressed against her abdomen to ward off nausea. When her appetite returned, she managed to consume several pieces of plain toast accompanied by tea with lime juice, a cure that she called "a help."

Although Harriet was hardy by nature, the first few days of her trip were off to a rocky start.

Over the next few weeks, the *Peru* traveled along the coast of California and then Mexico. The weather grew warmer, and Harriet was thrilled by the sight of flying fish, dolphins, water snakes, and whales off the shores of Acapulco. The landscape changed dramatically as they headed farther south, stopping in Guatemala and El Salvador. Harriet was quite taken with the cuisine in Central America, tasting everything she could and writing lists of dishes to sample in the future. Succulent fruits like mangoes and papaya and delicacies like chocolate, turtle eggs, and local cheeses were among her favorite foods.

From El Salvador, Harriet and Frank traveled to Panama City, which was in a swirl of change in the aftermath of a rebellion that had led to Panama's independence. Harriet's first view of the country's coast differed significantly from that of President Theodore Roosevelt, who would travel to Panama from Washington, DC, in 1906. Approaching from the Atlantic Ocean, he described how the "jungle-covered mountains looked clearer and clearer until we could see the surf beating on the shores, while there was hardly a sign of human habitation." On land, Roosevelt marveled at the tropical forests, with "palms and bananas, bread-fruit trees, bamboos, lofty ceibas, and gorgeous butterflies and brilliant colored birds fluttering among the orchids." Jaguars roamed the jungles, and crocodiles lurked beneath the turgid waters of the Chagres River. Roosevelt, an avid sportsman and naturalist, longed for the opportunity to go hunting and "collect specimens."

When Harriet reached what she called "Columbia [*sic*] or Panama Republic" on February 18, 1904, American battleships were visible near the shore. Although she enjoyed the view aboard the ship, she noted that it was "so different from what we imagined," since Panama was "not very tropical looking." American construction of the Panama Canal would not begin for two months. By the time Roosevelt toured the Canal Zone two years later, significant progress had been made.

In the Culebra Cut, just a mud-filled passage when Harriet visited it in 1904, "great work is being done," Roosevelt reported to his son Kermit in November 1906. "There the huge steam-shovels are hard at it; scooping huge masses of rock and gravel and dirt previously loosened by the drillers and dynamite blasters."

Racial inequality defined working conditions in the Canal Zone; white Americans employed there had brought Jim Crow discrimination with them. "With intense energy men and machines do their task, the white men supervising matters and handling the machines, while the tens of thousands of black men do the rough labor where it is not worth while to have machines do it," Roosevelt observed. Their collective work would permanently transform both the region and global trade. "It is an epic feat," Roosevelt concluded, "and one of immense significance."

Harriet was not able to witness the canal's completion on this journey, but she did arrive in time to see the inauguration of Panama's first president, Dr. Manuel Amador Guerrero, on February 20. At 4 p.m., she photographed the crowds gathered in Cathedral Square, a grand plaza dominated by a baroque cathedral constructed in 1796. During the inauguration, the head of ceremonies emphasized the legitimacy of Guerrero's election and Panama's recent revolution, the *New York Sun* reported, highlighting "the humiliations Panama had suffered from Colombia." President Guerrero outlined his plans for the new republic and declared his support for the canal, which, he argued, "would create an economic and industrial revolution" that would benefit Panamanians. And the United States, as well; during the inauguration, men from the US armed forces kept a watchful eye on the ceremonies.

On February 23, Harriet and Frank departed from Panama on the ship *Limari*, a vessel carrying American miners, South Americans, English military officers, and at least one shipbuilder. One week later, Harriet glimpsed the green mountains of Peru stretching high above the blue waters of the Pacific Ocean. Her first days in the city of Lima, which she reached by train, were defined by stretches of illness and punctuated with danger, including a violent earthquake that made her and Frank feel "way back in [their] hearts that it was the end." She was pleased when the time came to board the ship *Peru* three weeks later, leaving behind Lima's noisy streets and a climate that she found uncomfortable.

As autumn descended on South America in April 1904, Harriet and Frank reached Bolivia. They made their way inland to the high country by train, rapidly ascending to an altitude of fourteen thousand feet without allowing for adequate time to adjust. Sleep evaded Harriet; she felt "smothered," as if there were "a weight on [her] chest." Despite experiencing significant altitude sickness, Harriet felt vivified by the countryside and marveled at her new surroundings. "Wonderful lakes . . . llamas by hundreds, alpacas, [and] Indian huts," she recorded in her travel journal. After reaching La Paz, Bolivia, a "red-roofed city" nestled in the depths of a canyon high in the Altiplano, Harriet felt quite at home.

Thirteen thousand feet above sea level, La Paz was a "city of the clouds," Harriet later wrote, "protected from the icy blasts which sweep across the bleak plateau above." First inhabited by Bolivia's Indigenous Aymara people, La Paz had been colonized in the mid-sixteenth century by Spanish conquistadors in search of riches. Harriet admired La Paz's historicity but expressed her concern about the potentially damaging consequences of development, observing that La Paz had "not as yet been greatly marred by that buccaneer and despoiler of natural beauty: modern civilization." During her stay, she saw a growing number of foreigners, predominantly "American engineers, German merchants, and British capitalists," citizens of the three nations with some of the largest investments in South America at the turn of the twentieth century.

Keeping in mind the historically disruptive role of outsiders entering La Paz, Adams took on the role of ethnographer. She wrote down her observations of the Aymara people, whom she later described as "the earliest American aborigines known to us, the builders of Tiahuanaco," an empire "now in sand-swept ruins not far from Lake Titicaca."

Their appearance and way of life were of great interest to her, and she spared no details in recording their attire and occupations. They went barefoot and wore short pants, which enabled them to climb the local hills more easily. Rainbow-colored "woolen ponchos cover[ed] their broad shoulders," and their skin was "russet brown in shade, their hair straight and black." Of the Indigenous people she met not only in La Paz, but also during her lifelong travels, she later told a reporter, "In many of the ancient countries the so-called peasants have acquired so much wisdom and culture that has come down to them outside of books. . . . There's a world to learn of birds and plants and animals from primitive people."

Unlike most explorers of the nineteenth century, Harriet paid attention to how colonialism had harmed Indigenous peoples like the Aymara. Under Spanish rule, they had suffered terribly via a system called *repartimiento*, Harriet reported, which was a "source of oppression and fraud" relating to the Aymaras' obligation to "pay exorbitant prices for [Spanish] articles utterly useless to them." Worse was *mita*, which involved "forced labor in mines and plantations, where the poor Indians died by the thousands from over-exertion and ill-treatment." She lamented this history of oppression and was careful to take respectful photographs that captured twentieth-century Aymaras' community traditions and unique way of life. When it came time to leave, Harriet regretted departing from La Paz, which was so different from "our progressive America, where we rush and strive from morning till night, where all of the cities are alike, and every man, woman, and child dresses like every other."

Days spent in the saddle in the Sierra Nevada had prepared Harriet for an excursion from La Paz into the surrounding regions. On one memorable day, she rode a mule beyond the city's outermost limits, where the path "clung to the cliff and the canyon developed into a miniature Yellowstone in coloring." As their group proceeded, the "mountain walls grew more rugged, more picturesque in their rainbow attire." That night, she wrote in her journal of the dazzling landscape, calling the adventure "the most wonderful ride I've ever taken."

After a week in La Paz, Harriet and Frank departed for Tirapata, a rural town in Peru from which they would launch an expedition into the steep, rocky Andes in search of the Inca Mining and Rubber Company's gold mines. In Tirapata, they spent several days packing essential supplies and preparing for their departure. Harriet was "restless and anxious to start," anticipating that she would be "very tired and cold but that [the] scenery [would] be fine."

At 9:30 a.m. on April 19, 1904, after loading their mules with supplies, Harriet mounted her gaunt horse and nudged him into action. Over the next four days, she, Frank, and the other members of their group forded icy rivers and endured cold temperatures that decreased as they climbed the steep mountainsides, the snowy peaks sparkling in the sunlight. Covering nearly forty miles per day, Harriet spotted Incan ruins and abandoned towns. A violent snowstorm overtook the group on the third day of the journey, as they began their downward march from an altitude of about seventeen thousand feet. The following day provided a brief respite from inclement weather, but they were again pelted with precipitation the next evening. Despite the frigid conditions, Harriet recounted the event with typical stoicism. "Out on ledge in storm all night," she remarked in her diary. "The adventure I have wanted I guess and it's kind of exciting. . . . A long long night so wet and cold. No sleep or food." When morning came, she was relieved when the exhausted group set off again and finally reached the first mine in Santo Domingo, a moment that felt "like Heaven."

It took a week for Harriet to regain her strength at the mining site after the taxing journey. By April 30, she felt well enough to continue with Frank deeper into the interior of the continent. With horses and pack mules, their group set off into the mountains. The terrain changed markedly as they descended from the heights of the previous mining camp into a lush forest. Around them grew "royal palms—ferns and such climbers and creepers," while the forest's floor was crisscrossed by countless streams. Birds of all kinds twittered as they darted from trees dimly illuminated by the sunlight.

When night fell, however, the dangers of the forest became apparent. Vampire bats thrived in the humid environment. With large ears, porcine snouts, and teeth like razors, they roamed the forest by starlight in search of their next meal. As creatures that sustained themselves by drinking the blood of mammals, vampire bats were drawn to the human travelers as they slept. These bats were frequently infected with rabies. While rain fell on the campsite in the dark of night, a vampire bat swooped down and sank its teeth into Frank's ear. With no accessible treatment for rabies, all he and Harriet could do was wait to find out if he had been infected. Harriet remained stoic in the face of such uncertainty.

Other creatures enlivened subsequent days of travel. In the thick mist, Harriet spotted wild turkeys running across the ground and enormous monkeys climbing high into the forest canopy. At night, vermin and insects plagued the weary travelers, she remembered, "to say nothing to wild animals sneaking about the hut," making sleep elusive. Harriet even tried her hand at hunting, shooting "a brown monkey about 30 lbs. between La Union and Pampa, bringing him in with us."

Signs soon emerged that encroaching development endangered the wildness of the Andes. Harriet and Frank's group inspected a rubber-making site in May. The laborious process left behind "gashed trees" and "smoked rubber." One observer from the period compared it to making maple syrup. "The trees are tapped when the sap begins to run, and the milk, as they call it, is boiled in a big kettle until it is reduced to its proper consistency." At the time of her visit, Harriet seemed unaware of how the extraction of products like rubber and precious metals threatened to destroy the homeland of the Indigenous communities she admired.

Having completed their initial foray into the forests of the Andes, Harriet and Frank returned to the mines in Santo Domingo on

May 7. There, Harriet recuperated while Frank visited additional excavation sites nearby. Eleven days later, Harriet and Frank departed for the mining region of Potosí, Bolivia, a journey that would take them across barren landscapes at exceptionally high altitudes. Riding horses, and supported by mules carrying their supplies, they began their ascent into the mountains the following day. Harriet was immediately stricken with altitude sickness, and frigid weather compounded her illness. On the third night of the journey, after riding for twelve hours without stopping, she suddenly felt herself sliding out of her saddle. Like a rag doll, she flopped to the ground, overcome with exhaustion. Unable to move, she "lay down on [the] frosty ground by Indians with llamas," animals that may have helped provide warmth in the deadly cold. "Too ill to go on and no huts in sight or sounds," she lamented.

With no choice but to shelter in place, the group hunkered down for the long night at an altitude of fifteen thousand feet. The sky overhead was dotted with crystal stars. Around her Harriet could hear the breath of the llamas and the murmuring of the Indigenous caretakers. Refusing to sleep from fear that she would freeze to death, she lay awake for the entire night, shivering violently under layers of soft rabbit-skin blankets. "Tears came to my eyes, my lids froze together," she later remembered. When Harriet emerged from her makeshift bed the next morning, her "hair was covered in frost and the blanket over [her] head as well as all the others, soaked." The dampness compounded an already dangerous situation, as the human body loses heat faster when wet. Harriet gathered her last ounce of strength and rose stiffly from the hard ground to seek shelter.

After boiling hot water for tea as the sun came up, the group set off at about 8 a.m. Seven hours later, they finally caught sight of a small house above the town of Potosí. They were saved! Suffering from hypothermia, Harriet rejoiced at the sight of a small stove in her new lodgings. While the others hungrily sat down to dinner, Harriet was too sick to swallow even a morsel of food and promptly went to sleep.

It was a miracle she had survived the trek; she marveled at her good fortune and remained impressed by the stamina of the people who called this formidable landscape home.

DURING THE MONTHS ahead, Harriet and Frank explored additional mines throughout the Andes. Photographs of the mining operations in Santo Domingo from the period depict rickety wooden structures nestled into mountainsides, where young Indigenous men labored in cavernous tunnels. Piles of rocks and industrial debris littered barren landscapes, a sign of the environmental destruction wrought by foreign mining companies in their pursuit of profit. Two acquaintances of Harriet, Max T. Vargas and Martin Chambi, captured images of bricks of gold, stacked alongside black pistols and stamped with the name of the Inca Mining Company. These brazen displays of weapons and wealth likely called to mind her home state, California, and the lawlessness of the Gold Rush era.

Peru's ancient historical sites also thrilled Harriet. She reached Cuzco, the capital of the Incan Empire, in June 1904. Situated in the heights of the Andes, the city heaved with life, and architectural wonders abounded: Incan stone walls and the rich interiors of cathedrals equally fascinated her. "See many marks of the red Inca in foundations, doorways, etc.," she wrote. "In churches [there were] many gold and silver Inca relics." Lying in the hills above the city, the impressive fifteenth-century Incan ruins of Sacsayhuaman also shed light on the region's history. "Saw ancient aqueducts, seats of the Incas, sliding stones, and a number of new features." The walls of the fortress, built from massive toothlike blocks of stone, towered above her diminutive frame.

Whenever they traveled through particularly remote regions of South America, Harriet was conscious of the fact that she was likely the first non-Native woman to set foot there. On a pampa, or plain, outside Potosí, Harriet once recorded that she was a "great curiosity" to the local Indigenous people, who were perplexed by her skin color and appearance. Unlike some explorers of the prior century, however, Harriet remained keenly aware of the generations of Indigenous women who had called these lands home. "These women before me since prehistoric days have traveled the almost impregnable ways on foot, unprotected and unequipped for hardships," she would later tell a reporter. "My questioning thought as I meet one difficulty after another is always, 'How did those women meet the dangers? What were their thoughts?'" Harriet may have been the first white woman to traverse the furthest reaches of the Andes, but she fully recognized that she was not the first woman to do so and acknowledged with admiration the courageous people who had blazed trails in centuries past.

One of Harriet's greatest exploratory accomplishments in the remainder of her time in the Andes was her ascent of El Misti, a massive dormant volcano in southern Peru. At 19,200 feet, El Misti's snow-capped peak pierced the clouds above the town of Arequipa. The Inca people believed it was a sacred place and made *capococha*, or human sacrifices, at its base, following their conquest of the region in the 1500s. Reaching Arequipa on July 6, 1904, Harriet began preparing for the journey to the top of El Misti. If she felt nervous about the altitude, which was less than two thousand feet shy of the highest peak any person had scaled at that point, she did not reveal it in her travel log.

At 5 a.m. on July 10, Harriet awakened, drank coffee to fortify herself, and mounted her mule. Traveling in a group of seven people that included Frank, Harriet steeled herself against the cold that enveloped them as they ascended the mountain. El Misti's barren landscape offered little shelter from the elements, and the hut where the group took cover was pounded with wind throughout the first night. Flashes of lightning illuminated the bleak landscape, sending rats scurrying to the shadowy corners of their primitive lodgings.

But when the sun rose the next day, Harriet was dazzled by the view of "a mountain of gold." The group mounted their mules and proceeded four thousand feet on a "trail in lava," she wrote, which led them "up to [the] summit." It was a terribly taxing ascent, causing her heart to ache with pain. At last, the travelers reached El Misti's great peak. There, Harriet remembered, "we had the most wonderful and comprehensive of all of the views of our lives." It was a moment she would treasure forever and an accomplishment that would place her among the ranks of the world's greatest explorers.

IN THE MONTHS that followed, she and Frank continued on their journey, visiting Chile before rounding the Strait of Magellan and heading north. The next year they traveled through Argentina, Paraguay, Uruguay, Brazil, Venezuela, Guyana, French Guiana, Suriname, Curaçao, and Colombia. From Cartagena, their final destination in South America, she and Frank began the trip back to the United States. They returned by way of Panama before sailing to New York City on the *Orinoco*. When Harriet finally sighted land, she rejoiced. "A Red Letter day," she wrote in her journal of her arrival in the bustling metropolis on a spring day. After more than two years of travel by boat, mule, train, and horse, she was ready to enjoy the comforts of home.

After regaining her strength and organizing her notes, Adams began to share her experiences with the public through lectures to enthusiastic audiences. Journalists immediately recognized that her explorations had made her an expert on South American geography and anthropology, two areas of study that had been established at the university level only in the previous two decades. Newspapers described her as "intimately acquainted with the interesting features of the southern continent" and acknowledged her "extraordinary" exploratory contributions in "penetrating to the furthermost frontier points," where she lived among the Indigenous peoples and documented their activities. Her straightforward empathetic descriptions often differed from those of late nineteenth-century western anthropologists, many of whom viewed anything non-European as inferior. By contrast, Adams praised as "advanced" the "ancient American civilizations" that predated European contact and recognized the negative impact of

European colonialism, or what she called the "Spanish invasion," on Indigenous peoples of South America.

Shortly after returning from abroad, Adams made contact with a magazine established less than two decades earlier. It was called *National Geographic*. The members of the society that published it recognized that they lived in an exciting age of global exploration and argued for their country's place in the race to make new discoveries. In the magazine's inaugural issue, National Geographic Society President Gardiner Hubbard claimed that "America refuse[d] to be left in the rear" and "already her explorers are in every land and on every sea." He praised American adventurers' contributions to geographic knowledge in the realms of sea and air, thanks to new technologies that facilitated scientific observation and data collection. Furthering the world's understanding of its own conditions would be an ongoing project that "demand[ed] the best efforts of their countrymen to encourage and support," Hubbard concluded.

Adams became one of *National Geographic*'s first female contributors. She was thrilled to share her discoveries through the platform, publishing exciting stories and poignant images of the people she met during her travels over the next few decades. One of the magazine's most popular lecturers, she captivated public audiences of up to 1,500 people with images and was the first American to present color photography in such a setting. Her first article, "Picturesque Paramaribo," appeared in June 1907 and described her journey on the Suriname River in what was then a Dutch colony. In the years that followed, Adams wrote twenty additional essays, often featuring her stunning photographs of landscapes and Indigenous people in their native dress. With her descriptions of places ranging from the Incan ruins of Peru and Bolivia to the mountains of Chile and Argentina, Adams transported readers to far-flung locales. Her discussions of Indigenous people revealed their labor conditions, religious practices, diet, and attire. For the most part, Adams wrote of her subjects with respect, although her descriptions of people of African descent in South America were sometimes tinged with the racist stereotypes of the Jim Crow era.

Chester Harding painted this portrait of Daniel Boone in 1820. It is the only known portrait of Boone made during his lifetime. The frontiersman died soon after its completion. *National Portrait Gallery*

Boone gazes expectantly upon the Kentucky frontier from an overlook. Alfred Jones created this etching of William Ranney's painting, *Daniel Boone's First View of Kentucky*, in 1850. *Indianapolis Museum of Art at Newfields*

Daniel Boone's popularity reached new heights in the 1960s, when the actor Fess Parker reenacted his life on the eponymous television show. This lunchbox and thermos feature images of Parker wearing a coonskin cap, though Boone likely never wore one. *The Daniel Boone Memorabilia Collection, Fort Boonesborough*

In 1832, artist George Catlin visited the Hidatsa village where Sacagawea had lived when she joined the Corps of Discovery in 1804. He named the painting *Hidatsa Village, Earth-Covered Lodges, on the Knife River, 1810 Miles Above St. Louis. Smithsonian American Art Museum*

Members of the Corps of Discovery went over the Continental Divide via Lemhi Pass on August 12, 1805. This pass is in the Beaverhead Mountains and separates the states of Idaho and Montana. *Photograph by Amanda Bellows, 2021*

Sacagawea's unmarked grave is most likely here, at the reconstructed Fort Manuel Lisa, where she probably died in December of 1812. Fort Manuel Lisa is located in Kenel, South Dakota, on the Standing Rock Reservation. *Photograph by Amanda Bellows, 2021*

The United States Postal Service created this stamp depicting Sacagawea in 1994 as part of its Legends of the West series. Over the centuries, artists have had to imagine her likeness because there are no known portraits of Sacagawea from her lifetime. *Smithsonian National Postal Museum*

The US Mint issued a golden dollar coin in 2000 that depicted Sacagawea and her infant son, Jean Baptiste Charbonneau. *United States Mint*

Currier and Ives produced this lithograph depicting California gold miners in 1871, more than two decades after the Gold Rush began. The negative environmental consequences of mining, such as soil erosion and logging, are evident in this representation. *Library of Congress*

James Beckwourth found a pass through the Sierra Nevada in 1850. Today, a marker commemorates his discovery. *Photograph by Amanda Bellows, 2021*

An African American miner searches for gold in California in 1852. *California History Room, California State Library*

Laura Ingalls Wilder lived in the "surveyors' house," located in De Smet, South Dakota, from 1879 to 1880. *Photograph by Amanda Bellows, 2021*

Spirit Mound is located more than one hundred miles south of De Smet, South Dakota. This rock formation emerged 13,000 years ago and is a spiritual site for the Yankton, Mandan, and Lakota people. The Corps of Discovery reached Spirit Mound in 1804. *Photograph by Amanda Bellows, 2021*

Laura Ingalls Wilder signs books for her young readers, 1930s or '40s. *Laura Ingalls Wilder Memorial Society*

Heavy snowfall prevented trains from reaching Dakota Territory settlers like Laura Ingalls Wilder during the winter of 1880–81. *Minnesota Historical Society*

This statue of John Muir by Valentin Znoba stands near Muir's childhood home in Dunbar, Scotland. *Photograph by Amanda Bellows, 2022*

Muir was born in this house in 1838. *Photograph by Amanda Bellows, 2022*

Muir played on the beaches of Dunbar as a child. *Photograph by Amanda Bellows, 2022*

John Muir and President Theodore Roosevelt stand at Glacier Point in Yosemite National Park in 1903 during a three-day camping trip. *Library of Congress*

Tunnel View, Yosemite National Park. *Photograph by Amanda Bellows, 2021*

Florence Merriam Bailey at her
campsite near Queens, New Mexico,
1901. *American Heritage Center,
University of Wyoming*

Florence Merriam Bailey with
seagulls at the beach. *Lewis County
Historical Society*

Demand for feathers in the name of women's fashion fueled an ecological crisis
at the turn of the twentieth century. Gordon Ross, "The Woman Behind the
Gun," *Puck*, 1911. *Library of Congress*

William H. Sheppard as a young man in the late nineteenth century. *Presbyterian Historical Society*

William Sheppard plays the banjo for Congolese children, 1900. The banjo, an instrument with West African roots, emerged among communities of enslaved people in North America and the Caribbean during the seventeenth century. *Presbyterian Historical Society*

William Sheppard stands with Chief Maxamalinge, king of the Bakuba, 1900. *Presbyterian Historical Society*

Harriet Chalmers Adams stands next to a llama at a zoo in 1912. Llamas likely once provided life-saving heat to Adams during a 1904 excursion in Bolivia. *Library of Congress*

The fifteenth-century ruins of the Incan fortress Sacsayhuaman towered above curious travelers during the early twentieth century. This photograph, circa 1905–06, was likely taken by Max T. Vargas and Martin Chambi, men who traveled with Harriet through parts of South America. *University of California–Los Angeles Library*

A display of Inca Mining and Rubber Company gold bars and pistols recalled the lawlessness of the California Gold Rush era sixty years earlier. This photograph was also likely taken by Vargas and Chambi. *University of California–Los Angeles Library*

Above: Possible self-portrait by Matthew Henson, *Arctic Fever Notes*, 1903–05. *National Archives*

Left: Matthew Henson (*center*), Ooqueah, Ootah, Egingwah, and Seegloo at the North Pole, April 7, 1909. *National Archives*

Left: The cover of Matthew Henson's autobiography, *A Negro Explorer at the North Pole*, 1912. *Photographed at the Earl S. Richardson Library, Morgan State University by Amanda Bellows, 2021*

Bottom: Matthew Henson holds a photograph of Robert Peary in 1953. This picture was taken by Roger Higgins. *Library of Congress*

Amelia Earhart's birthplace and childhood home in Atchison, Kansas. *Photograph by Amanda Bellows, 2021*

From her front porch, Amelia Earhart could view the Missouri River. Less than a century before her birth, the Corps of Discovery passed through this stretch of the river during their journey to the Pacific Ocean. *Photograph by Amanda Bellows, 2021*

At the White House, President Herbert Hoover gave Amelia Earhart a National Geographic Society Gold Medal in honor of her successful solo transatlantic flight, 1932. *Library of Congress*

In 1933, Amelia Earhart received the first Society of Woman Geographers Gold Medal from explorer Harriet Chalmers Adams for her solo transatlantic flight. *Library of Congress*

Amelia Earhart opens the door of a Stearman-Hammond Y-1 monoplane, 1936. *Library of Congress*

Sally Ride obtained this Soviet coin depicting cosmonaut Valentina Tereshkova. The coin was minted in the Soviet Union on the twentieth anniversary of the 1963 *Vostok VI* mission, when Tereshkova became the first woman in space. In 1983, Sally Ride became America's first woman in space after the completion of the STS-7 mission. *National Air and Space Museum*

Five of the first six female astronaut candidates conduct water training exercises near Homestead Air Force Base in Florida, 1978. *Left to right:* Sally K. Ride, Judith A. Resnik, Anna L. Fisher, Kathryn D. Sullivan, and M. Rhea Seddon. *NASA*

Official NASA portrait of Sally Ride, 1984. *NASA*

Liftoff of the space shuttle *Challenger* on June 18, 1983. Sally Ride became the first American woman in space during this mission. *NASA*

Summer on the east coast of Greenland. *Photograph by Bathsheba Demuth, 20*

The Yellowknife Bay Formation on Mars, photographed by NASA's *Curiosity* rover, a mobile spacecraft, 2013. *NASA*

Adams maintained her professional relationship with *National Geographic Magazine* throughout her career but regretted that the organization failed to fund her research trips while providing thousands of dollars to male explorers. For instance, in 1915 National Geographic refused to grant her money for an excursion to Africa because of the organization's limited resources. But that same year, Frank Chapman, an ornithologist and peer of Florence Merriam Bailey, went on a National Geographic–funded trip to the Urubamba Valley in Peru. He returned with no fewer than 744 bird carcasses to study. Adams found ways to continue exploring, mostly on her own while Frank pursued a career at Pan American Union between 1907 and 1933. She made trips to the Caribbean, Central Asia, and the Philippines before the outbreak of World War I, during which she journeyed to France to work as a war correspondent. After the war ended, Adams visited Indian reservations across the United States before returning to South America and Europe.

Although she became a fellow of England's Royal Geographical Society, which also counted African American missionary William Sheppard among its ranks, Adams was excluded from countless explorers' clubs, which provided comradeship and financial support for men in the United States. Undeterred, Adams helped build one of her own, becoming the first president of the Society of Woman Geographers. Established in 1925 in Washington, DC, the society shattered barriers and challenged sexist views like those of a journalist who remarked that its existence proved that "a woman can be charming, wear pretty clothes, and do a man-sized piece of work in primitive, uncivilized waster of territory."

Through her leadership, Adams helped the organization connect female explorers around the world and advanced geographic knowledge. Mary Vaux Walcott, who succeeded Adams as president, later praised her predecessor for the "countless hours" she spent "in drafting the hundreds of letters in our files, and in sending them forth with their messages of encouragement to women in all parts of the world." The Society of Woman Geographers soon included among its ranks the anthropologist Margaret Mead and the aviator Amelia Earhart, who received the society's first medal for her path-breaking flight across the Atlantic Ocean in 1932.

Adams spent most of her life fearlessly traveling through remote regions of the world even after sustaining serious injuries in 1926 while on a climbing expedition in Spain's Balearic Isles. As she approached old age, she journeyed to Africa for the first time, before finally retiring with Frank in Europe, where she died in 1937. Over decades of exploration, she helped transform the field by publicizing her experiences as a female adventurer. In 1912, she "wondered why men [had] so absolutely monopolized the field of exploration" and why "women never go to the Arctic, try for one pole or the other, or invade Africa, Tibet, or unknown wildernesses?"

8

Matthew Henson

THE ARCTIC EXPLORER

When Matthew Henson awoke on March 1, 1909, the sun had not yet risen. Through the thick, cold walls of his igloo, he could hear the fierce wind sweep across the forbidding landscape of Cape Columbia, the northernmost point of Canada. It was a place of black headlands and frozen plains, frigid seas and floating ice. In the darkness of the Arctic morning, Henson contemplated the journey ahead. Today he would embark upon a 413-mile trek by dogsled to reach the North Pole.

The past week at the campsite on Cape Columbia had been difficult. Henson, who was African American, and a group of seventeen white and Inuit men, led by Commander Robert Peary, endured gale-force winds, blinding snow, and temperatures that descended to fifty-seven degrees below zero Fahrenheit. By six o'clock in the morning, the men had packed up their belongings and were ready to depart. As gusts of wind swirled about them, they stood on their sledges, listening for Peary's command of "Forward, March!" During the final phase of their expedition, Henson knew that he would have to use all of his physical and mental strength to persevere.

The North Pole had long represented an elusive prize to intrepid explorers, who wondered what icy worlds lay unknown at the top of the earth. In the sixteenth century, the pursuit was largely commercial— European adventurers searched for an Arctic trade route to connect the Atlantic and Pacific Oceans. But by the nineteenth century, polar explorers' objectives began to change. Now they strove to discover unknown regions for the sake of science, as well as personal and national triumph. To the British explorer Sir William Edward Parry, who led the first attempted voyage to the North Pole in 1827, the Arctic was a place of "floating mountains of ice," a region "rude and colossal" with countless dangers that imperiled even expert navigators. Although Parry's effort to find the North Pole was unsuccessful, his expedition inspired subsequent generations of British, Norwegian, Swedish, and American adventurers.

The historian Edward Larson described the nineteenth-century pursuit of the North Pole as "a fundamentally romantic goal promising glory to anyone who could achieve it," guaranteeing the explorer both fame and fortune through lecture fees and book sales. Robert Peary, the American leader of the 1908–9 expedition that Matthew Henson joined, called it the "prize of the centuries." Reaching the North Pole was a dream that demanded the utmost of those who sought to realize it: relentless physical training, mental toughness, navigational knowledge, and fluency in the Inuit language. Did the 1908 team have the skills and strength required to successfully complete such a journey together?

MATTHEW HENSON'S PATH from his birthplace in rural Maryland to uncharted Arctic terrain was a remarkable one. In 1866, just one year after slavery was abolished in the United States, Matthew was born to Caroline and Lemuel Henson on a farm in Charles County.

Although Abraham Lincoln's Emancipation Proclamation of 1863 freed enslaved people living in the Confederate states during the Civil War, it did not apply to slaves living in Maryland, a border state that had remained part of the Union during the Civil War. Not wanting to lose slaveholding Marylanders' support during the fight to crush the white Southern rebellion and preserve the Union, Lincoln exempted from his proclamation Maryland and three additional Union states where slavery was legal. On November 1, 1864, however, just months before the war's conclusion and about two years before Matthew's birth, Maryland adopted a new state constitution that outlawed slavery. The tens of thousands of African American Marylanders who remained enslaved at the beginning of the war were finally free.

Matthew's parents, who were born free, spent the 1860s as sharecroppers on a farm at the site of a former slave market. Caroline and Lemuel Henson lived near the sweeping Potomac River, which enriched the soil of the region's tobacco plantations. During the century that preceded the Civil War, tobacco was the predominant crop in Maryland, and white plantation owners relied heavily on slave labor. Just over half of Black Marylanders were enslaved during the nineteenth century; the free Black population resided primarily in Baltimore.

The war's end in 1865 did little to improve the Hensons' situation, nor that of other free African Americans in rural Charles County. Like Caroline and Lemuel, millions of freedpeople who took on work as sharecroppers after emancipation found themselves entangled in an agricultural system that the historian Henry Louis Gates Jr. has described as "neo-slavery." They grew crops on land owned by others in return for a portion of the annual harvest. Most Black sharecroppers faced significant economic hardship, limited geographic mobility, and racism; without funds to purchase land of their own or migrate to cities to work as wage laborers, they were unable to escape rural poverty.

Somehow Caroline and Lemuel found a way out of sharecropping. Before Matthew's seventh birthday, they moved to the District of Columbia, likely in search of industrial work. Matthew was left behind, facing not only an unsettled childhood but also the dangers of white supremacist violence in Reconstruction-era Maryland. One of Henson's first biographers recounted how terrorist groups threatened the lives of the African American residents of Charles County, recording that Matthew "knew the sounds of the night riders of the Ku Klux Klan. Hidden in bushes, he had witnessed the sadism, the obscenities committed by these white men."

Before Reconstruction's end, Henson's parents died, leaving him an orphan with an uncertain future. He too would leave Charles County for the District of Columbia, where his uncle lived and where he enrolled at the N Street School. Henson joined the more than twenty-five thousand African Americans who migrated to Washington between 1861 and 1877. Together they helped build a thriving Black community, establishing schools, churches, and community organizations. By spending part of his childhood in a cosmopolitan setting far from the Maryland countryside, Henson was exposed to a diverse range of individuals engaged in careers beyond farming, inspiring him to think more expansively about his own possibilities.

At about the age of eleven, Henson ran away from his uncle's home and headed north to Baltimore, a city with a bustling seaport where travelers came and went from all over the globe. He soon met a kind woman who took an interest in the itinerant young man and employed him at her restaurant, where he met Baltimore Jack, a former seaman. Jack's thrilling stories of life as a sailor "whetted the youngster's appetite for the sea," one journalist wrote, and Henson soon found work at Baltimore's harbor.

The young adventurer set off on his first voyage in 1879 aboard the *Katie Hines*, a steam-powered merchant vessel bound for Hong Kong. As a cabin boy, he worked under the command of Captain Childs, who gave him his first lessons in seafaring. Henson recalled that he took naturally to a life of travel. "After my first voyage I became an able-bodied seaman, and for four years followed the sea in that capacity, sailing to China, Japan, Manila, North Africa, Spain, France, and through the Black Sea to Southern Russia," he later wrote. These far-flung journeys expanded his horizons and prepared him for future excursions to the inhospitable Arctic.

IT WAS A day like any other in 1888 when Matthew Henson encountered a person with an enticing proposal. Henson, by then an athletic young man with piercing dark eyes, was working at B. H. Stinemetz, a hat shop in Washington, DC. Henson started there as a clerk after retiring from service on the *Katie Hines* when Captain Childs died. Stocking the hats kept him occupied, although the pace of life was certainly slower than it used to be. But on that day, a six-foot-tall individual with gray eyes and a long red mustache entered the store. His name was Robert Peary, and he was a man of grand ambitions.

Henson's senior by ten years, Peary was born in the rural town of Cresson, Pennsylvania, a mere two hundred miles from Henson's birthplace. He later moved to Maine, where he led an active childhood outdoors, making the most of the state's frigid winters and brisk summers. Peary graduated from Bowdoin College with a degree in civil engineering and joined the Civil Engineer Corps of the US Navy in 1881. Three years later, he was chosen to head an expedition to Nicaragua to survey a potential waterway. Peary's experience in the humid jungles of Central America heightened his interest in exploration and readied him for more dangerous excursions in the Arctic. He took his first trip to Greenland, then largely unexplored and unmapped, in 1886. With the Danish assistant governor Christian Maigaard, Peary traversed one hundred miles of icy interior terrain by dogsled before heading home after running out of provisions. It was a taxing, exhilarating excursion that left him eager to return—with a partner who could help him go even farther.

Peary was preparing for a second trip to Nicaragua to conduct additional canal surveys when he walked into B. H. Stinemetz. When Peary revealed that he was looking for not only a hat but also a valet for his upcoming trip, Henson's boss recommended him for the job. When Peary offered it to him, Henson readily accepted. It was a gamble to travel with a stranger in a foreign land. But Peary and Henson got along well in Nicaragua. Together they trekked through dense forests and humid jungles where cougars and jaguars wandered. Little did Henson know that this journey would be only the first of many with Peary.

The connection they formed during their trip in 1888 would endure for the next two decades. After they returned to the United States, Henson went back to the hat shop, but Peary soon secured him a position as his messenger at the League Island Navy Yard in Philadelphia. From there, they began planning a daring new expedition back to Greenland. As the story goes, Henson's colleague at the League Island Navy Yard doubted he could withstand the harsh climate of the frozen north. He wagered $100 that Henson would lose at least one of his digits to frostbite during the trip. Henson was determined to prove him wrong.

Both Peary and Henson knew that time was of the essence because the competition was becoming fierce. While the two men were in Nicaragua, the Norwegian voyager Fridtjof Nansen skied across Greenland's ice cap, the first person to do so, one-upping Peary's attempt during his 1886 trip.

Reflecting on what compelled him to turn his attention to the region, Peary recalled that a kind of "Arctic fever" overcame him in the 1880s. A mental preoccupation with the frozen north, Arctic fever afflicted ambitious young men like Peary due to what one journalist called the "magnetic mystery which hovers over a large portion of these unexplored seas and lands." Its pull on adventurers was irresistible, he continued, for "no fear of suffering is sufficient to subdue the desire to solve the great problem. The only cure for the arctic fever is the discovery of the North Pole." Arctic fever gripped not only Peary but also Henson, who was similarly prepared to endure the most extreme weather and physical deprivation in the quest to reach the top of the world.

Imperialist impulses also fueled the frenzy among adventurers to be the first to set foot on the North Pole. US victory in the Spanish-American War led to a surge of interest in further extending American power through global exploration. An American explorer who reached the North Pole could claim glory not only for himself but also for his nation. Unlike the government-sponsored excursions of the early nineteenth century, undertaken by the Corps of Discovery and similar groups, most global polar expeditions in the late nineteenth and early twentieth centuries were privately funded, sometimes by financiers of the Gilded Age or new institutions like the National Geographic Society and the American Museum of Natural History. The new generation of explorers sought not to conquer or claim northern territory for themselves or their country of origin but rather to study its geography.

Peary and Henson recognized that they were in a race. Their preparations for a new investigative trip to Greenland proceeded rapidly. Peary assembled a team of experts and secured funding for a voyage that would begin in the summer of 1891. As commander of the expedition, Peary proposed to map the northern coast of Greenland. At that time, there was a dispute between scientists and geographers as to whether Greenland was an island or a continent. By charting its contours, Peary hoped to discover the most efficient means of accessing the world's most northern regions. Of the journey's central purpose, he said, "The vexed question of the finding of the North Pole will have been answered in the affirmative, for it will be necessary, in order to reach the Pole, only to penetrate further and further into the frozen country along the line of the western coast of Greenland." The plan sounded so simple.

On June 6, 1891, Henson and Peary, along with Peary's wife, Josephine, and four other members of the "North Greenland Expedition" departed from New York City on the *Kite*, a humble eighteen-year-old Norwegian vessel that had once been used for seal fishing. The journey north began inauspiciously when an unsecured ship's tiller broke the two bones in Peary's lower leg. Once the adventurers reached Greenland, they hunkered down for the frigid Arctic winter in a makeshift wooden house on the banks of Smith Sound among the members of the local Inuit community.

The following May, Peary and Eivind Astrup, one of his team members, began a twelve-hundred-mile expedition to the northeast coast of Greenland. An enormous ice cap, spanning hundreds of miles and stretching thousands of feet into the sky, stood between them and their destination. By sledge, they traveled for weeks over what Peary described as "a barren waste of snow" where no man had previously set foot. The wind ceaselessly swept across the ice cap during their journey, offering no respite to the weary travelers. Peary later recalled that when the winds built up to a furious speed, they created a "hissing white torrent of blinding [snow] drift" that made it nearly impossible to find their way.

Finally, the scenery changed dramatically: piles of reddish-brown rocks seemed to emerge from the snow. "Leaving the sledge and our supplies at the very edge of the rocks, leading our dogs, and with a few days' supplies upon our backs, Astrup and myself started on over this strange land, bound for the coast, which we knew could not be far distant," Peary remembered. After walking for four long days over hazardous terrain, Peary and Astrup reached the northern coast of Greenland. "We came out at last upon the summit of a towering cliff, about 3,500 feet high," Peary recorded, and then took in the view.

Before them lay a great bay leading to the Arctic Sea, proof at last that Greenland was an island. As they "gazed from the summit of this bronze cliff, with the most brilliant sunshine all about us . . . and a herd of musk-oxen in the valley behind," they found it "almost impossible . . . to believe that [they] were standing upon the northern shore of Greenland." The exhausted men spent a full week there, feasting on musk-ox meat and enjoying the blooms of yellow poppies before beginning the arduous journey back to camp.

Once reunited with the rest of their party, they made their way to the United States. Henson and Peary survived the return trip. Their colleague John Verhoeff, a mineralogist from Kentucky, did not; he had tumbled into an inaccessible crevasse. His death deeply distressed Henson, who mourned the fact that Verhoeff's body was never found. The perils of the Arctic were abundantly clear.

Back home, Henson found himself permanently changed by his first Arctic adventure. Up north, he had learned new skills from the members of the Inuit community who joined or supported the North

Greenland Expedition. Because their ancestors had lived in the Arctic for hundreds of years, the Inuit knew how to endure extreme weather, hunt, and forage for food in a land of scarce resources. Photographs from Peary and Henson's expeditions offer a glimpse into Inuit lives. Men and women alike wore their hair long and dressed in clothing made from thick animal skins and furs. They navigated the icy waters of the Arctic in slim wooden canoes and used spears to slay seals, walruses, and beluga whales.

Henson adapted to this way of life more easily than Peary did. He recorded that he began "dressing in the same kind of clothes, living in the same kind of dens, eating the same food, enjoying their pleasures, and frequently sharing their griefs." In subsequent forays to the Arctic, he would become increasingly adept at constructing and driving dogsleds, building the igloos that were critical to survival, hunting reindeer and musk oxen, and navigating what Josephine Peary called the "white Sahara." He also gained fluency in the Inuit language.

Living in close proximity with the Inuit people, Henson developed important relationships that he would keep for the rest of his life. As a testament to the strength of these friendships, Henson affirmed: "I have come to love these people. I know every man, woman, and child in their tribe. They are my friends and they regard me as theirs." Thanks to the Inuit people's generosity toward Henson and the other members of the North Greenland Expedition, they were now better equipped for the journeys ahead.

The 1891 expedition had also solidified Peary and Henson's relationship. They would make six subsequent treks to the Arctic in the two decades that followed. Of his decision to ask Henson to become his assistant, Peary explained, "He has shared all the physical hardships of my arctic work. . . . [He] can handle a sledge better, and is probably a better dog-driver, than any other man living, except some of the best of the [Inuit] hunters themselves." During each of their adventures, Peary and Henson acquired additional knowledge about polar conditions and geography as they pushed farther north. Both men also fathered Inuit children. Anaukaq, the son of an Inuit woman named Akatingwah, was born in 1906. Peary's son Karree, also born in 1906 to an Inuit woman called Aleqasina, survived to adulthood and visited the United States with Anaukaq in 1987.

Although Peary relied on Henson's facility in traversing treacherous terrain and communicating with the Inuits, they were not equal partners. Whereas the Presbyterian missionaries William Sheppard and Samuel Lapsley became as close as brothers in the Congo, Peary viewed Henson as his inferior due to the color of his skin. Hungry for fame, he may even have chosen Henson over a white partner because he wanted to receive all the accolades. Jim Crow America would be less likely to bestow such recognition on Henson.

The years spent in the Arctic took a toll on both men. Frigid weather devastated their bodies. Henson barely survived his months-long trek across the barren stretches of North Greenland in the spring of 1895, and Peary lost most of his toes to frostbite during the winter of 1899. Pangs of hunger distracted them during what Henson called their "starvation expeditions," when the men were forced to supplement meager rations of beans and biscuits with wild game or dog meat. Still, Henson returned with his digits intact, so he could claim his winnings from his League Navy Yard colleague.

Between arduous journeys to the Arctic, Henson and Peary not only sought to regain their strength but also to gather supplies and raise funds for future trips north. As the leader of these expeditions, and as a white man, Peary gained a degree of fame. He lectured widely, published accounts of his travels, and received prestigious awards. To finance his Arctic journeys, in 1898 he helped organize the Peary Arctic Club, a group of wealthy men who helped Peary raise more than $100,000 over the course of his career. Hard-driving and egocentric, Peary claimed top media attention for himself. In advance of at least one Arctic expedition, he had his team members sign contracts, ensuring that Peary had the first chance to publish his version of events or deliver lectures. Henson, by contrast, received far less attention for his contributions to the success of their polar treks. Despite his unusual position as one of the nation's only Black explorers, his accomplishments, like those of many other African American men and women living in the age of Jim Crow, were often minimized—or ignored—by the media.

In the late nineteenth century, the achievements of African Americans were frequently overlooked, and their opportunities for advancement were significantly limited by legalized segregation, discrimination, and white supremacist violence. The Supreme Court's "separate but equal" ruling in *Plessy v. Ferguson* (1896) paved the way for the passage of inequitable state laws that prevented African Americans from attending schools with white children, riding in the same train cars as white passengers, and sharing public facilities. In the South, where Democrats controlled the statehouses, African Americans were disfranchised through rigged literacy tests or poll taxes; their public schools were defunded, and they were barred from attending state institutions of higher education.

White supremacists enacted campaigns of terror against African Americans who challenged the political, economic, and social order. Lynching rates rose dramatically during this period and peaked in the 1890s, when more than a thousand African Americans were murdered. The journalist and activist Ida B. Wells-Barnett, whose efforts brought lynching to the forefront of national attention, bravely condemned its "awful death-roll." She documented and publicized episodes of murder around the country, arguing that the phenomenon of lynching was "appalling, not only because of the lives it [took], the rank cruelty and outrage to the victims, but because of the prejudice it foster[ed]" against the Black community.

African American communities were also targeted during race riots that erupted in cities across the country at the turn of the twentieth century. In Wilmington, North Carolina; New York City; New Orleans; and Atlanta, racism fueled the destruction of Black neighborhoods and acts of violence against African Americans. In 1908, the year that Henson and Peary sought to reach the North Pole, white mobs in Springfield, Illinois, rampaged through Black neighborhoods, destroying $200,000 worth of property, assaulting and killing African Americans, and forcing two thousand Black residents to flee.

In the face of discrimination and racial terror, African Americans pushed for personal and collective advancement. They established churches, civic organizations, and clubs that supported the community. African Americans also founded successful businesses that made up significant sections of cities across the South and the Midwest. Entrepreneurs like Madam C. J. Walker, the first female self-made millionaire, created business empires by offering products and services to the growing Black middle class. African Americans organized to protest housing discrimination; they unionized to improve unfair labor conditions on Southern farms; and they boycotted segregated public transportation. In short, across the nation they challenged racism and created meaningful, successful lives for themselves in a country that treated them as second-class citizens.

At the turn of the twentieth century, few white Americans thought of exploration as a field in which African Americans could succeed. Racial theorists advanced pseudoscientific arguments purporting that Black Americans were physiologically inferior to white Americans and therefore incapable of succeeding as explorers, particularly in the Arctic. Some posited that African Americans were better suited to warm weather, an argument that gained popularity in the decade before Henson reached the North Pole.

From his earliest days as an adventurer, Henson encountered whites who believed African Americans lacked the "grit" to withstand tough conditions and doubted his abilities to survive the Arctic climate. He remembered one conversation during which he was told that he "couldn't stand the cold—that no black man could." When confronted with such assertions, Henson would respond that he was indeed capable of succeeding as a polar explorer—and was "willing to die if necessary to show them."

BY THE SUMMER of 1908, the moment had come for Henson and Peary to set off on the ultimate mission: a year-long journey to find the yet undiscovered North Pole. Months of fundraising in 1906 and 1907 had secured $75,000 from Peary Arctic Club members and small donors across the nation to purchase equipment and boilers for the *Roosevelt*. Peary ensured that, for their most ambitious journey yet, the ship was constructed of wood harvested from US forests, forged from metal from US mines, and built according to designs created by American engineers. In an age of intense nationalistic competition, Peary wanted everything about the expedition to project American ingenuity and power.

At 1 p.m. on Monday, July 6, 1908, the SS *Roosevelt* pushed back from a pier at East Twenty-Fourth Street in New York City with Henson and the other twenty-one members of the expedition. Crowds of people shouted with excitement on the bright midsummer's day as the 182-foot vessel made its way up the East River, pulled by the navy tugboat *Narkeeta*. Boats of all shapes and sizes saluted joyfully. One newspaper reported that all of the whistling and tooting created "such a din as . . . hasn't been heard in those parts for some time," a cacophony that left the *Roosevelt*'s crew with "a roaring in their ears when it was over."

The ship sailed to Oyster Bay, Long Island, where President Theodore Roosevelt, wearing a white cotton suit and perspiring profusely in the July heat, inspected the boat and greeted its crew. A keen proponent of exploration and an adventurer in his own right, Roosevelt offered words of encouragement to each man, telling one, "I'm glad to meet you and I hope you'll reach the pole." But to Henson the stop seemed little more than an unwelcome delay. At the age of forty-one, he was eager to get underway on the most thrilling adventure of his life. "I am waiting for the command to attack the savage ice- and rock-bound fortress of the North, and here instead we are at anchor in the neighborhood of sheep grazing in green fields," he remarked of their sojourn on Long Island. Finally, they set sail for Greenland. On July 29, the *Roosevelt* slipped between the imposing icebergs of the Labrador Sea as it approached the coastline.

While the *Roosevelt*'s crew focused on the journey ahead in July 1908, newspapers commented amply upon the expedition. Journalists lavished praise on Peary, but they largely ignored Henson. When he was mentioned, it was often in less than flattering terms. For instance, Richmond's *Times Dispatch* referred to him by the wrong name, calling him "Dave Henson" and describing him as "the negro cook." Meanwhile the *Evening Star* called him "the negro from the eastern shore of Maryland, who has made three journeys into the icy regions with [Peary]" and who, according to Peary, was "faithful to the core and of inestimable value in such work." Such paternalistic descriptions utterly failed to acknowledge the extent to which Henson's survival skills, dogsledding prowess, and language proficiency were critical to Peary's survival.

The *Roosevelt* pushed farther north in the months that followed, finally arriving at Cape Sheridan on September 5, 1908. On this barren site, located at the northeastern corner of Ellesmere Island, Canada, the expedition established its camp for the next five months. The long, dark Arctic winter loomed before them, one final hurdle to overcome before they could set out for the North Pole. There was much to do to prepare for their attempt to reach the top of the world. At night, the men slept aboard the *Roosevelt,* but they spent the diminishing daylight hours on land. Henson recorded in his diary that he spent his time "carpentering . . . interpreting, barbering, tailoring, dog-training," and striving to achieve good relations with the members of the Inuit community at the camp. He made frequent use of two prized tools: a handsome wooden pocketknife and large foot-long saw.

As autumn advanced, the bleak weather and lack of sunlight began to diminish Henson's spirits. His diary notes the constancy of "the black darkness of the sky, the stars twinkling above, and hour after hour going by with no sunlight," a phenomenon that, coupled with a lack of game despite many hours spent hunting, left him "tired, sick, sore, and discouraged." To pass the time, Henson attempted to lose himself in books from the communal library: Dickens, Kipling, and Thomas Hood. Henson's Bible also offered comfort during the gloomiest months of the year. He drew strength from his Christian faith, believing, he later said, that God gave a man "vision and . . . inspiration [to] lift him above level of himself and send him forth against all opposition or any discouragement to do and to dare and to accomplish wonderful and great things for the world and for humanity."

When winter arrived at Cape Sheridan, temperatures plummeted and dreadful storms besieged the campsite. Henson did not underestimate their power, having witnessed gale-force winds drive an eighty-four-pound boulder into one of his Inuit companions. Such storms produced in him a sensation of "abject physical terror, due to the realization of perfect helplessness" in the face of such force. During short days and endless nights, Henson tolerated these terrible tempests and put his energies into completing the arrangements for the expedition's final advance.

Working alongside him were many of the thirty-nine Inuit men, women, and children who had come to Cape Sheridan aboard the *Roosevelt*. "Miss Bill" sewed his fur clothing for his trip to the North Pole, while the Inuit boys helped with smaller tasks aboard the ship. Henson admired these close-knit families, noting that there was "a great deal of affection among them." But like Harriet Chalmers Adams, he noticed the negative impact of development on their community. "It is sad to think of the fate of my friends who live in what was once a land of plenty, but which is, through the greed of the commercial hunter, becoming a land of frigid desolation. The seals are practically gone, and the walrus are being quickly exterminated," he reflected. "It is my conviction that the life of this little tribe is doomed."

By February 1909, spring's approach signaled that the time had come for Henson and the other members of the group to leave the comparative safety of the *Roosevelt* and begin their northward trek along the seventieth meridian of the Western Hemisphere. They targeted Cape Columbia, some ninety-three miles northwest of Cape Sheridan. From Cape Columbia, they would traverse 413 nautical miles to reach the North Pole, where Peary would measure the elevation of the sun at noon to confirm the location. In total, seven American and nineteen Inuit expedition members, driving twenty-eight sledges pulled by 140 dogs, would support the advance to the North Pole.

Henson set off from Cape Sheridan on February 18 at 9 a.m., shortly after a swift-moving storm swept across the camp. "The time to strike had come," he remembered, as the group left behind the *Roosevelt* on "what might be a returnless journey." The route to Cape Columbia was one of "somber magnificence," taking them past "huge beetling cliffs" and "dark savage headlands" that "project[ed] out into the ice-covered waters of the ocean."

It was a taxing trek for Henson, his three Inuit assistants, and the two dozen dogs that dragged their four heavy sledges across the packed snow and rough ice. Ravenous from their daily exertions, the men feasted on pemmican, a nutrient-dense type of dried meat traditionally prepared by Indigenous peoples, and drank warm tea or sweet condensed milk. Although they yearned to eat the abundant snow to quench their thirst, they knew that doing so would fatally lower their body temperature. Henson tried to keep warm in bearskin clothes, *kamiks* (sealskin boots), and mittens made of sealskin and polar bear fur, but the "awful cold" froze his face and nose. When one of the Inuit men, Ootah, suffered from frostbite on his toe, Henson helped as best he could. Ootah removed his boot and put "his freezing foot under my bearskin shirt," Henson remembered, "the heat of my body thawing out the frozen member."

After four days of frigid temperatures and high winds, they reached Cape Columbia, where they joyfully reunited with the other members of the expedition who had reinforced their advance. Henson celebrated their arrival but knew that the most difficult part of the journey still lay ahead. Hundreds of miles of snow and ice remained between them and the North Pole. As dawn broke on March 1, the men prepared for the departure that represented what Peary called "the drawing of the string to launch the last arrow in [his] quiver." They would move in separate groups, with different tasks to accomplish along the way. While some would break the trail, others would construct igloos or supply the main members of the expedition with food and fuel. At 6 a.m. the men set off across the ice-covered polar sea in search of the North Pole.

Henson's team led the way across rough terrain, using pickaxes to carve a route through densely packed ice. During the days ahead, they made slow progress across the frozen sea, encountering snowstorms, shifting ice, fog, and subzero temperatures. In some places, the very ice they stood upon threatened to break off and leave them stranded in open water. Sometimes the sled dogs resisted the drivers' orders, a protest against the treacherous conditions. Of the bone-chilling temperatures, Henson remembered that "the wind would find the tiniest opening in our clothing and pierce us with the force of driving needles. Our hoods froze to our growing beards and when we halted we had to break away the ice that had been formed by the congealing of our breaths and from the moisture of perspiration exhaled by our bodies." When he lay down to rest at night, he found little respite from the cold; only brutal fatigue enabled him to sleep in his igloo.

One hundred and thirty-three nautical miles lay between the explorers and the Pole. The supporting parties departed, and now only Henson, Peary, and four Inuit men remained. In this final phase, Henson and Peary would lead two separate teams whom Peary had selected because of their abilities as sledge drivers and navigators. Henson was joined by Ooqueah, who was twenty years of age, and Ootah, a thirty-four-year-old man whom Henson called the "most experienced" sledge driver; he had accompanied them on an Arctic journey in 1906. In Henson's eyes, Ootah was "the best all around member of the tribe, a great hunter, a kind father, and a good provider." Peary's group included Seegloo, who was twenty-four years old, and Egingwah, "a big chap" of twenty-six years who weighed 175 pounds.

But Henson felt most keenly the strength of his bond with Peary, a relationship they had built during their eight expeditions together. As midnight on April 2—the hour of departure—approached, Henson gazed at Peary. In Henson's words: "We knew without speaking that the time had come for us to demonstrate that we were the men who, it had been ordained, should unlock the door which held the mystery of the Arctic." The next four days were "a memory of toil, fatigue, and exhaustion" during which they slept little and sought to make constant progress amid hazardous conditions. Henson bore the brunt of the work as trailbreaker, relying on the compasses attached to his sled and wrist, as well as dead reckoning, to navigate their route. To calculate the time during backbreaking marches, he checked his round nickel pocket watch, its glass face flashing in the Arctic sun.

As the days passed, the North Pole seemed within reach. But the men's exhaustion made them vulnerable to the Arctic's numerous dangers. Just miles from their goal, disaster struck as they were attempting to traverse a series of ice floes that bobbed dangerously above the water. In what Henson called "an instant of hideous horror," he lost his footing and tumbled into the sea. "I tore my hood from off my head and struggled frantically," he remembered. Floundering in waters of about thirty-five degrees Fahrenheit, he tried to grip the ice, but his gloved hands kept slipping back into the water.

Moments from death, he felt a firm hand on the back of his neck. It was Ootah. Risking his own life, he yanked Henson out of the sea and pulled him onto the ice. As Sacagawea did for the Corps of Discovery, Ootah also rescued the tools aboard the sledge, which Henson knew were "the essential portion of the scientific part of the expedition . . . the Commander's sextant, the mercury, and the coils of piano-wire." After stripping off his drenched boots and beating his trousers dry, Henson rejoined the broader group, only to find Peary similarly sopping wet from his own unexpected plunge into the Arctic Sea.

On April 6, thirty-seven days after their departure from Cape Columbia, the two teams suspected they had finally reached the North Pole. To this day, however, it remains a matter of dispute which man arrived first. Peary unloaded his scientific instruments, which included a chronometer, a sextant, and an artificial horizon. Calculating the angular distance between the sun and the artificial horizon, he sought to confirm their position: the coordinates were 89° 57. Henson was overjoyed. "A thrill of patriotism ran through me," he remembered. "And I raised my voice to cheer the starry emblem of my native land" as the men planted an American flag. Then the group let "three hearty cheers [ring] out on the still, frosty air." Peary walked among his comrades, shaking the hand of each one to congratulate him.

But the time for celebrating was short. The exhausted men needed to set up camp for their sojourn at the North Pole. First, they named the site Morris K. Jessup, after one of the Peary Arctic Club's primary financial supporters. Then Henson, Ootah, Ooqueah, Seegloo, and Egingwah began repairing their sledges for the return journey and unpacked the supplies. Peary fell into what he called "absolutely fatigue-compelled sleep" while the others worked. A few hours of rest revived him. The next day, he took more measurements several miles from the camp's vicinity and again at the actual site. Satisfied that they had truly reached the North Pole, the team planted another American flag atop what Henson called "a huge paleocrystic floeberg" that towered above them. At long last, they had realized their dream of standing at the top of the world.

BUT THE TIME to revel in victory was short. The six men still faced a dangerous journey home. Intending to retrace their footsteps for a quick return, they hoped to arrive at Cape Columbia before changing tidal conditions made their route impassable or storms covered their trail. They surely were haunted by the thought that if an accident befell them on their return, the world would never know of their incredible discovery. Henson's worst fears were nearly realized during the seventeen-day trek back to camp at Cape Columbia. Unable to fully capture the experience in words, he recorded that the trip was "a horrid nightmare" defined by "haste, toil, and misery as cannot be comprehended by the mind." Pushing themselves to their physical limits to execute a speedy return, the men encountered strong winds that crushed the ice, obscuring the tracks left on their outbound route. Utterly exhausted from the stress of the advance to the North Pole, Peary rode on a sledge for most of the journey home as "dead weight," in Henson's words.

Finally, on the morning of April 23, 1909, the expedition's members reached the camp at Cape Columbia. Exhaustion overcame them, and for two straight days, they slept without a care for the future. When he awoke, Henson discreetly assessed the visages of the other members of the expedition, finding them "gaunt . . . seamed and wrinkled," evidence of the journey's hardships. Later, turning his gaze to himself, he saw in the mirror "the pinched and wrinkled visage of an old man" who had undergone an enormous amount of stress and lost twenty pounds. Even the dogs were, Peary observed, "simply lifeless with fatigue."

Their period of rest strengthened them for the final phase of their Arctic journey. Henson, Peary, and the Inuit men departed from Cape Columbia and crossed more than ninety miles to reach the *Roosevelt*, the ship that had brought them from New York to the Arctic. There, the six men reunited with the supporting members of the expedition, but their joy at returning safely was dampened upon learning of the death of their partner Ross Marvin days earlier, when he likely fell through the ice.

On board the ship, Henson slowly recovered from the ordeal. While awaiting warmer weather that would release the icebound *Roosevelt*, he regained his strength. By mid-July, shifting tides and ample sunshine had begun to melt the enormous chunks of ice that imprisoned the *Roosevelt* at Cape Sheridan. A path southward gradually opened, and Henson rejoiced when they finally departed for New York City on July 17, 1909.

Impatient to share the news of their discovery of the North Pole with the world, the men anxiously anticipated their arrival at the closest telegraph station, in Indian Harbor, Newfoundland and Labrador. Before they reached it, however, the expedition received shocking news after coming ashore at Cape Saumarez and at Etah in Greenland. They learned that Frederick Cook, a member of Peary's 1891–92 Greenland expedition, had outrageously claimed to have reached the North Pole the previous year.

Doubts quickly arose, and the team sought to investigate Cook's claims. In his diary, Henson recorded that he and Donald MacMillan, another member of their expedition, spoke with the two teenage Inuit members of Cook's team in Etah. The young men confessed that they had "not been any distance north," and the expedition members concluded that there was "no foundation in fact for such a statement [that Cook had reached the North Pole]." Furthermore, Henson and Peary believed that Cook "was never good for a hard day's work; in fact he was not up to the average, and he [was] no hand at all in making the most of his resources." But Henson and Peary did not know Cook's current location or whether he had already announced his supposed discovery to the rest of the world. Could they reach the telegraph station at Indian Harbor before Cook publicly proclaimed himself the winner in the race to the North Pole?

Just miles from its destination, the *Roosevelt* was unexpectedly waylaid for three long days near Cape York due to a violent storm. Henson recalled how "the gale increased so considerably that the *Roosevelt* was forced to lay to under reefed foresail, in the lee of the middle pack, until the 29th [of August], when the storm subsided and the ship got under way again." They came into port at Indian Harbor on September 5, 1909. But by then, it was too late. Four days earlier, Cook had arrived at the Shetland Islands, where he wired the tale of his trek to the North Pole to the *New York Herald*. Now all Peary could do was telegraph his challenge to Cook's claims in a message to the Peary Arctic Club, which read, "Stars and Stripes nailed to the North Pole." In the weeks that followed, the race to the Pole transformed into an all-out public relations blitz. The *New York Times* championed the Peary expedition while the *New York Herald* backed Cook, but most of the public ultimately sided with Peary, who would ardently seek to dispel critics' doubts for the rest of his life.

FOR PEARY, THE journey represented his greatest achievement as an explorer. The day he announced his expedition's success at Indian Harbor symbolized what he later called "the victory of twenty-three of the best years of a man's life," a time of "brute hard labor, cold, darkness, and hunger . . . such days of physical hell and nights of agony of disappointment and deferred hope, as few can realize or imagine."

Reaching the North Pole also represented a moment of triumph for the country of Peary's birth, the United States. His public telegraph "marked the cap, the climax, the finish, the closing of the book on four hundred years of splendid sacrifices and struggles and heroism—a book some of the brilliant pages of which have been written by almost every civilized nation on the globe, and the final chapter and the last word of which has been written by the United States." In the years that followed, Peary took much of the glory, including a National Geographic Society Hubbard Medal and media attention for the expedition's collective accomplishment.

For Henson, the discovery of the North Pole also symbolized more than a personal achievement. Reflecting on the meaning of his success in a period of anti-Black discrimination and segregation, Henson viewed the moment as an example of interracial cooperation. In his autobiography, Henson framed the discovery of the North Pole as an example of "another world's accomplishment . . . done and finished, and as in the past, from the beginning of history, wherever the world's work was done by a white man, he had been accompanied by a colored man." Using self-deprecating language likely intended to appeal to white readers, Henson continued: "I felt all that it was possible for me to feel, that it was I, a lowly member of my race, who had been chosen by fate to represent it, at this, almost the last of the world's great *work*."

But later, when speaking directly to the Black members of the Harriet Tubman Community Club of Hempstead, New York, Henson emphasized how his efforts also challenged discrimination. "Whatever I have done [has] not been for myself alone but to help up lift the Negro standard to a higher sphere in the eye of the civilized world," he said. Ultimately, Henson saw himself as a kind of spokesperson or ambassador for people of African descent, recognizing that his participation in the expedition to the North Pole could serve as a global symbol of Black advancement.

But back in the United States, Henson received little attention for the critical role he played in the expedition. At home, newspapers targeting predominantly white audiences overlooked or minimized Henson's role. For example, on September 7, 1909, the front page of the *New York Times* trumpeted news of the expedition's accomplishment: "Peary Discovers the North Pole After Eight Trials in 23 Years." Headlines such as this one directed all the attention to Peary, ignoring the fact that Henson had led one of the two teams of Inuit men that reached the North Pole. Nine days later, the *Times* finally ran a story that presented Henson's perspective but referred to him as Peary's "colored lieutenant." In this account, which was reprinted in several newspapers, Henson corrected an initial report that suggested only Peary had reached the North Pole.

Other journalists framed Henson's participation in paternalistic or racist terms that played down his intelligence and his navigational skills. For instance, several newspapers described Henson in the language of popular turn-of-the-century "plantation literature" about faithful slaves and devoted freedpeople. The *Tacoma Times* called him Peary's "black bodyguard," a man whose "devotion to his master [was] unbounded," while the *Nashville Banner* commented on Henson's "attachment to Commander Peary," which it characterized as "another instance of the faithful Negro servant often evidenced in the ante-cival [*sic*] war period of the South, and still more emphasized, perhaps, by those who followed their masters into the Confederate service."

Newspapers also referenced common racist pseudoscientific theories about African American physiognomy, noting their surprise that Henson had reached the North Pole at all. For instance, one journalist wrote that Henson's success refuted "the general supposition that the negro can't stand cold weather and is a warm climate person only." Reflecting on Henson's capacity to bear conditions in the Arctic, another reporter similarly mused that "the tropical origin of the Negro race and its long habitation in sunny climes, it might be supposed, has rendered its members unfit to endure extreme cold, and that is the prevalent opinion concerning them." In the words of a journalist writing for the Black-owned *Nashville Globe,* such reports strove to "modify or detract from Henson's work and worth" when in fact Henson's victory showed that "there is no place on all this mundane sphere where a member of any other race can go that a member of Henson's race cannot put his foot as far as he who goes farthest."

Peary's own account of the expedition downplayed Henson's contributions. In "The Discovery of the North Pole," published in *Hampton's Magazine*, Peary wrote that he allowed Henson to join him on the expedition as a reward for his "many years of faithful service." While he acknowledged that Henson possessed important skills and was "useful," he also made the astonishing claim that Henson "had not as a racial inheritance the daring and initiative of my Anglo-Saxon friends." Using this racist reasoning, Peary concluded that it was safer to bring Henson along with him to the North Pole than to have him serve as a supporting member of the expedition, which would "subject him to dangers and responsibilities with which he was temperamentally unable to cope."

By contrast, newspapers and periodicals targeting a Black readership wholeheartedly celebrated Henson's success and enumerated the ways in which he helped the expedition's members reach the North Pole. West Virginia's *Advocate* emphasized Henson's previous Arctic experience, while the *Colored American Magazine* published a poetic tribute to Henson, which praised his Inuit language abilities, prowess in guiding sledges, and hunting skills. In the months that followed Henson's return to the United States, he traveled around the country, speaking to packed audiences in New York City, Washington, DC, Baltimore, and Saint Louis, thrilling them with stories and images from his polar adventures. Ultimately, Henson's story as a man who had overcome childhood poverty during Reconstruction to explore the farthest regions of the earth served as a source of inspiration and pride for the African American community.

Matthew Henson lived almost fifty more years after he returned from the North Pole. But he and Robert Peary, partners in exploration for so many years on their Arctic quests, ultimately fell out. They saw each other only twice in the months after they returned, before parting ways forever in 1910. Henson's wife, Lucy, whom he married in 1907, publicly criticized Peary for his mistreatment of her husband. She excoriated Peary in the *San Francisco Chronicle* in 1910: "Since they returned from the North Pole, Peary has dropped Matt entirely, and has held no communication with him nor done a single thing in recognition of his twenty-three years of service. . . . So far as Peary knows, or cares, for all the interest he has shown, Matt might be starving to death. Such ingratitude is pretty hard." When Henson published his autobiography, *A Negro Explorer at the North Pole* (1912), Peary wrote a brief foreword that only begrudgingly acknowledged Henson's "participation in the final victory."

Lucy rightly recognized the precariousness of their situation and Peary's hardheartedness. After Henson's lecture tours ended, he struggled to find a new career that would support his family financially. At one point, he worked as a maintenance man in a Brooklyn garage for $16 a week. He finally found a measure of stability in 1913, when President Taft appointed him to the clerk office of the US Custom House in New York.

Only in Henson's later life did he receive fuller recognition for his achievements. The exclusive Explorers Club, headquartered in New York City, accepted him into its ranks in 1937. Congress recognized Henson's co-discovery of the North Pole in 1944, when, along with other members of the expedition, he was given a Peary Polar Expedition Medal, a gesture with a certain irony, given Henson's role in the mission's success. President Dwight Eisenhower hosted Henson and his wife, Lucy, at the White House in 1954, where he received a commendation for his courage as an Arctic explorer. Matthew Henson died at St. Clare's Hospital in New York on March 9, 1955, and was buried in Woodlawn Cemetery. More than thirty years after his death, his remains were reinterred, with honors, in Arlington National Cemetery alongside those of his wife, near the graves of Josephine and Robert Peary.

9

Amelia Earhart

THE AVIATOR

The propeller of the cherry-red plane began spinning slowly. Its golden trim gleamed in the evening light as it sped down the runway. A crowd of people watched with anticipation as the single-engine Lockheed Vega 5B ascended into the sky above Harbor Grace, Newfoundland. It was 7:12 p.m. on May 20, 1932, and Amelia Earhart had begun her attempt to cross the Atlantic Ocean. Five years to the day had passed since Charles Lindbergh completed the first solo nonstop transatlantic flight in his Ryan M-2 monoplane, the *Spirit of St. Louis*. Since his daring feat, other pilots had tried and failed to fly the Atlantic. Now Earhart would attempt to outdo them.

Most Americans thought that flying was a man's sport. Women, the conventional wisdom ran, lacked the temperament, stamina, or intelligence to pilot their own aircraft. To prove them wrong, Earhart spent years honing her skills. She bought her own airplanes, acquired a pilot's license, and flew in airshows and as a private citizen. For months, she covertly prepared for her dangerous transatlantic crossing. Finally the day had arrived. Earhart had not planned where to land, but she was confident in her ability to traverse the two thousand miles of cold, empty sea that separated Newfoundland and Europe.

ADVANCES IN THE field of aviation had dramatically expanded the possibilities for adventurers in the twenty-three years since Matthew Henson stood at the top of the world. After Wilbur and Orville Wright successfully flew an engine-powered aircraft for the first time at Kitty Hawk, North Carolina, in 1903, aviators raced to push the limits of flight. Just eight days after Henson set sail for New York City from Cape Sheridan after reaching the North Pole in 1909, the Frenchman Louis Blériot flew from France to England over the English Channel, a journey that showed the world how distance could be compressed through flight. But the hazards of aviation remained; the *New York Times* reported that gusts of wind spun Blériot's wooden-framed monoplane "in two complete circles" during a harrowing descent, which ended with "a rather severe bump which broke the propellor." Nonetheless, Blériot's triumph inspired other pilots to compete for fame by flying farther, higher, and faster than all others before them.

Between 1910 and the advent of World War I in 1914, courageous aviators from around the world tested their skills and their airplanes' capacities in public competitions. Setting new records for distance and altitude, pilots dazzled audiences with impressive feats. But flying remained a deadly pastime in those years. The year after Blériot crossed the English Channel, thirty-two aviators perished in accidents—a stunning 5 percent of all pilots globally. Nonetheless, crashes failed to dampen the enthusiasm of those who foresaw the potential of airplanes.

Some proponents of air travel argued that airplanes could allow people to reach previously inaccessible regions and efficiently transport goods. Others predicted their usefulness as tools of war. Pilots could conduct reconnaissance, for instance, by flying across enemy lines to survey troop buildup or weaponry placement. Planes might one day take part in combat or drop bombs. As these military functions became the focus between 1911 and 1914, European governments and private financiers invested more heavily in aircraft development. In response to growing demand, factory owners ramped up production.

The outbreak of World War I in 1914 made it clear that a nation's fleet of aircraft would play a decisive role in its geopolitical power. In 1915, engineers began equipping airplanes with rapid-fire guns that had greater reach than handguns or long guns. Combat between pilots represented a new phase of warcraft, demanding that fliers develop new skills in attack and evasion. The most powerful fighter was the Fokker Eindecker, a German-produced airplane from 1915. Realizing the threat posed by Axis aerial dominance, the Allied forces stepped up their own production of innovative new fighter planes during the remaining years of the war. The American lieutenant J. A. Armstrong described airplanes as an integral part of the experience of battle after 1917, calling them "fierce destroyers of the sky" that flew like eagles above the carnage of the battlefield. By 1918, people around the world recognized that aerial combat had become a permanent element of modern warfare.

World War I also ushered in a new and terrifying era of aerial bombing via massive, hydrogen-filled airships called zeppelins, which targeted military bases and civilians alike. On January 19, 1915, a German Zeppelin L-3 traveled to Great Yarmouth, a coastal town in eastern England, where it released thirteen bombs. The surprise raid, which occurred under the cover of night, took fewer than six minutes. It nonetheless generated "considerable excitement" as Great Yarmouth's residents "rushed into the streets" and "children scream[ed]" in fear, the *New York Times* reported.

Inside nearby buildings, people felt tremors like those of an earthquake as the bombs released their fury. The unprecedented attack, the first of its kind, sent shock waves across the globe. The *Times* of London told its infuriated readers that Americans were similarly outraged, with US newspapers calling the bombing "a disgrace to civilization." Germany, however, defended the raid, issuing a statement declaring that "air war is acknowledged to be a means of modern warfare as long as it is carried out within the rules of international law."

Whereas World War I prompted destructive new uses of aircraft, the Allied victory in 1918 ushered in a period of peacetime aeronautic advancement. Polar exploration was one of the most exciting areas of activity. In 1925, the Norwegian polar adventurer Roald Amundsen set off for the Arctic by airplane with Lincoln Ellsworth, an American with wartime pilot training who had gone on to survey the Andes by plane. Though inspired by the explorers of the Arctic who preceded them, Amundsen and Ellsworth believed that airplanes could transform the field entirely. "The possibilities were great," Amundsen later reflected. While explorers had, for centuries, "worked with his primitive means, the dog—the sledge—" under grueling conditions, "the complete and troublesome journey should be changed now to a speedy flight."

Sixteen years earlier, reaching the North Pole by aircraft would have seemed an impossible dream to the American explorers Matthew Henson and Robert Peary. Their journey to the North Pole had required trekking across fields of snow and ice by dogsled for weeks on end. But the rapid aeronautical advances of subsequent years made the goal of reaching the top of the world by aircraft achievable. Ellsworth even remembered discussing the prospect with Peary before he passed away in 1920, concluding that the aging explorer "was enthusiastic about the project."

In two Dornier Wal aircraft, Amundsen and Ellsworth planned to complete a round-trip flight from Svalbard, Norway, to the North Pole, which has a latitude of 90° north. They took off on May 21, 1925, and were thrilled when they first glimpsed the "entire panorama of Polar ice [that] stretched away before [their] eyes, the most spectacular sheet of snow and ice ever seen by man from an aerial perspective," Amundsen remembered. After about nine hours of flying, they landed their planes in order to discern their location and learned that they stood just shy of the North Pole, at a latitude of 87°44' north. It was a remarkable moment, one that made clear how twentieth-century technological advances could power exploration in new ways. As he looked out across the frozen landscape, Ellsworth recalled thinking how "we had left civilization, and eight hours later we were able to view the earth within ninety miles of the goal that it had taken Peary twenty-three years to reach."

The descent damaged one of their airplanes, however, and the men ultimately failed to reach the North Pole. The following year, they tried again, this time, in a race against competitors. Richard Byrd and Floyd Bennett, both World War I navy pilots, left Spitsbergen in a Fokker airplane on May 9, 1926. Upon their return, they claimed that they had flown over the North Pole. In recognition of this accomplishment, President Calvin Coolidge bestowed the US Congressional Medal of Honor on the two men the following year. Decades later, however, analysis of Byrd's notebook suggested a failed endeavor.

Two days after Byrd and Bennett returned from their mission, Amundsen and Ellsworth made their second journey, this time in a massive airship called *Norge* (*Way to the North*), designed by the Italian engineer Umberto Nobile. After departing from Svalbard en route to Alaska on May 11, Amundsen, Ellsworth, Nobile, and their crew approached the North Pole the following day. The heavy fog through which they had been traveling cleared just in time for them to assess the barren landscape below. "The ice was much broken up at the Pole and a mass of small ice-floes was observable," they recorded. At 1:25 a.m. Greenwich Mean Time, the *Norge* passed over the top of the world. "Now we are there," declared their Norwegian navigator, Hjalmar Riiser-Larsen, as he confirmed their location with a sextant through one of the airship's small windows. Looking down from the *Norge*, the men dropped a Norwegian flag, an Italian flag, and an American flag onto the ice. For Ellsworth, who turned forty-six that day, the moment was one to remember forever. "It was with an extraordinary, quite indescribable feeling that Ellsworth undertook this task," the men later wrote, for "when again will a man plant the flag of his country at the Pole on his birthday?"

Their accomplishment drew immediate international acclaim. Upon hearing the news via radio communication of the expedition's success, Coolidge expressed his "hearty congratulations" to the explorers. In Rome, the *Daily Mail* reported, there were "delirious demonstrations of joy and relief" as the bell rang in the Piazza del Campidoglio. In addition to the expedition's demonstration of bravery, it answered a scientific question. Amundsen, Ellsworth, and Nobile saw no solid land covering the North Pole, a finding that challenged the contemporary theory that an Arctic continent rested at the top of the world. Airplanes, it was becoming clear, offered new opportunities of discovery that the explorers of bygone eras could have scarcely imagined.

SIX YEARS AFTER the airship *Norge* soared over the North Pole, Amelia Earhart prepared to set a record as the first woman to cross the Atlantic Ocean on an eastbound flight. The dangerous route had claimed the lives of six men between 1919 and 1927 as they competed to claim the Orteig Prize of $25,000 for completing the first transatlantic flight. Lindbergh had ultimately won, but no woman had yet flown solo from North America to Europe. The world watched anxiously, wondering: Could she make the journey?

Amelia Earhart had grown up during the golden age of aviation. But little about her childhood in a small Kansas town foreshadowed her career as a fearless explorer of the skies. Born in 1897, Earhart lived for the better part of each year in the house of her maternal grandparents in Atchison. It was a charming neo-Gothic residence, perched high on a grassy hill, with striking arched windows through which Earhart could survey the verdant landscape. Below the house flowed the powerful Missouri River, where, less than a century earlier, the members of the Corps of Discovery had made their way across the continent to the Pacific Ocean.

The Kansas of Earhart's youth, however, was quite different from the scene that William Clark described in his journal in 1806, when the region was populated by bold Kaw warriors who robbed boats traveling on the Missouri River. Earhart was aware of Kansas's history as a frontier territory in the early nineteenth century, but, she lamented in her autobiography, *The Fun of It*, "There were no Indians around when I arrived, though I hoped for many a day some would turn up." Settlers had not only displaced Native Americans but also disrupted the state's ecosystem through overhunting and farming. Whereas the members of the Corps of Discovery encountered deer and bison during their journey through the Midwest, by Earhart's day, "the nearest [she] got to buffaloes was the discovery of an old fur robe rotting away in the barn."

From age three, Amelia lived in Atchison with her grandparents, Amelia and Alfred Otis; her parents, Amelia Otis and Edwin Stanton Earhart, resided fifty miles away in Kansas City, Missouri. Her mother came for extended stays with Amelia's younger sister, Muriel, but their father remained in Kansas City, where he worked as a lawyer for a railroad company. Earhart's bedroom was modest but cozy, with delicate floral wallpaper and a comfortable bed covered with quilts. As a girl who was "exceedingly fond of reading," she later recalled, she felt quite at home in a house filled with books. The works of Charles Dickens, Sir Walter Scott, George Eliot, and others stimulated her imagination and opened her mind to the world beyond Atchison.

Amelia's girlhood was unusual in that she was permitted to freely exercise her body. In grade school, she was "fond of basketball, bicycling, [and tennis]" and never turned down the opportunity to play "any and all strenuous games." Although she grew up during the Progressive Era, a time when rigid societal expectations for women began loosening, Earhart lamented that "though reading was considered proper, many of [her] outdoor exercises were not." But she pushed against these expectations. One favorite pastime of hers was to explore the region's geologic wonders. With her school friends, she participated in excursions "up and down the bluffs of the Missouri River," where "the few sandstone caves in that part of the country added so much to our fervor that exploring became a rage." Perhaps Amelia and her friends were inspired by the fictional adventures of Mark Twain's character Tom Sawyer, whose adventures in a Missouri cave, which Twain had called "a vast labyrinth of crooked aisles that ran into each other and out again and led nowhere," thrilled young readers at the turn of the twentieth century.

As Amelia grew older, her family was uprooted more than once, which ultimately prepared her for a career as an aviator constantly on the move. In 1908, Amelia and her sister, Muriel, relocated with their parents to Des Moines, Iowa, where Edwin had taken on a new role at the Chicago, Rock Island and Pacific Railroad Company. Before long, however, his drinking spiraled out of control—and the lives of his family along with it. Bouts of drunkenness cost Edwin his well-paying job in Des Moines, and the family was forced to move to Saint Paul, Minnesota, when he agreed to work as a clerk for lower pay at the Great Northern Railroad Company. There, Amelia attended Central High School, where she earned high marks in English, physics, mathematics, and German. In 1914, however, the family moved again, to Chicago, this time leaving Edwin behind. Amelia finished coursework for her diploma at Hyde Park High School in 1915.

The United States' entry into World War I in 1917 changed the course of Amelia's life. Although she had anticipated enrolling at a college such as Bryn Mawr, following preparatory work at the Ogontz School in Pennsylvania, she decided instead to devote herself to service. During a trip to Toronto, where her sister, Muriel, attended St. Margaret's College, Amelia remembered how the sight of wounded Canadian soldiers, "men without arms and legs, men who were paralyzed, and men who were blind," revealed "for the first time . . . what the World War meant." Intending to become a nurse's aide with the American Red Cross, Earhart found herself stranded in Toronto due to administrative issues and spent the remainder of the war there, volunteering at the Spadina Military Convalescent Hospital. Laboring for ten hours each day, she "did everything from scrubbing floors to playing tennis with convalescing patients." The experience was a far cry from the peaceful days of her youth in Atchison; now she was a witness to the trauma caused by modern battle. World War I "so completely changed the direction of my own footsteps," Amelia remembered, "that the details of those days remain indelible in my memory."

Airfields used by Canada's Royal Flying Corps, later called the Royal Air Force, enlivened the city. The University of Toronto even offered an innovative training program for male pilots to prepare them for combat in Europe. During the scant hours she spent outside of the hospital, Earhart found herself drawn to these sites of aeronautic activity. "I believe it was during the winter of 1918 that I became interested in airplanes," Earhart later wrote. Alighting on the airfields were "full-sized birds," robust planes that "slid on the hard-packed snow and rose into the air with an extra roar that echoed from the evergreens that banked the edge of the field." Although she regretted that "no civilian had opportunity of going up," despite her request, she still "hung around in spare time and absorbed all [she] could." One powerful sensation, "the sting of the snow on [her] face as it was blown back from the propellors" during take-off, remained impressed on her memory. Many of Earhart's patients at the hospital served the Royal Flying Corps as aviators, and she listened eagerly to their stories of flight.

Although Earhart then felt a "sense of inevitability of flying" one day, she did not yet contemplate a career as a pilot. There were several prominent female aviators prior to the onset of World War I, but American women were banned from serving as military pilots during the conflict. Instead, Earhart turned her attention to medicine. After the armistice, she enrolled at Columbia University in New York City but soon learned that she "probably should not make the ideal physician," she remembered ruefully. Medical research interested her, but patient care was less enticing. When her father, now sober, asked her to move to Los Angeles, California, where he had begun practicing law,

Earhart readily agreed to leave Columbia and head west to a state that had long represented opportunity and adventure.

THE GOLDEN STATE stood at the forefront of innovation in the field of aviation when Earhart arrived in 1920. World War I had fueled industrial production in California, where factories churned out supplies to support the military efforts abroad. Businessmen also opened manufacturing companies that designed and constructed airplanes for the United States. Loening Aeronautical, founded in 1917, produced early fighter planes while the Loughead Aircraft Manufacturing Company constructed amphibious planes that could land on water. In total, factories in Southern California made seventeen thousand airplanes during World War I, cementing its status as a production hub.

After the war ended, aviation became even more popular among Californians. The Golden State's good weather created favorable conditions for flying, and airfields multiplied near cities like Los Angeles, where Amelia joined her reunited parents, who were trying to rebuild their marriage in a rented house at 1334 West Fourth Street. For Earhart, a tall, slim, natural athlete who enjoyed riding horses and "almost anything else that [was] active and carried out in the open," she recalled, "it was a short step from such interests to aviation" in a region that was "particularly active in air matters." Intrigued by the pilots whose daring stunts thrilled audiences during weekend events, Amelia became a regular attendee of local airshows, often accompanied by her skeptical father. Amelia "felt in [her] bones that a hop would come soon," sensing the irresistible, magnetic pull of flight.

Rather than try to dissuade her from falling in love with one of the most dangerous activities of their day, a profession that claimed the lives of skilled aviators when engines failed or parts broke, Edwin agreed to his daughter's request to fly as a passenger. He paid ten dollars to the pilot Frank Hawks, a handsome World War I veteran with piercing eyes and a strong chin, to take her up in his biplane above the sprawling city. Departing from Wilshire Boulevard, Hawks and Earhart soared into the sky on a flight that would convince her of her destiny as an aviator. "As soon as we left the ground, I knew I myself had to fly," wrote Earhart in her autobiography. High above Los Angeles, she glimpsed the magnificent deep blue of the Pacific Ocean and peered down at Beverly Hills.

When Earhart returned to earth, she was transformed, willing to do whatever it took to become a pilot. It was a heady moment, one that empowered her to gain command of a life that had often been dictated by circumstances beyond her control. When she returned home, she calmly told her parents, "I think I'd like to learn to fly." But her insides were aflutter with the knowledge that she knew "full well [she'd] die if [she] didn't." Edwin and Amy did not withhold permission, but her father explained that he could not pay for the expensive lessons.

So Earhart got a job at a telephone company. From Monday through Friday, she toiled away to earn the money that she would spend on weekends learning how to fly. Joining the workforce was a decision that set her apart from other women of her station. Only one out of five workers were female in 1920, but the establishment of the US Department of Labor's Women's Bureau that year signaled a new trend: women would increasingly work outside the home in the decades that followed. Women also made strides in politics in 1920, when nearly a century of activism led to the ratification of the Nineteenth Amendment, which ensured citizens' ability to vote would not be "denied or abridged . . . on account of sex."

Earhart remained keenly aware of the unorthodox nature of her decision to partake in traditionally male-dominated activities like working and flying. It was with hesitancy that she began to embrace her status as an independent woman who defied traditional gender roles. Perhaps taking her cue from the "flappers" of the twenties, women who bobbed their hair, smoked, drank, and wore less restrictive clothing, Earhart slowly adopted a more liberated lifestyle. When she started taking flying lessons, she began wearing "breeks and a leather coat," a "semi-military outfit" that protected pilots from the elements. Amelia also remembered how "up to that time [she] had been snipping inches off [her] hair secretly, but [she] had not bobbed it lest people think [her] eccentric." The approval of others mattered to Earhart, and she "tried to remain as normal as possible in looks, in order to offset the usual criticism of [her] behavior" during an era when it was "very odd indeed for a woman to fly." In time, however, she would become more comfortable with the notion of feminism and showed the world what women were capable of through her own successes as a pilot.

Despite the headwinds she faced in a decade of cultural upheaval, Earhart rapidly gained confidence under the tutelage of Anita Snook, known as Neta. Female instructors were exceedingly rare in the 1920s, when many pilots were male veterans of World War I. But young women like Earhart looked up to the examples set by pioneering female aviators of the preceding decade, such as Harriet Quimby, the first female American licensed pilot who flew across the English Channel in 1912, and Alys McKey, a record-setting aviator who lobbied to serve in a combat role during World War I.

Neta Snook had an impressive list of her own accomplishments, which included supporting the Allied aeronautic war effort and performing aerial stunts at airshows. At Kinner Field in Los Angeles, she taught Earhart how airplanes functioned and how to properly take off and land. They practiced in a Curtiss Canuck, which Earhart described as similar to "the famous Jenny of wartime memory." Snook initially thought Earhart "would make an excellent student," but she found herself "really provoked" on occasion by Earhart's admitted habit of "daydreaming" in the cockpit.

On one memorable day, Earhart failed to check the half-empty gas tank before taking off on a flight with Snook, a mistake that could have proved deadly and prompted Snook to wonder if she had "misjudged [Earhart's] abilities." What her student may have lacked in natural talent and attention to detail, however, she made up for with passion. Before long, Earhart had successfully purchased her first airplane, a sunny yellow Kinner that she named *The Canary*, with financial help from her mother and sister. She continued to log hours in the air with Snook, gaining confidence in her abilities as a pilot.

In 1924, after the breakdown of her parents' marriage, Earhart and her mother headed east from California by car, stopping at national parks along the way: Yosemite, Sequoia, Crater Lake, and Yellowstone. Marveling at the magnificent scenery in landscapes that had been long inhabited by Indigenous people, Earhart connected what had once been America's wilderness territory with new frontiers in aviation. "As thrilling to me as the national parks were the long stretches of open country dotted with air mail beacons," she remembered.

Founded by the US Postal Service just six years earlier, the groundbreaking Transcontinental Airway System comprised a series of gas lights between New York and San Francisco, which guided courageous pilots as they delivered mail from coast to coast. Much of the western portion took pilots through the "spacious skies" over the landmarks of Katharine Lee Bates's famous song: the "fruited plains" of Nebraska and the "purple mountain majesties" of the Colorado Rockies. The airway system was a technological wonder, one that prompted Earhart to write, "It is such things which make the real thrills in aviation." Before long, she would delight observers around the world through her own efforts to advance aviation's frontiers.

FAMILY DIFFICULTIES AND financial hardships kept Earhart grounded between 1924 and 1927. She moved to Boston, Massachusetts, in 1925. There, opportunities to fly were few and far between, and as an unmarried woman, she had to earn a living. Earhart found employment as a social worker at Denison House, a Progressive Era settlement house for Chinese and Syrian immigrants. In her spare time, Earhart helped run the National Aeronautic Association's local chapter, which apprised her of aviation developments and connected her to a network of like-minded enthusiasts.

While overseeing the children at Denison House one day in 1928, Earhart received an unexpected phone call. When she held the receiver to her ear, she heard a male voice say, "Hello. You don't know me, but my name is Railey—Captain H. H. Railey." It was the man who was then responsible for managing the financial aspects of Richard Byrd's upcoming aerial expeditions over the continent of Antarctica. "He asked me if I should be interested in doing something for aviation which might be hazardous," she later wrote. But Railey would not reveal what the project would entail. Intrigued, Amelia agreed to meet him that night to discuss his proposal.

With anticipation, she headed over to Hilton Howell Railey's office. A slender man, thirty-two years of age, Railey wasted no time getting to the point. "How would you like to be the first woman to fly the Atlantic?" he asked Earhart. She maintained her composure, but Railey noticed a spark of interest in her eyes. Earhart pressed him for more information. Two men were preparing to fly across the Atlantic Ocean in a Fokker airplane once owned by Byrd, he explained. Their flight would be financed by an American woman living with her British husband in London. Her name was Amy Phipps Guest, and she believed that the transatlantic journey would represent to the world the warm relations between the United States and England in the postwar era. Guest had wanted to take the ride herself, but her family refused to let her undertake such a hazardous journey. Railey wondered: Would Earhart be interested in accompanying the male pilot and mechanic as their passenger?

It was a tantalizing proposition, but one that carried significant risks. Earhart saw her chance to make history and agreed under two conditions. First, she wanted to inspect the airplane's equipment and become acquainted with the pilot. Second, Earhart wanted to fly the plane for part of the journey. Her first request was easy to meet. To her disappointment, however, Earhart discovered that she was unable to fly the plane. "Despite my intentions," she wrote, "the weather encountered necessitated instrument flying, a type of specialized flying in which I had not had any experience."

During the weeks that followed, the pilot, Wilmer Stultz, and the mechanic, Louis Gordon, prepared for the team's covert flight on a bright orange monoplane called *Friendship*. Hiding their plans from the press, they sought to avoid generating publicity until the big day. First, they installed new parts in the *Friendship*, including pontoons and three 225-horsepower Wright Whirlwind engines. Next, they gathered data on weather conditions and packed their supplies.

Finally, Earhart wrote her will. She may have felt that her adventure was akin to an exploratory excursion, perhaps like the one Richard Byrd was busily planning. Shortly before her departure, she visited Byrd and his wife in their house, which she remembered was "just then bursting with the preparations for his Antarctic expedition—a place of tents and furs, specially devised instruments, concentrated foodstuffs, and all the rest of the paraphernalia which makes the practical, and sometimes the picturesque, background of a great expedition."

To prepare for her own journey, Earhart donned clothes that she compared to those needed for "a strenuous camping trip," when a person "couldn't tell what might happen." Over a delicate white silk blouse, she layered a warm sweater and a long, heavy leather coat to protect her from the cold at high altitudes. Her curly cropped hair was covered with a leather hood that buckled under the chin, and her determined gray eyes were protected with oversized goggles.

Like the members of the Corps of Discovery, who carried their belongings in army-issued knapsacks, Earhart similarly tucked her personal items into a military pack. But while the men of the early nineteenth-century brought with them razor blades and hard soaps, she chose to bring a toothbrush, a comb, handkerchiefs, and a small container of cold cream, a twentieth-century luxury, for the journey. Her publicist, George Palmer Putnam, loaned her field glasses that he had used on a prior expedition to the Arctic, and Earhart carried Richard Byrd's book, *Skyward*, as a gift for Amy Guest, the flight's sponsor. Inside the book, the explorer had written a note to Guest that began, "I am sending you this copy of my first book by the first girl to cross the Atlantic Ocean by air—the very brave Miss Earhart."

On June 3, 1928, the *Friendship* took off from Boston. As the plane headed up the East Coast toward Newfoundland, Earhart admired the terrain below. Over Nova Scotia, she saw a "jagged coast" with "few roads" and "many little houses nestle[d] in the woods seemingly out of communication with anything for miles." Perhaps most striking, however, was her sense of a connection to the aerial explorers who had preceded her over the same foreboding landscape as they tested the limits of flight. "During the last two years this remote country has had many visitors from the air," she later mused. "There have been Lindbergh and Byrd," who completed transatlantic flights one month apart in 1927, and "in the old days, the N.C. 4s, disregarding the incidental flights which doubtless have winged over this territory." Now it was Earhart's turn to make history.

After stops in Halifax in Nova Scotia and Trepassey in Newfoundland, the *Friendship* returned to the air on June 17 to cross the Atlantic. Unfortunately, their view throughout most of the journey was obscured by what Earhart described as "a sea of clouds." Earhart, Stultz, and Gordon endured frigid temperatures and what Earhart called the "heaviest storm I have ever been in, in the air, and had to go through," one that violently shook the *Friendship* from nose to tail. Probably due to nerves, she lost her appetite during the flight and ate only three oranges. When the time for the expected sighting of land came and went, the trio began to worry. Their fuel was quickly running out, but they clung to the hope that "land—some land—must be near."

Suddenly, a "blue shadow" emerged from the gloom. "*It was land!*" Shouting for joy, the crew celebrated, and Stultz brought the plane down on the coast of what was, they soon found out, Burry Port, Wales. June 18, 1928, was a day of firsts—not only their first "crossing of the Atlantic" by air but also their first time in Europe, or what Earhart jokingly referred to as their "introduction to the Old World." The explorers of the preceding century had pushed west toward new frontiers, but the aerial pioneers of the twentieth century were now forging east, powered by exciting new technologies.

The days that followed their almost twenty-one-hour flight were a blur of activity. Much of the initial media attention was directed toward Earhart because, although she had not flown the plane, she was the first woman to have made such a journey. The lead story in the *Boston Globe* had this headline: "Miss Earhart's Plane Comes Down in Wales: Tanks Almost Empty After Stormy Passage"; it entirely ignored the roles of Stultz and Gordon. In England, the *Guardian* similarly touted Earhart as the "First Woman to Succeed in Atlantic Flight," calling her "Lady Lindy" and explaining to readers how she had "captured the imagination of America during the few days preceding her successful flight from Trepassey, Newfoundland to Llanelly, Wales."

After traveling to London, where she stayed with Amy Guest and participated in publicity events related to the flight, Earhart returned to the United States. Over the following weeks, Putnam took advantage of national interest and secured engagements for her in cities around the country. As her fame grew, thanks in large part to her publicist's maneuverings, the pair grew closer. Although Putnam was married, he and Earhart began a romantic relationship that spurred Putnam's wife of almost two decades, Dorothy, to file for divorce in 1929. Earhart later confessed in an interview that she and Putnam "came to depend on each other, yet it was only friendship between us, or so—at least—I thought at first." She continued, "At least I didn't admit even to myself that I was in love." But with the passage of time, Earhart realized: "I could deceive myself no longer. . . . I knew I had found the one person who could put up with me."

Putnam eagerly sought to win Earhart's heart following his divorce, by some accounts proposing to her no fewer than six times between 1929 and 1931 because of what she explained to him was her "reluctance to marry." Earhart finally capitulated on February 7, 1931, when she wed Putnam in a short ceremony that was not attended by anyone from her family. But in a letter to Putnam the night before their wedding, she insisted that he "let [her] go in a year" if they were unhappy as husband and wife. Aviation remained of the utmost importance to Earhart, who feared that marriage to Putnam would prevent her from pursuing her dreams. Happily, however, she soon discovered that Put-

nam would throw his support behind her newest, most dangerous adventure yet: a solo flight across the Atlantic Ocean.

WHILE READING THE newspaper at the breakfast table with
George one morning, Amelia summoned up her courage. The time
had come to pose a question that had been weighing on her mind.
She put down the paper and looked him in the eye. "Would you *mind*
if I flew the Atlantic?" Amelia asked in a steady, measured voice. In
that instant, George felt a powerful mix of emotions, which he later
described as "the fusion of a clutch at the heart, and something akin
to elation, in the presence of so adventurous a spirit." He must have
realized then that marrying Amelia meant coming to terms with her
insatiable urge to push the limits of flight, a desire that could put her
life at risk.

What prompted Earhart to contemplate such a feat in 1932? Much
had changed since her flight with Stultz and Gordon in 1928. During
those four years, Putnam later reflected, widening aviation experiences "had crystallized the purposes and personality of AE," who
gave "slavish attention to becoming expert in the air" over hundreds
of hours of training. Pure enjoyment primarily inspired Earhart to
attempt to fly across the Atlantic, she explained in her autobiography.
"It was clear in my mind that I [would undertake] the flight because I
loved flying," she wrote. "I chose to fly the Atlantic because I wanted
to. It was, in a measure, a self-justification—a proving to me, and to
anyone else interested, that a woman with adequate experience could
do it."

Additional factors may also have influenced her decision. Since her
successful transatlantic journey on the *Friendship*, Earhart had tried to
maintain her place in the public eye by participating in a transcontinental air derby for female pilots, the first of its kind, and by attempting to break airspeed records. But she faced growing competition from
other outstanding women aviators, including Elinor Smith and Ruth
Nichols, both of whom intended to try to cross the Atlantic Ocean. If

Earhart could beat them to it, she would secure for herself a special place in history.

And of course, there was the matter of money. The Great Depression gripped the nation in 1932. The economic downturn began in October 1929, with a dramatic stock-market crash; four years later the trend showed little sign of relenting. One in four Americans was unemployed in 1932, while millions more scraped by on the limited income from the meager jobs they could find. Neither the budding aviation industry nor celebrities like Earhart were immune to these effects. Earhart was also financially supporting her mother, who needed assistance after the death of her husband, Edwin, in 1930 from cancer. By replicating Charles Lindbergh's eastbound transatlantic solo flight, Earhart could recapture the world's attention and raise much-needed funds.

As her publicity manager, Putnam could help her capitalize on the attention that would surely follow such a feat by booking speaking engagements and other public appearances. But the risks were tremendous. Since Earhart's flight on the *Friendship*, eight pilots had attempted to cross the Atlantic. Six had died. But having secured Putnam's approval that morning, she quietly began planning her journey.

Preparations began in earnest that spring. Bernt Balchen, a Norwegian aviator who had participated in Arctic expeditions with Amundsen, Ellsworth, and Byrd, agreed to get Earhart's plane into shape. With the mechanic Edward Gorski and the aeronautical engineering expert Edwin Aldrin, whose son Buzz would be the second man to set foot on the moon, Balchen oversaw improvements to Earhart's Lockheed Vega. They repaired its fuselage, installed a five-hundred-horsepower motor, and added fuel tanks that could carry enough gasoline and oil to power the Vega at a cruising speed of 140 miles per hour across eighteen hundred miles of open sea. Waiting for a period of good visibility, Earhart assessed weather reports provided by the US Weather Bureau and mentally prepared for the possibility of departing at a moment's notice.

For his part, Putnam urged Earhart to set off on the five-year anniversary of Lindbergh's solo flight, recognizing that the timing would make a good story in the press. Meanwhile, Earhart avoided generating what she called "advance publicity—the questions of reporters, the protesting and/or congratulatory messages of friends, [and] commercial overtures," which "would have been unimportant and exhausting." Instead, she sought to "pursue [her] daily life normally, fit, happy, and rested, physically and mentally," and to "concentrate on the essentials."

On May 19, 1932, Earhart conferred with Balchen and Gorskii at New Jersey's Teterboro Airport, where her Lockheed Vega was parked. The weather report showed reasonable flying conditions. Amelia decided it was time to embark upon her career-defining adventure across the Atlantic Ocean. She rushed home to the house she shared with Putnam in Rye, "changed into jodhpurs and windbreaker," she recorded, "gather[ed] up [her] leather flying suit, maps, and a few odds and ends and raced back to the [air]field." It was mid-afternoon when she, Gorskii, and Balchen set off for Harbor Grace, Newfoundland, the launching point for her flight to Europe. They stopped in Saint John, New Brunswick, where they spent the night before pushing onward to Harbor Grace the next morning.

Word began to spread of her impending flight, and newspapers began breathlessly reporting her movements on May 19. The following evening, Earhart was ready to take off on her own from Harbor Grace. Balchen, the experienced Arctic explorer, marveled at her serene demeanor as she bid him farewell before the mass of people who had come to watch her departure. "Do you think I can make it?" she asked him privately. "You bet," he responded with a big smile.

IT WAS FIVE years to the day since Charles Lindbergh attempted his transatlantic flight from New York to Paris. Now, it was Earhart's turn. At 7:12 p.m., she lined up her Vega on the runway and "gave her the gun," she later wrote. "The plane gathered speed, and despite the heavy load rose easily." Under her composed façade, Earhart likely struggled with the knowledge that the success of her journey was far from assured. Indeed, disaster struck soon after she took off from Harbor Grace. Soaring at twelve thousand feet above the sea, Earhart noticed her altimeter malfunctioning, its hands spinning "around the dial uselessly." All pilots relied upon the altimeter, and Earhart had never dealt with a technical failure of this kind. Now it would be impossible for her to gauge the distance between her plane and the dark, flat ocean in the black of night.

Not long after, a huge lightning storm pierced the skies and began to violently shake her plane from nose to tail. "I was considerably buffeted about," she remembered, "and with difficulty held my course." She squinted through the blinding flashes of light that penetrated the gloom for the next hour, as the Lockheed Vega was continually rocked by turbulence. When Earhart finally passed through the worst of the storm, she "climbed for half an hour" above the clouds before noticing the plane was "picking up ice." The frozen water ruined her tachometer, which showed her rate of travel, another critical piece of information. Without a sense of her speed or altitude, Earhart knew that she could not safely judge the distance between her Lockheed Vega and the wild waves below.

It was the dead of night, and land was hundreds of miles away. The moment was one of life or death. To avoid spiraling into a watery grave, Earhart had to safely descend to an altitude low enough to prevent the accumulation of ice. In the blackness, she brought the plane down. Fog enveloped her aircraft, and she did not know whether impact was imminent. As the seconds ticked by, the plane motored on through the mist. It seemed as though crisis had been averted.

Looking around the airplane, Earhart suddenly noticed that it was burning. Tongues of fire leaped through an opening in the exhaust manifold, a part of the airplane's equipment that contained the flames produced by combustion. Unsure whether the fire would stay within safe limits or if the airplane's metal parts would buckle from its heat, she summoned her last reserves of courage. For Earhart, there was no choice but to keep flying and hope for the best. Minutes turned into hours, and before long, rays of sunshine broke through the darkness.

Now Earhart could more clearly estimate her altitude. She scanned the great illuminated sea for signs of human activity that might indicate that land was near.

Time was running out. Her exhaust manifold showed signs of severe fatigue. Earhart knew that she "should come down at the very nearest place, wherever it was." At last, she saw Ireland's coastline beneath ominous thunderclouds. Beginning a violent descent, she questioned whether it was even possible to land on the rolling hills below. "Not having the altimeter and not knowing the country," she remembered, "I was afraid to plow through those lest I hit one of the mountains." She finally found a safer place to alight in Londonderry—it happened to be a pasture where cows grazed. "I succeeded in frightening all the cattle in the country, I think, as I came down low several times before finally landing in a long, sloping meadow," she later joked. But her accomplishment was nothing to laugh about. Any one of the misfortunes that befell her during the transatlantic flight could have downed her plane. But Earhart had survived. For the first time in history, a female pilot had flown across the Atlantic Ocean.

EARHART'S FLIGHT DREW widespread praise and confirmed for the world that Lindbergh's flight was no fluke: transatlantic travel by airplane would be the way of the future. She also set multiple new records, which affirmed her status not only as the foremost female aviator of her day—but as one of the greatest of pilots in the world. In 1932, no one but Earhart had survived two transatlantic crossings. Her 2,026-mile flight from Harbor Grace to Londonderry, completed in a record-shattering fourteen hours and fifty-four minutes, also represented the fastest flight across the Atlantic at the time.

When US president Herbert Hoover heard the incredible news of Earhart's arrival in Londonderry, he sent a cablegram that expressed the "pride of the nation." He congratulated her "upon achieving the splendid pioneer solo flight by a woman across the Atlantic Ocean." Choosing his words carefully, President Hoover connected her accomplishment to those of the explorers who preceded her. He lauded her "dauntless courage" and recognized that her flight had broken barriers in the fight for women's rights by demonstrating "the capacity of women to match the skill of men in carrying through the most difficult feats of high adventure." It was an accomplishment that secured Earhart's place as America's most famous female flier.

Others in Earhart's position might, after such a triumph, contemplate enjoying a quiet life, but Amelia was not ready to retire. When she returned to the United States, she received many awards, including a National Geographic Society bronze medal and a gold medal from Harriet Chalmers Adams on behalf of the Society of Woman Geographers, the first bestowed upon an explorer in the organization's history. Earhart smashed additional records in the years that followed, inspiring downtrodden Americans and pushing the frontiers of flight during the Depression era by flying new, risky routes from Hawaii to California and from Los Angeles to Newark.

During the summer of 1937, Earhart embarked upon her most ambitious trip to date: a solo circumnavigation of the globe beginning and ending in Oakland, California. No person had ever succeeded in such an attempt. Her trip began on May 20, when she departed from Oakland with Putnam; Ruckins McKneely Jr., her mechanic; and Fred Noonan, her navigator, in a Lockheed Electra 10E monoplane. After stopping in Miami for repairs, Earhart and Noonan set off on June 1 to complete their trip around the world. Almost daily, Earhart reached a new country: Venezuela, Dutch Guiana, Brazil, Senegal, India, and Australia. Newspapers followed her every move, reporting on the distances flown and the types of weather encountered. Before long, Earhart reached the most dangerous leg of the journey: between Darwin, Australia, and Oakland, California. Flying over thousands of miles of open water in the Pacific Ocean, she and Noonan would have to land in Lae, New Guinea; Howland Island; and Honolulu, Hawaii.

On June 29, Earhart successfully reached Lae. Two days later, she and Noonan began the 2,570-mile trip to Howland Island, a stretch that one newspaper ominously warned was "the most hazardous flight on her leisurely journey around the world." Howland Island was an atoll set in the midst of the vast Pacific Ocean, an extremely difficult location to reach relying primarily on celestial navigation and radio contact with stations below. The flight was expected to take more than eighteen hours, and Earhart planned to broadcast her location every thirty minutes. Earhart expertly steered the heavy plane, weighed down by numerous gallons of fuel, along the runway and into the air. The skies were calm and visibility was good.

At Howland Island, the crew of the US Coast Guard ship *Itasca* waited impatiently to hear from Earhart. At 2:15 p.m. Greenwich Mean Time, her steady voice briefly broke through over the radio. But when the hour of her expected arrival approached, there was no sign of the Electra. Then, at 7:12 p.m., they heard a panicked message: "KHAQQ calling *Itasca* we must be on you but cannot see you but gas is running low been unable to reach you by radio we are flying at altitude 1000 feet." Earhart and Noonan were likely just ten miles from Howland Island, but for unknown reasons, they could not locate it. During these fraught minutes, the *Itasca* tried to reach Earhart. Due to radio transmission problems, Earhart could not receive their messages so that she could better discern her location. Circling the skies in search of Howland Island, Earhart sent her last message to the *Itasca*: "We are on the line of the position 157–337," she reported. "We are running north and south."

On July 2, 1937, the Associated Press broadcast the tragic news that "America's foremost woman flier and her navigator, Fred Noonan, were lost at sea today in the desolate wastes of the central Pacific after apparently overshooting tiny Howland Island." Neither the Lockheed Electra nor the bodies of Earhart and Noonan were ever found, but their last received radio transmission indicated that they had not found a place to land. It was a tragedy that shocked the world and generated significant speculation during the century that followed. Amelia Earhart had been, in the words of the *New York Times*, a woman whose passion for life drove her to "explore [her] way through it to that dark forest, that tumbling mass of driven water, where its boundaries lie."

10

Sally Ride

THE ASTRONAUT

The sky above Cape Canaveral, Florida, was black during the early hours of the morning on June 18, 1983. In the salt marshes near Kennedy Space Center, run by the National Aeronautics and Space Administration (NASA), herons, egrets, and a host of other wildlife slept peacefully. But the stillness before dawn was broken by sounds emanating from Pad 39A. There, NASA workers busily prepared the space shuttle *Challenger* for departure, inspecting its complex parts and pouring hundreds of thousands of gallons of liquefied gases into its external tank. In front of the Operations and Checkout Building, a white RV called the "Astrovan" waited to bring the *Challenger*'s crew members to the launch complex. As the doors of the O & C Building opened, the blinding flash of photographers' cameras pierced the darkness. Four men and one woman, all of them dressed in blue jumpsuits, confidently exited and strode toward the van.

It was an auspicious day, one that represented a milestone in the

history of American exploration. For the first time, an American female astronaut prepared to launch into space, and the world watched.
Her name was Sally Kristen Ride, and she had trained for this moment
for the past five years. In good spirits, Ride smiled and chatted with
NASA workers as her crewmates suited up and climbed aboard the
Challenger. Then Ride covered her bouncy brown hair with her black
communication cap, put on her launch-and-entry helmet, and entered
the cavelike cockpit of the space shuttle, which was illuminated by the
tiny lights of dozens of switches. It was almost time. Soon powerful
rockets would ignite beneath her, propelling the spacecraft beyond the
Kármán line, which demarcates the end of the earth's atmosphere. In
that moment, Sally Ride would cross a new frontier and become the
first American woman in space.

JUST OVER FIFTY years earlier, Amelia Earhart became the first
female aviator to complete a solo transatlantic flight. After Earhart's
death while trying to circumnavigate the globe, the American busi-
nessman Howard Hughes successfully flew around the world in a sleek
silver Lockheed Super Electra Special 14-N2 monoplane in fewer
than four days. His accomplishment impressed onlookers and prompted
many to consider, as the journalist Kirke Simpson then mused, that
"however earth-bound [a person's] own trudge through life . . . time
and space no longer mean much to humanity." In the aftermath of
Hughes's journey, however, worries about national security emerged.
"Amid the patriotic thrills there lurks also grave concern for the people
not only of this country but all of the Western Hemisphere," reflected
Simpson, where "security . . . behind its ocean ramparts, is not so certain as it was before Hughes took to the air four days ago." Such fears
were not unfounded.

America's golden age of aviation coincided with the alarming rise
of fascism in parts of western Europe. During the 1920s and 1930s,
the Italian National Fascist Party founder, Benito Mussolini, and the
German Nazi chancellor, Adolf Hitler, eliminated political opposition,

established dictatorships, and rapidly strengthened their military forces. Meanwhile, Japan's emperor, Michinomiya Hirohito, similarly built up his country's armies with an eye to territorial expansion. Between 1931 and 1939, each ruler invaded nearby sovereign countries, including Ethiopia, China, Austria, and Czechoslovakia, brazen acts that flouted international law.

Communism also posed a growing threat to security around the world. In the Soviet Union (USSR), the dictator Joseph Stalin clung to power, murdering his opponents by the millions through starvation and political purges. In accordance with an agreement with Stalin, Hitler's troops attacked western Poland in September 1939, an action that triggered declarations of war from Great Britain and France. Germany's strategy relied heavily on aircraft: its military was supported by thirteen hundred fighter planes and bomber planes. Allied forces quickly recognized the German Luftwaffe as the most advanced flying corps of its day. Its tactics, including reliance on the element of surprise, spawned the creation of a new word: *blitzkrieg*, or "lightning war."

The sudden outbreak of World War II in 1939, just one year after Hughes touched down at Floyd Bennett Field in Brooklyn, pushed aviation to greater heights. Air power became an even more critical component of modern warfare, enabling nations to conduct surveillance, drop bombs on unsuspecting targets, and engage in battles. Aerial combat between Allied and Axis powers, which began in 1939, resulted in shocking numbers of military and civilian casualties over the next six years, as the world became entangled in the deadliest international conflict to date. By 1946, up to seventy-eight million people, or 3 percent of the global population, had perished as a result of the war.

It was a surprise air strike that ultimately dragged an isolationist United States into the conflict. On December 7, 1941, 353 Japanese dive-bombers and torpedo bombers attacked the headquarters of the US Pacific Fleet at Pearl Harbor, Hawaii. They destroyed battleships and airplanes in a devastating assault that killed 2,404 people. In his somber address to Congress the following afternoon, President Franklin Delano Roosevelt famously called December 7 a "date which will live in infamy" and rallied the nation to join the fight against the Axis powers.

To meet the military challenges ahead, the United States began recruiting soldiers and ramping up industrial production, a task managed by the War Production Board. Factories churned out planes at impressive rates that reached ninety-six thousand aircraft in 1944 alone, when the Allied forces' D-Day invasion turned the tide of war in Europe in their favor. Six million American women played a critical role in the war effort as industrial workers. Jobs in production facilities offered women upward mobility. For example, Rose Widmer, the daughter of Swiss immigrants in rural Oregon, became a real-life "Rosie the Riveter," making $39 a week at the Douglas Aircraft Company in Santa Monica. Factories also produced heavy weaponry, naval craft, and tanks, which were sent to the European and Pacific theaters of war.

Two technological advances of the World War II era revolutionized humankind's relationship with the skies. The first was rocketry, a scientific field with roots in the 1920s. During World War II, German scientists began using the Vergeltungswaffen-2 (V-2), a deadly ballistic missile fueled by a rocket engine. After the Allied victory over Germany, the US military studied V-2s and worked with German scientists to construct ballistic missiles for America. Unlike planes that use oxygen from the environment, V-2 rockets carry the fuel and oxygen needed for combustion in environments with little to no oxygen. After the war, leaders in the United States and the Soviet Union quickly recognized rockets' incredible potential to facilitate transportation, communication, and military dominance. Advances in rocketry ultimately opened outer space as the world's new frontier.

The second major technological advance of the World War II era was America's detonation of the first atomic bomb, on July 16, 1945, in Jornada del Muerto, a desert in New Mexico. Three weeks later, the United States sent a B-29 Superfortress bomber to release a powerful atomic bomb over the Japanese city of Hiroshima. After a second atomic bomb was dropped over Nagasaki, Japan surrendered. As many as 246,000 people, most of them civilians, perished as a result of the two blasts. World War II finally came to an end, but peacetime would be short-lived. Tensions ran high among the former Allied powers in the new nuclear age, when the skies served as a potential pathway for weapons of mass destruction.

Although the USSR joined forces with the Allies in 1941 to help defeat Hitler after he invaded Soviet territory, the United States remained wary of the Communist nation. When Germany surrendered in May 1945, the USSR refused to relinquish control of territory it had annexed during the war in eastern Europe, including much of Poland. The Russian bear's actions set the stage for the Cold War, a period of extraordinarily tense relations between the United States and its former ally.

In response to Stalin's ambition to spread communism throughout eastern Europe and expand the borders of the Soviet Empire, the United States formulated a new policy: containment. President Harry Truman described it to Congress in 1947 as providing "financial and economic assistance" to "help free peoples to maintain their free institutions and their national integrity against aggressive movements that seek to impose upon them totalitarian regimes." America's commitment to defending democratic nations was more than rhetorical bluster. Congress immediately approved $400 million in anti-Communist aid and passed the National Security Act (1947), which strengthened the US military by establishing the Department of Defense, the Central Intelligence Agency, and the National Security Council. The Soviet threat seemed more dire than ever two years later, when the USSR detonated its first atomic bomb at a test site in the northeastern region of modern-day Kazakhstan. The short period of US nuclear supremacy had ended, and a new era defined by the threat of mutual assured destruction (MAD) had begun.

THIS WAS THE world into which Sally Ride was born. In May 1951, the month of her birth, war raged on the Korean Peninsula. The

Chinese People's Volunteer Army and Korean People's Army, both trained or supplied by the Soviet Union, had been fighting against US-backed South Korean soldiers, US soldiers, and UN soldiers for control of the region after North Korea's invasion in 1950. As a proxy war within the greater Cold War, the conflict was a major test of the United States' commitment to containing communism and a catalyst for technological and military development in what would become known as the space race.

Little did Sally Ride's parents know that their infant daughter would one day advance US military and scientific goals in the broader effort to secure dominance of the skies. Her early childhood in the suburbs of Los Angeles, California, offered little suggestion of her future career. Sally's father, Dale, was a World War II veteran turned professor, who taught political science at Santa Monica Community College. Her mother, Joyce Anderson, was a homemaker with progressive political views who undertook charitable work in women's prisons. Neither Dale nor Joyce harbored a love of science, but Dale emphasized to Sally and her younger sister, Karen, "the importance of education to get ahead in the world," Ride remembered. Ultimately, she felt free to follow her varied passions: "Anytime I wanted to pursue something that they weren't familiar with, that was not part of their lifestyle, they let me go ahead and do it."

Perhaps more than anything else, as a child Sally loved spending time outside. Like fellow explorers Harriet Chalmers Adams and Amelia Earhart, women who spent their formative years in the Golden State, Sally excelled in athletics and embraced California's outdoor lifestyle. In an interview with Ride's biographer and friend, the journalist Lynn Sherr, Karen told how her sister searched for "any way she could get out and about." From launching herself into the sky on the family's trampoline to kayaking, Sally strengthened her body and developed her stamina.

As an adolescent, Sally became hooked on tennis. An outstanding player, she was soon traveling to tournaments across the state with her wooden Dunlop Maxply racket. Her sister recounted that "when the kids played baseball or football out in the streets, Sally was always the best. When they chose up sides, Sally was always the first to be chosen. She was the only girl who was acceptable to the boys." Like Amelia Earhart, Sally also remained sensitive to others' impressions of her femininity, or lack of it. When a reporter later referred to her as a tomboy, she promptly retorted, "I really don't like that term. Tomboy, when applied to a girl, means a girl acting like a boy. As opposed to a girl acting like a girl." To Sally, athleticism and girlhood were wholly compatible.

As Sally grew older, glimmers of interest in science emerged. In junior high school, she learned about physics and began subscribing to *Scientific American*, a magazine that illuminated the intricacies of outer space, the natural world, and the human body. Her parents' gift of a silver Bushnell Sky Rover telescope also piqued her curiosity about the stars. In the California night sky, then largely untouched by light pollution, Sally spotted glowing planets and dazzling constellations, which heightened her sense of wonder about the vast universe.

But it was only as a high schooler at the private all-girls West-lake School, where Sally matriculated in 1965 on a tennis scholarship, that she fully realized her passion for science. Her classmate and friend Susan Okie described the teenage Sally as a "fleet-footed 14-year-old with keen blue eyes, a self-confident grin, and long, straight hair that perpetually flopped forward over her face." At Westlake, teachers introduced Sally to physics, physiology, and mathematics, topics that engaged her quick mind. English and Spanish language classes bored Sally to tears, Okie remembered, but "she had a bottomless reservoir of patience" when it came to the "mental gymnastics" required of students of science and math. One teacher in particular, Dr. Elizabeth Mommaerts, made a strong impression on her as a scientist and mentor. Her home country, Hungary, had unsuccessfully revolted against expansionist Soviet rule a decade earlier, in a bloody conflict that prompted terrified freedom-seekers to leave the country by the thousands. Thanks to the influence of Mommaerts and others, by her senior year, Sally decided that she wanted to major in astrophysics as a college student, a field then dominated by men.

The nation desperately needed more young scientists like Sally Ride. In the space race against the Soviet Union, the United States lagged behind its ferocious competitor. In October 1957, when Sally was six, the Communist nation shocked the world with its surprise launch of *Sputnik*, the first man-made satellite to orbit the earth. Rocket power sent the nearly two-hundred-pound object into space, an achievement that seemed to pave the way for the launching of animals, people, or more ominously, weapons into space. It was a clear victory for the ascendent Soviet Union.

The Communist Party's press mouthpiece, *Pravda*, enthusiastically spread the news among Soviet readers, with headlines like "The World's First Artificial Satellite of the Earth: Created in the Soviet State!" and "Triumph of Soviet Science and Technology." President Dwight Eisenhower's administration quickly assessed the public response and came to a troubling conclusion: "American prestige" had "sustained a severe blow," and citizens in "friendly countries" feared that "the balance of military power" was tipping toward the Soviet Union. To make matters worse, the USSR launched a second satellite, *Sputnik 2*, just one month later.

Seeking to score points against the USSR, the United States blasted its own satellite, *Explorer 1*, into orbit in January 1958. A Redstone rocket propelled the thirty-pound satellite into orbit as a group of reporters watched; they had sworn to keep any failure a secret. News of *Explorer*'s first successful orbit thrilled its architects. With *Explorer 1*, the two *Sputnik*s, "after circling the earth for a few months in solitude, had received a brother in orbit," the German American NASA rocket scientist Ernst Stuhlinger later said. It was a minor victory but one that helped the United States recapture "some of the prestige it lost to Russia last October," one journalist wrote.

To continue to successfully compete against the Soviet Union, the United States needed to increase funding for scientific research and better educate its youngest citizens. One year after the launch of *Sputnik 1* captivated the world, Congress passed a law to found NASA. This government agency would revamp and expand the role of its predecessor, the National Advisory Committee for Aeronautics (NACA). President Eisenhower publicly expressed his hope that the new organization would energize US space exploration and bolster American military capacities.

America needed more scientists too, especially physicists. After *Sputnik*'s launch, the National Science Board argued that the Soviet feat was "an impressive demonstration of the strong position of Russian science and education" and consequently "urged that both short and long range steps be taken continually to improve [America's] scientific position." To counter Soviet ambitions of space dominance, Congress also passed the National Defense Education Act in 1958. The law helped triple federal funding at the university level, particularly for mathematics and science. This effort targeted Sally Ride's generation, which was soon to begin entering college.

The space race had officially begun and there was no turning back. During the next decade, while a deeply unpopular proxy war raged in Vietnam, the USSR and the United States competed to gain bragging rights on a range of "firsts." They tested the abilities of living creatures to withstand the hardships of space travel, launching dogs and chimpanzees into orbit. The Soviet Union notched a major victory by sending the first human, the cosmonaut Yuri Gagarin, into space in 1961.

Sally Ride was just nine years old, but she may have glimpsed the headlines of the *Los Angeles Times* that day, warning readers that Gagarin's flight "could bring nearer the time when hydrogen bombing by satellite is possible" and reporting Congress's panicked call for the United States to "match the Russian man-in-space achievement in a full-speed-ahead program." Overton Brooks, the chair of the House Space Committee, publicly lamented, "We are in conflict, a contest to forge to the forefront in the exploration of space, man's last great frontier." Five days after Gagarin's return to earth, geopolitical tensions between the United States and the Soviet Union rose. A failed

US-Cuban plot to overthrow the country's Communist-backed leader, Fidel Castro, exposed US ineptitude on the world stage. How could the United States restore its public image and gain the upper hand against the Soviet Union?

THE ANSWER: BECOME the first nation to send a man to the surface of the moon. The celestial object nearest to the earth, the moon had long intrigued astronomers and amateur observers alike. To begin to truly unlock its mysteries, it would be necessary to walk upon its surface. "Uncharted areas of science and space," John F. Kennedy said in his acceptance speech as presidential nominee at the Democratic National Convention in July 1960, represented one critical aspect of the so-called New Frontier. While the "pioneers of old" relinquished "their safety, their comfort, and sometimes their lives to build a new world here in the West," he told his audience in Los Angeles, California, the time had come to take on new challenges.

One year later, Kennedy made explicit the goal of sending American astronauts to the moon. In an address to Congress on May 25, 1961, President Kennedy declared his belief that "this nation should commit itself to achieving the goal, before this decade is out, of landing a man on the moon and returning him safely to the earth." He argued that "no single space project in this period [would] be more impressive to mankind, or more important for the long-range exploration of space." But to achieve this goal, more money was needed. Kennedy asked Congress to authorize billions of dollars in additional spending. This huge sum signaled the depth of the US commitment to the mission to the moon. Congress ultimately approved Kennedy's request, recognizing the dual nature of the goal: the exploration of space would increase knowledge and enhance US security. Being the first to achieve what Kennedy called a "mastery of space" was now a top priority.

Although President Kennedy had stipulated that the US government planned to land a *man* on the moon, young women around the country harbored dreams of becoming astronauts. But the odds were against them. NASA specified that applicants had to have experienced wartime combat, or served as test pilots, or worked on submarines, or survived harsh conditions as Arctic or Antarctic explorers. Back then, women were prohibited from occupying the specified military roles, and opportunities for far-flung exploratory trips were few and far between, a situation that all but disqualified women from the NASA program that was meant to test a person's ability to withstand a lunar journey.

Still, a small group of women, many of whom were recruited by Amelia Earhart's organization of female pilots, the Ninety-Nines, held out hope that one of them might be among the first to set foot on the moon. In 1961, more than a dozen underwent grueling test regimens, led privately in Albuquerque, New Mexico by the NASA Special Life Sciences Committee Chair W. Randolph Lovelace, to assess their physical and mental fitness. Although more than half of the women passed, the trial was terminated when the Navy denied the use of its facilities for further testing. At the time, some Americans believed that women were unable to endure the demands of space travel, which included g-forces during liftoff, radiation, and isolation. The next year, NASA sent a now infamous letter to an American student, Linda Halpern, expressing the belief that they had "no present plans to employ women on space flights because of the degree of scientific and flight training, and the physical characteristics, which are required." That policy remained in place for the next fifteen years, depriving capable American women of the chance to use their skills in the service of their country.

The Soviet Union had no such hesitations. On June 16, 1963, they sent the first woman, Valentina Tereshkova, into orbit around the earth in a spacecraft called *Vostok VI*. A milestone in the global women's rights movement, it proved that sex was no barrier to surviving in space. *Pravda* crowed that Tereshkova's accomplishment represented a "new triumph in space exploration," calling her the "world's first woman-cosmonaut," one who successfully piloted her vessel around the planet. By radio telephone, the Soviet leader Nikita Khrushchev reportedly said to Tereshkova, "Now you see what women are capable of." The comment made it resoundingly clear that women possessed the mettle to survive in space. Sally Ride was only twelve at the time of Tereshkova's groundbreaking flight, but later she would acquire a Soviet coin commemorating the event's twentieth anniversary. Still, the Soviets had gotten no closer to the great prize: the moon.

By 1969, the United States prepared to cross a frontier that had long been the stuff of science fiction. Ten Apollo missions, testing the capacities of the rockets, spacecraft, and astronauts alike, had preceded this moment. In July, NASA determined that the time had come to send astronauts to land on the surface of the moon. Neil Armstrong, a Korean War veteran and former navy pilot, was selected as commander of the Apollo 11 mission. His background as an NACA test pilot, who designed and flew rocket-powered aircraft that reached exceptionally high speeds and altitudes, had helped him qualify for NASA's astronaut training program in 1962.

The week-long journey from the earth to the moon and back would test the willpower and fortitude of Armstrong and fellow astronauts Buzz Aldrin and Michael Collins. With millions around the world watching live on television, the group departed from Cape Canaveral, Florida, on July 16. While Collins orbited the moon in the Command Module "Columbia," Armstrong and Aldrin descended in the Lunar Module "Eagle," safely landing on the surface of the moon on July 20. During their unprecedented journey, the men completed a series of complex maneuvers, which included exiting earth's orbit and entering the moon's, as well as landing on its uncharted terrain. Armstrong and Aldrin collected lunar samples and performed experiments before returning with Collins to earth, as the world looked on in awe. The nearly flawless mission affirmed US strength in the space race and elicited worldwide wonder at what humankind could achieve through exploration.

Sally Ride, then a sophomore at Swarthmore College, was transfixed by the Apollo 11 mission. "I was playing in a national junior tennis tournament that day," she later remembered, and "convinced other players to watch the landing." As Armstrong bravely pressed his foot into the rocky lunar surface and said, "That's one small step for [a] man, one giant leap for mankind," Ride felt that "he became [her] hero." At the time, Ride was still wavering about whether to pursue science or tennis as a career. But the moon landing made up her mind. Suddenly, she "knew that [she] would become a scientist," she recalled, "and that led to [her] training as an astronaut." But at the time, she believed that the chances of actually becoming an astronaut were slim. "I just assumed there would never be a place for women," Ride lamented, since the Apollo astronauts all possessed combat or test-pilot experience that she could never attain. Little did Ride know that NASA would soon unexpectedly open its doors to young women like her.

THE OPPORTUNITY PRACTICALLY fell into her lap. It was 1977, the height of the US feminist movement that pushed for women's equality. Ride was a graduate student at Stanford University. She had transferred there in 1970 to complete her undergraduate work in physics and stayed on to get her doctorate in astrophysics. After picking up a copy of the *Stanford Daily* one day, she marveled at a front-page headline: "NASA to Recruit Women." Ride later said, "The ad made it clear that NASA was looking for scientists and engineers, and it also made it clear that they were going to accept women into the astronaut corps." NASA had had a change of heart following what Ride's biographer, Lynn Sherr, described as a "twenty-year effort to integrate the astronaut corps and [a] six-month campaign to attract female scientists." Ride was largely ignorant of these endeavors.

Intrigued, Ride recalled "whipp[ing] off a little handwritten note asking for more information" from the Johnson Space Center in Houston, Texas. Perhaps an undergraduate colloquium she aced, "Life Science Problems in Space Exploration," would come in handy after all. She filled out the paperwork, crossed her fingers, and waited. More than eight thousand others also submitted their applications to NASA. But Ride, a strong candidate thanks to her background in astrophysics, made it to the final round. A grueling series of stress tests and interviews in Houston preceded the selection committee's final decision. In the wee hours of the morning on January 16, 1978, a phone call shattered the silence of Ride's house. The voice on the line delivered life-changing news: she, at the age of twenty-six, had been accepted into NASA's astronaut training program. "[I] couldn't believe it was Houston," she later said. "I thought maybe I was dreaming."

During this same period, the agency was preparing for a new phase of exploration. After the United States won the race to the moon, NASA's objectives began to shift. Now, the agency focused on building an innovative space shuttle, considered by NASA to be "humankind's first re-usable spacecraft." Rather than create a new vehicle for each launch, NASA wanted to construct a spacecraft that could reach space using rocket power, orbit the earth independently, and land on its own wheels after the completion of each mission. Transporting a crew of up to seven passengers at a time, the space shuttle could also move materials from earth to a permanent, occupied space station, which NASA hoped to construct in the decades ahead. Its final purpose was covert; NASA planned for the space shuttle to facilitate reconnaissance by launching military satellites to gather critical intelligence about the Soviet Union.

Twenty-nine men and six women made up NASA's 1978 class of astronaut candidates. The group reflected NASA's new commitment to develop a more diverse team. Ride's fellow classmate Judith Resnik decisively told the press, after the women were selected, "There is no reason women can't perform as well as men in space." The new space shuttle program offered the "thirty-five new guys," as they called themselves, the opportunity to reach new heights, both literally and professionally. But who would be chosen to travel into space?

Dr. Sally Ride, having completed her dissertation, moved to Houston and threw herself into NASA's intensive training program. Her primary goal, "to fly in space," she later said, remained at the forefront of her mind. But to prove herself to NASA administrators, Sally knew that she had to "master the technical subjects . . . basically to do everything that [she] needed to do to put [herself] in a position to be selected for a flight." At the Johnson Space Center, scientific seminars engaged Ride as she devoured content on avionics and aerodynamics.

One of the few civilian astronaut candidates, Ride also reveled in the chance to learn to fly Northrop T-38 supersonic trainer jets. This activity was "completely new to [her]," and "an awful lot of fun!" she later said. Ride's training notes for aerobatics, or complex aircraft maneuvers, reveal her prudent approach as a pilot, one that would serve her well as an astronaut facing the risks of space travel. "Be alert to any hazardous condition that may develop and know how to get out of it safely," she reminded herself.

Ride also grappled with her new position as a public figure. A naturally reserved person, she closely guarded her personal life from the press. Her disposition may have developed during childhood, when displays of affection were few and far between in the Ride household, and vocal expressions of love were largely absent. Although Ride felt secure and protected by her parents as a young girl, she grew up to become what her friend Susan Okie described as "protective of her own emotions." Ride "learned early how to preserve her own privacy," Okie wrote, donning "psychic camouflage that made her an elusive character even in high school."

One aspect of her identity that she shielded from others, including many of her closest friends, was her sexual orientation. During college, Ride dated John Tompkins and also enjoyed a romantic relationship with a woman named Molly Tyson. When Ride moved to Houston in 1978, she did so with her boyfriend, Bill Colson, a physicist. As an astronaut candidate, she dated the test pilot Robert Gibson and married fellow astronaut Steve Hawley in 1982. During her marriage, however, Ride fell in love with Tam O'Shaughnessy, a female friend from her junior high school tennis circle. In 1987, Ride and Hawley divorced; O'Shaughnessy and Ride moved in together two years later. Their partnership would be the longest and most fulfilling of Ride's life.

And yet it was not until after Ride's death that the relationship was made public. Tam would later say that Ride "didn't want to be defined by the lesbian/gay label just as she didn't want to be defined by a gender label." Perhaps Ride feared that homophobia would tarnish her reputation or cause controversy at NASA during an era when same-sex marriage was illegal. One thing, however, was certain: Ride hoped to be recognized primarily for her myriad professional accomplishments as an astronaut and scientist.

IN 1982, SALLY RIDE caught her big break. NASA was preparing to send the space shuttle *Challenger* on the agency's seventh Space Transportation System mission (STS-7). Since the launch of the *Columbia* (STS-1) on April 12, 1981, the twentieth anniversary of Gagarin's flight, NASA had continued to test its shuttles' abilities. The space shuttle program gained significant support during Ride's early years at NASA under President Ronald Reagan, who wrote in his executive diary that "space truly is the last frontier." Reagan also viewed the USSR as a continued threat to US security and oversaw NASA's budget growth from $5.5 billion in 1981 to $9 billion in 1988. Seeking to keep up with the United States, the USSR constructed its own version of a reusable spacecraft, *Buran*, but it reached orbit only once, in 1988, on an uncrewed flight.

A little over a year before the *Challenger* was due to depart on STS-7, Johnson Space Center's director of flight crew operations, George Abbey, summoned Ride to his office. He informed her that she had been chosen to be the first American woman to fly into space. It was the chance of a lifetime, a history-making opportunity. It would demand intense training in advance of the mission, no easy task under the press scrutiny that would undoubtedly follow. Nonetheless, Ride felt reassured that NASA would "shiel[d her] from the media so that [she] could train with the rest of the crew and not be singled out," she later said. What had led NASA to pick her instead of one of the other female candidates? Reflecting on the path that led to her selection, Sally recalled that her philosophy had been to "just do as good a job as you can on the assignments you are given and try to make sure that the community . . . knows that you are working hard." Her strategy had been a winning one.

There was much to do to get ready. During the previous year, Ride had honed her skills from mission control as capsule communicator for STS-2 and STS-3, proving through effective dialogue with crew members via radio that she possessed the cool demeanor and technical mastery required of a space shuttle crew member. For STS-7, her first trip to space, Ride would serve as a mission specialist. Robert "Bob" Crippen, a former naval aviator, would act as commander, Frederick Hauck as pilot, and John Fabian and Norman Thagard as the other mission specialists. Four of the five crew members had never before traveled to space; only Crippen was a spaceflight veteran. Ride's primary responsibility was the operation of the *Challenger*'s robotic arm, which would remove satellites from the orbiter's cargo bay and deploy them into orbit during the course of their six days in space. She had mastered this task at NASA, where she worked with engineers to refine the arm's remote manipulator system.

Ride would later remember how "the training really accelerated and intensified during that two months before the flight," when she spent "virtually all [her] time trying to learn things . . . practice and just stuff that one last fact into [her] brain." In a black-and-white ninety-six-page notebook, she neatly jotted down her notes, first in pencil. Then she added important details in bright green ink. Ride studied the procedures for operating the payload assist module in transporting satellites ("launch with sunshield open") and for performing tests using the shuttle pallet satellite system, which was made in Germany.

The STS-7 crew's daily schedule, finalized three weeks before lift-off, gave Ride a sense of how every hour would be spent aboard the *Challenger*. Between periods of sleep and eating prepackaged meals, the astronauts would assess the spacecraft's orientation, deploy satellites, and conduct experiments. To ready her body for the lack of gravity she would experience in orbit, Ride floated around the back of a diving KC-135 airplane and swam in NASA's water environment training facility.

Press attention was an unwanted distraction. Reporters' sexist questions were irritating. The *Washington Post* reporter Judy Mann summarized the host of queries Ride had to "put up with from the press" before the *Challenger*'s launch. They ranged from "Would the flight affect her reproductive organs?" and "Did she intend to become a mother after her flight?" to "Does she weep when things go wrong in flight simulations?" and "Does she think women ought to be astronauts?" She also endured the slight when journalists and interviewers young and old refused to use her title, "Dr.," and when they referred to her familiarly by her first name, while addressing men by surname. Despite being treated as an object of curiosity rather than a skilled professional, Ride, according to one journalist, "remain[ed] calm, unrattled and as laconic as the lean, tough fighter jockeys who surround[ed] her."

Finally, after months of preparation and anticipation, the day of the *Challenger*'s launch from Cape Canaveral arrived. It was June 18, 1983, and Sally Ride was ready. Up at 3:13 a.m. after a restful night, thanks to the sleeping pill provided by NASA, Sally ate a large breakfast with her crewmates. Wearing her blue in-flight jumpsuit, she headed to Pad 39A and boarded the *Challenger*. At 7:33 a.m., the orbiter's three main engines started up followed by ignition of the two solid rocket boosters (SRBs) and slowly lifted the spacecraft into the air. The thick orange flames from the SRBs burned against the backdrop of the sun breaking through the morning clouds. "Lift-off, lift-off of STS-7 and America's first woman astronaut," cried a reporter who narrated the launch's video broadcast for excited viewers.

As the *Challenger* ascended, Ride was shaken violently from gravitational forces reaching three times the force of gravity (or 3Gs). Milky gray smoke trailed behind the rocket boosters, obscuring the view of the spacecraft from the ground. For two long minutes, the space shuttle climbed through the atmosphere. Then, tens of thousands of feet above the earth, the *Challenger* shed its rocket boosters, which began hurtling downward. But the orbiter continued its heavenward rise, carrying, as President Ronald Reagan had remarked earlier that month at a press conference, "a little bit of every American . . . [the nation's] pride and our prayers."

For Ride, the fear was palpable as her body hurtled toward the blackness of space and entered earth orbit. "It's an emotionally and psychologically overwhelming experience," she reflected. "[Your feelings] are fueled by the realization that you're not in the simulator, you're sitting on top of tons of rocket fuel and it's basically exploding underneath you." She experienced a tremendous range of emotions in the span of a few seconds, calling liftoff "exhilarating, terrifying, and overwhelming all at the same time." While her seat rattled forcefully seven seconds after ignition, Ride struggled to execute the task of reading aloud the first item on the crew's checklist. "I'll guarantee that those were the hardest words I ever had to get out of my mouth," she later said.

Ride felt a little more at ease once the *Challenger* reached outer space. The sensation of buoyancy in an environment devoid of gravity particularly appealed to her. "I really enjoyed being weightless," she said. "Although it took an hour or so to get used to moving around, I adapted to it pretty quickly." The other members of the STS-7 crew also adjusted well to the physical demands of space travel aboard the *Challenger*. Over the course of the next week, the *Challenger* traversed about 2.5 million miles as it circled the earth. What might Amelia Earhart, who died while attempting to fly around the world half a century earlier, have thought of Ride's chance to span the globe sixteen times in a single day?

When the time came to complete her assignments, Ride carried the mental weight of wanting to perform her work flawlessly while the world watched. One of the most challenging jobs was using the robotic arm to release and capture one of Germany's satellites, an unprecedented feat set to take place on June 22. As Sally prepared to snatch the metal object from space while the *Challenger* zoomed along at about seventeen thousand miles per hour, she thought to herself, "Oh, my gosh. This is real metal that will hit real metal if I miss."

But staying calm in high-pressure situations was one of Sally's gifts.

Despite her worries, she deftly maneuvered the arm to successfully complete what a *New York Times* journalist later called "one of the most significant tasks of the space age," one that "mark[ed] a new stage in the taming of the high frontier." It indicated that, in the "not-so-distant future . . . it could be routine to grasp satellites, mine asteroids, build space stations—in short, to clutch and shape instead of just to pass through space as an awestruck visitor."

After six days aboard the orbiter, the STS-7 crew prepared to return to earth. Inclement weather thwarted their plans to land at the Kennedy Space Center, where they would have made history as the first crew to do so. "I remember being disappointed that we weren't going to land in Florida," Ride said. Although she was pleased that their backup option was California's Edwards Air Force Base, where the weather was more predictable and which "almost felt like a second home," thanks to hours spent training there, she was still a bit disheartened. "There weren't many people there waiting for us!" she remembered, since the STS-7 crew's family and friends and the press had been expecting the *Challenger* to touch down in Florida.

On June 24, the *Challenger* exited its orbit and reentered the earth's atmosphere. A video camera captured the black-nosed, hulking craft gliding down toward the runway of Edwards Air Force Base against the backdrop of a cloudless blue sky. Without incident, the *Challenger's* landing gear descended, and the massive spacecraft slammed into the sandy earth, kicking up swirls of dust as it cruised down ten thousand feet of runway. A little over a minute later, the *Challenger* finally rolled to a stop. Following Commander Crippen, Ride and the rest of the crew calmly exited the spacecraft and conducted a postflight inspection. Even at the end of such a surreal journey, they remained consummate professionals.

LIKE SO MANY explorers before her, Sally Ride now faced a moment of significant transition. Her mission was accomplished, and it was time to deal with the inevitable onslaught of publicity. For an introvert like Ride, it was a challenge to balance her need for privacy with her desire to inspire others through her story. Public duties started shortly after a postflight medical checkup. At a press conference that day, Ride told reporters, in words that echoed those of Earhart, "The thing that I'll remember most about that flight is that it was fun, and in fact I'm sure it's the most fun I'll ever have in my life."

After their interviews concluded, NASA flew the STS-7 crew to Houston for an event at the Johnson Space Center. The mood was celebratory, but then Ride miscalculated the consequences of rejecting a bouquet of flowers from a NASA staffer during a ceremony in the crew's honor. The act generated negative news headlines. Although the wives of male crew members received roses, none of the male crew members were given flowers, and Ride had said she "wanted to be treated no differently than her four crewmates," one journalist reported. Had she wanted to make a show of shattering gendered stereotypes? Ride explained that she had a more practical reason: "I just wanted my hands free."

A telephone call from President Reagan proved a more positive moment during Ride's whirlwind of a day. From a yellow wingback chair in the White House Diplomatic Room, Reagan welcomed the crew home. "All of America was watching you and what you were accomplishing up there," he told them warmly. Singling out Ride, Reagan praised her for adeptly maneuvering the *Challenger*'s robotic arm, assuring Sally that her abilities rather than the fact of her sex made her the "best person for the job." Furthermore, he recognized the impact of Ride's achievement in terms of the women's rights movement, arguing that she had set the standard for how people would be chosen by NASA for future assignments.

Some observers linked Ride's accomplishment to those of the female explorers who preceded her. "Sally Ride Earns Place in History Next to Earhart," the *Miami Herald* attested, calling Ride's trip to space a milestone in aviation history as well as "the history of the women's rights movement." In Earhart's hometown, members of the Ninety-Nines proudly showed off a granite marker bearing Ride's name amid a forest containing stones commemorating other female aviators. Secretary of Transportation Elizabeth Dole connected Earhart's legacy to that of women like Ride, who "continue to break new barriers and reach higher levels of achievement," she said at the event.

What did Ride's success mean in the context of the Cold War, now in its fourth decade? The United States had secured a victory both symbolic and tangible, demonstrating its commitment to women's rights and continued prowess in operating a reusable spacecraft. But questions still remained as to which nation possessed the advantage in space. A little over a week after Ride returned to earth, a reporter for the *Atlanta Constitution* declared that "no winner [was] apparent" in the ongoing push for space dominance. While the United States had proved the efficacy of its space shuttle system, the Soviet Union's efforts to establish a "permanent, manned Soviet station in space" far exceeded NASA's slow movement in that direction.

FOR SALLY RIDE, STS-7's conclusion marked the beginning of her journey as an astronaut and NASA expert. One year later, she returned to space on the *Challenger,* alongside Kathryn D. Sullivan, on STS-41-G. During their week-long mission, the crew launched another satellite and conducted additional experiments. Sullivan even completed a space walk, becoming the first American woman to do so. But in January 1986, tragedy struck when the *Challenger* exploded during the launch of STS-51-L. The event was broadcast live on television. All seven crew members, including the first teacher-in-space, Christa McAuliffe, perished. Ride was devastated to lose four of her classmates, individuals who had become good friends over the past eight years. She narrowly dodged death herself, having recently been assigned to an upcoming flight on *Challenger.*

The explosion shook the nation's faith in NASA and the space program. To determine what had happened, President Reagan established a commission of experts that included Ride. She lent her knowledge to the investigation but retired from NASA about one year after the commission delivered its findings. For the rest of her life, Ride worked in academia and the private sector. There, she championed STEM learning through publications and her own company, Sally Ride Science. In 2012, she passed away at the age of sixty-one after a fierce battle against pancreatic cancer.

Following in the footsteps of men and women who challenged discrimination and other obstacles, Ride blazed a wide path into the frontier of space. Her achievements as an astronaut strengthened the position of the United States during the Cold War. Ride also made strides for women's rights by defying critics who believed women were unfit for space travel. Reflecting on her life in 2001, she marveled that twenty-first-century children assumed women could be astronauts without fully appreciating their long struggle for acceptance and their quiet contributions along the way. As the first American female astronaut, Ride inspired future generations to continue pushing the limits of human exploration.

The Future of American Exploration

America has always possessed an urge to explore, compete, and conquer; but it also seeks to preserve, conserve, and replenish. Which impulse will prevail in the twenty-first century, as challenging frontiers emerge? Will future explorers find ways to protect our planet and space, rather than exploit them? What technologies will fuel exploration, allowing adventurers to reach new regions of the earth or the planets beyond it? Finally, might innovative inventions allow explorers to glimpse the past, uncovering secrets that have unfolded over the course of billions of years?

Climate change has recently transformed the earth's most contested region: the Arctic. Over the past 140 years, the planet's average temperature has risen by nearly two degrees Fahrenheit. Rates of warming have been concentrated in the Arctic due to the amplified effects of melting sea ice and the subsequent release of heat from its waters. About 267 billion tons of glacial ice also vanished with each passing year since 2000, exposing lands holding huge amounts of oil, gas, and mineral deposits. *National Geographic* reports that the Arctic, "once considered a frozen wasteland," is today considered "an emerging frontier" because of the rich resources that lie buried beneath the thawing earth. It remains to be seen whether Arctic exploration will lead to the region's conservation through international agreement or whether nations will soon find themselves engaged in a "new Cold War" as they seek to profit from access to this polar region.

If warming continues, the Arctic will be more accessible than ever. Countless adventurers spent time and treasure over the past few centuries, searching for nautical routes through its frigid waters. Many gave their lives to the cause. But with less sea ice, ships will be able to rapidly circumnavigate the globe, since routes through the Arctic are up to 50 percent faster than those through the Suez or Panama Canals. Tourism to the Arctic has also exploded in recent years, with more than 106 trips offered to northern polar regions in 2022 alone. In 2019, ships with a passenger capacity of nearly 100,000 people visited the Arctic, making exploration of the inhospitable region possible for a growing number of people. But as human activity increases, so does the risk of environmental disruption.

While tourists might head to the Arctic for a glimpse of caribou, seals, or whales, archaeologists who venture there do so in search of history. At the Yana "Rhinoceros Horn Site," located north of the Arctic Circle in Siberia, researchers have discovered evidence of human settlement from thirty thousand years ago. Tools sculpted from woolly mammoth tusks, weapons, and sewing needles lay buried beneath the sediment, relics of a community from long ago. Genetic testing on excavated human teeth points to distant links between the people of ancient North Siberia and the earliest migrants to the Americas, who were explorers in their own right. While much remains unknown about the earth's diverse peoples, discoveries like these help scholars come closer to solving the complex mysteries of human history.

In the twenty-first century, other adventurers have turned to space in search of an alternative habitat. Some consider Mars, the fourth planet from the sun, to be the next frontier and a possible site of future human settlement. With two moons, a climate that might support life, and gravitational forces that are 62 percent less than those of the earth, Mars appears to be an attractive option for space-faring explorers. The SpaceX company, founded by Elon Musk, hopes to send a manned mission to the Red Planet in the next decade. Having developed an innovative refuelable spacecraft, *Starship*, SpaceX believes that a round-trip journey to Mars is almost within reach. Through missions to Mars, SpaceX ultimately hopes to make "humanity multiplanetary."

Further exploration of the moon also seems to be on the horizon, despite the fact that fifty years have passed since the last lunar mission. On December 11, 1972, Eugene Cernan and Harrison Schmitt became the eleventh and twelfth Americans to walk on the moon's rocky gray surface. As part of the Apollo 17 mission, they ventured to a previously unexplored area of the moon, collecting unusual rocks and testing a lunar roving vehicle. But one year later, with an economic recession looming, the share of federal budget dollars apportioned for NASA shrank. With limited political appetite for financing expensive missions to the moon, NASA focused on ramping up its new space shuttle program instead. During subsequent decades, not a single manned mission to the moon occurred, and the Apollo program became an artifact of history.

By the mid-2020s, however, NASA predicts that Americans will once again walk on the moon. Its new Artemis program, which takes its name from the Greek god Apollo's twin sister, will send the first female astronaut and first person of color to the moon's surface. To facilitate research efforts longer in duration than the Apollo missions, the agency plans to set up a permanent, solar-powered base camp near the moon's south pole. From there, astronauts could search for water to support the base camp and further explore the moon's surface via lunar terrain vehicles that would negate the need for cumbersome spacesuits.

At a moment characterized by difficult economic conditions similar to those of the 1970s, NASA has touted the fiscal advantages of the Artemis program. "Artemis missions enable a growing lunar economy by fueling new industries, supporting job growth, and furthering the demand for a skilled workforce," NASA proclaims on its Artemis website. As in the Cold War era, geopolitical concerns also remain front and center. Although the Soviet Union collapsed in 1991, Russia continues to pose a significant threat to world stability, most recently through its invasion of Ukraine in February 2022. Its ally, Communist China, also threatens to upend global peace in southeast Asia, where it has aggressively expanded its presence in the South China Sea.

To remain competitive militarily, the United States must maintain its ability to freely operate satellites and deploy weapons. John Raymond, the first chief of space operations in the US Space Force, warned in December 2020 that "Russia and China have made obvious their intention to challenge American preeminence in commercial and military space and to prevent the US from using its space capabilities in crisis and conflict, raising the prospect of war beginning in, or extending into, space." The US Space Force, a branch of the military founded in 2019, represents part of the nation's effort to counter these challenges, and the Artemis program is another critical tool.

A final, and novel, area of American exploration in the twenty-first century relates not to humankind's future in space but to the origins of the universe. For centuries, technological advances have propelled exploration, be they the navigational tools of the Corps of Discovery, the first airships and airplanes, or the rocket-powered spacecraft of the late twentieth century. As of 2021, NASA's James Webb Space Telescope enables explorers of the stars to look farther back in time than ever before. The universe began expanding 13.8 billion years ago, following the big bang, and the James Webb Space Telescope can show us the universe as it existed just 250 million years after that tremendous life-generating explosion.

Launched atop an Ariane 5 rocket on December 25, 2021, from French Guiana, the massive telescope began its unprecedented trek across one million miles of space after successfully separating from its rocket boosters. It was a historic moment, one that US president Joseph Biden described as "a new window into the history of our universe." Awestruck by the sight of the telescope's first images, which depicted starlight from billions of years earlier, Biden marveled at how the photos would "remind the American people, especially our children, that there's nothing beyond our capacity. . . . We can go places no one has ever gone before." Inventions like the James Webb Space Telescope force us to redefine exploration, expanding the concept to include not only ventures into new geographic regions but also those looking back in time.

President Ronald Reagan may have called space "the last frontier," but American adventurers of the twenty-first century have set their sights on tremendously varied terrain. From the melting Arctic and the dusty moon to the Red Planet and even the starlight of long ago, opportunities for exploration continue to beckon those of great fortitude. Using new forms of technology and innovative modes of transportation, they will make daring strides into unknown regions. If they successfully take up the mantle of the explorers of the past who forged paths, climbed mountains, and crossed seas, they too will expand human knowledge about the world and universe in which we live.

ACKNOWLEDGMENTS

It was a joy to study and write about the American explorers whose discoveries shaped the course of US history. I conducted much of the research for *The Explorers* during the COVID-19 pandemic, when travel opportunities were few and far between. At that time, it was especially thrilling to read about the far-flung adventures of women and men who climbed mountains and trekked through jungles in centuries past.

In the summer of 2021, however, I was able to travel across the United States on a research road trip to many of the places featured in *The Explorers*. Following the path of the Corps of Discovery, I departed by car from Saint Louis, Missouri, and journeyed to Fort Clatsop near the Oregon coast. On the way, I visited many museums and historic sites ranging from the Knife River Indian Villages in North Dakota, where Sacagawea lived among the Mandan and Hidatsa peoples, to Lemhi Pass on the border of Idaho and Montana, where the Corps of Discovery crossed the Continental Divide.

During this journey, I also went to Fort Boonesborough in Kentucky, Amelia Earhart's house in Kansas, Laura Ingalls Wilder's homestead in South Dakota, and the pass through the Sierra Nevada that James Beckwourth discovered. In California, I traveled to Coloma, where John Marshall first discovered gold on the banks of the

South Fork of the American River, and to John Muir's treasured Yosemite Valley. In the fall of 2022, I traveled to Dunbar, Scotland, to see Muir's birthplace and hometown. The endnotes of *The Explorers* contain a full accounting of historic sites visited.

Research for this book was supported by expert archivists, librarians, museum staff, and scholars who helped me acquire materials both in person and through digital delivery. I am deeply grateful to Kathryn Davis, who generously shared with me her transcriptions of Harriet Chalmers Adams's travel journals and offered insightful analyses about key questions pertaining to her life. Lacey Flint at the Explorers Club, Jessica LaBozetta at the American Heritage Center at the University of Wyoming, Sara Cordes at the California State Library, Eleanor Affleck at John Muir's Birthplace, and Jacqueline Lapsley at Union Presbyterian Seminary also assisted me in acquiring or locating materials.

I would also like to thank the staff members of the following institutions for their help: the Beulah M. Davis Special Collections at Morgan State University; the Bancroft Library at the University of California, Berkeley; Special Collections at the UCLA Library; the Smithsonian Institution; the Lewis County Historical Society; the Sacajawea Interpretive, Cultural, and Education Center; the Missouri River Basin Lewis and Clark Center; the Knife River Indian Villages National Historic Site; Fort Mandan State Historic Site; the Lewis and Clark Interpretive Center at Cape Disappointment; the Fort Clatsop visitor center; the Marshall Gold Discovery State Historic Park; the Museum at the Gateway Arch; the Little Bighorn Battlefield National Monument; the Laura Ingalls Wilder Memorial Society; the Amelia Earhart Birthplace Museum; the National World War II Museum; the *Intrepid* Sea, Air, and Space Museum; and the Kennedy Space Center.

I benefited greatly from scholars who offered sage advice at different stages of the writing process. In 2022, I presented chapter excerpts at the annual conference of the British American Nineteenth-Century Historians at the University of Leicester, where I received extremely helpful feedback. In 2023, I shared material with the participants of the Southern Intellectual History Colloquium at the University of Miami and received excellent advice. The members of the Brick Church Men's Association also offered thoughtful comments following a presentation of my research in 2023.

Ten experts in their respective fields kindly agreed to review different chapters in *The Explorers*. I sincerely appreciate their editorial suggestions and critiques, which made the book much stronger. My gratitude goes to Joseph Brentano, James Campbell, Kathryn Davis, Amy Foster, Doug Friedli, Jim Hardee, Edward Larson, Cheryl Palmlund, Susan Fox Rogers, and Donald Worster.

One of the great delights of working as a professor is the fellowship of the academic community. I am grateful for the scholarly insights and friendship of James Basker, Catherine Bateson, Brandon Byrd, Fitz Brundage, Elaine Abelson, Elizabeth Ellis, Thomas Evans, Oz Frankel, Federico Finchelstein, Tim Galsworthy, Aston Gonzalez, Hilary Green, Peter Kolchin, Sarah McNamara, Louise McReynolds, Natalia Mehlman Petrzela, Claire Potter, Julia Ott, Holly Pinheiro, Stephen Riegg, Amity Shlaes, David Silkenat, Gyorgy Toth, Jeremy Varon, Kevin Waite, and Eli Zaretsky.

I want to extend my heartfelt appreciation to my outstanding editor at William Morrow, Nick Amphlett, whose sharp mind, keen eye, and wise words have shaped this book from beginning to end. I would also like to thank my exceptional literary agent, Elias Altman, who shared my vision for this project from its earliest days and who has offered unwavering support and expert guidance through the publishing process. It has been an absolute privilege to learn from Nick and Elias and work with them over the past few years. I am also indebted to my attentive copy editor Susanna Brougham, production editor Laura Brady, cartographer Nick Springer, cover designer Paul Miele-Herndon, interior designer Elina Cohen, publicist Martin Wilson, and marketing director Kasey Feather.

Lastly, I would like to thank my family members—Missye and Brian, Matt and Briana, Charles and Erica, Maureen and Gary,

Charles and Deborah, Mary and Helen—for their unwavering support over the past few years. I am grateful for the love of my husband, Marcus, and our three children, who are so dear to me. This book is dedicated to my parents, Anita and Mark, who led me on my first explorations of the American landscape as a child and whose love of history and the environment has been a constant source of inspiration.

NOTES

PROLOGUE

1   *a small wooden fort:* Roy Edgar Appleman, "Historic Sites in the South," *Regional Review* 2, no. 5 (1939), accessed at https://www.nps.gov/parkhistory /online_books/regional_review/vol2-5c.htm on September 1, 2022; "Boone's Fort," drawing by James Reeve Stuart, 1903, Wisconsin Historical Society, accessed at https://www.wisconsinhistory.org/Records/Image/IM24939 on October 13, 2023; and Bobbi Dawn Rightmyer, "James Harrod and Daniel Boone: Early Frontiersmen," *Kentucky Monthly*, November 3, 2021, accessed at http:// www.kentuckymonthly.com/magazine/kentucky-explorer/james-harrod-and -daniel-boone-early-frontiersmen/ on September 19, 2022.

1   *built near a salt lick:* Daniel Boone to John Filson, "Colonel Boone's Autobiography," 1784, accessed at https://www.danielboone.org/p/boones-autobiography -following-pages.html on September 1, 2022; and "National Register of Historic Places Registration Form for Fort Boonesborough Townsite Historic District," National Park Service, March 1994, page 8, accessed at https://npgallery .nps.gov/NRHP/GetAsset/NRHP/94000303_text on September 6, 2022.

1   *place of abundance:* Ellen Eslinger, ed., *Running Mad for Kentucky: Frontier Travel Accounts* (Lexington: The University Press of Kentucky, 2004), 3.

1   *prized by the Shawnee:* Eslinger, ed., *Running Mad for Kentucky*, 3.

1   *historic buffalo traces:* Eslinger, ed., *Running Mad for Kentucky*, 6, 8; and Rickie Longfellow, "The Cumberland Gap," US Department of Transportation, accessed at https://www.fhwa.dot.gov/infrastructure/back0204.cfm on September 19, 2022.

2   *an illegal transaction:* Henderson signed the Treaty of Sycamore Shoals with a group of Cherokee men in 1775. Eslinger, ed., *Running Mad for Kentucky*, 6, 8; and Michael A. Lofaro, *Daniel Boone: An American Life* (Lexington: The University Press of Kentucky, 2010), 50–51.

2   *a failed siege:* Boone to Filson, "Colonel Boone's Autobiography," and John Mack Faragher, *Daniel Boone: The Life and Legend of an American Pioneer* (New York: Henry Holt and Company, 1993), 189.

2   *thousands of settlers:* Longfellow, "The Cumberland Gap."

2   *"howling wilderness":* Boone to Filson, "Colonel Boone's Autobiography."

2   *"the prototype and epitome":* Lofaro, *Daniel Boone*, ix.

2   *bitter winters:* Boone to Filson, "Colonel Boone's Autobiography."

2   *dramatic rescue operation:* For a full account of this event, see Bob Drury and Tom Clavin, *Blood and Treasure: Daniel Boone and the Fight for America's First Frontier* (New York: St. Martin's Press, 2021).

2   *new book on Kentucky:* Michael A. Lofaro, "The Eighteenth-Century 'Autobi-
    ographies' of Daniel Boone," *The Register of the Kentucky Historical Society* 76,
    no. 2 (1978): 86.

2   *a suspected attack:* Lofaro, "The Eighteenth-Century 'Autobiographies' of Dan-
    iel Boone," 86; "History: The Filson Historical Society," accessed at https://
    filsonhistorical.org/about-us/history/ on February 27, 2023; and Jeff Suess,
    "How Losantiville Became Cincinnati and Why a Founding Father Never
    Even Saw the City," *Cincinnati Enquirer,* January 31, 2021, accessed at https://
    www.cincinnati.com/story/news/2021/01/31/cincinnati-lost-founding-father
    -john-filson-losantiville/4216024001/ on February 27, 2023.

3   *A heavily edited version:* The original title, altered for readability, is spelled "The
    Adventures of Col. Daniel Boon." John Trumbull, *The Adventures of Colonel
    Daniel Boon, One of the First Settlers at Kentucke* (Norwich: Printed by John
    Trumbull, 1786).

3   *reprinted twelve times:* Lofaro, "The Eighteenth Century 'Autobiographies' of
    Daniel Boone," 86, 92–93.

3   *multiple runs:* Lofaro, "The Eighteenth Century 'Autobiographies' of Daniel
    Boone," 91–92.

3   *"great . . . back-woodsman":* Lord Byron, *Don Juan,* 1837, accessed at https://
    www.gutenberg.org/files/21700/21700-h/21700-h.htm on February 27, 2023,
    referenced in "Daniel Boone," *Encyclopaedia Britannica,* accessed at https://
    www.britannica.com/biography/Daniel-Boone on February 27, 2023.

3   *fallen into disrepair:* "National Register of Historic Places Registration Form
    for Fort Boonesborough Townsite Historic District," National Park Service,
    March 1994, page 7, accessed at https://npgallery.nps.gov/NRHP/GetAsset
    /NRHP/94000303_text on September 6, 2022.

3   *"The fame of Boone":* Clarence Walworth Alvord, "The Daniel Boone Myth,"
    Journal of the Illinois State Historical Society 19, no. 1/2 (1926): 16.

4   *"the most celebrated":* Stephen Aron, "The Legacy of Daniel Boone: Three Gen-
    erations of Boones and the History of Indian-White Relations," *The Register of
    the Kentucky Historical Society* 95, no. 3 (1997): 221.

4   *The first explorers:* Simon Worrall, "When, How Did the First Americans Ar-
    rive? It's Complicated," *National Geographic,* June 9, 2018, accessed at https://
    www.nationalgeographic.com/science/article/when-and-how-did-the-first
    -americans-arrive—its-complicated on August 29, 2022.

5   *stone tools and weaponry:* Worrall, "When, How Did the First Americans Ar-
    rive?"

5   *Nearly four million:* This number from the 1790 census did not include Na-
    tive Americans. "Censuses of American Indians," *United States Census Bureau,*
    accessed at https://www.census.gov/history/www/genealogy/decennial_census
    _records/censuses_of_american_indians.html on July 26, 2023.

I. SACAGAWEA: THE NAVIGATOR

13 *In 1800:* Meriwether Lewis, July 28, 1805, in *Journals of Lewis and Clark Ex-
    pedition,* accessed at https://lewisandclarkjournals.unl.edu/item/lc.jrn.1805
    -07-28#lc.jrn.1805-07-28.02 on November 8, 2022; site visit, Sacajawea Inter-
    pretive, Cultural, and Education Center, Salmon, Idaho, July 31, 2021.

13   *Her homeland:* Sally McBeth, "Memory, History, and Contested Pasts: Re-imagining Sacagawea/Sacajawea," *American Indian Culture and Research Journal* 27, no. 1 (2003): 4; site visit, Sacajawea Interpretive, Cultural, and Educational Center, July 31, 2021; and John W. W. Mann, *Sacajawea's People: The Lemhi Shoshones and the Salmon River Country* (Lincoln: University of Nebraska Press, 2004), 2–3.

13   *provided nourishment:* Site visit, Sacajawea Interpretive, Cultural, and Educational Center, July 31, 2021; and April R. Summit, *Sacagawea: A Biography* (Westport: Greenwood Press, 2008), 4.

13   *silvery spotted salmon:* Summit, *Sacagawea*, 2; site visit, the Missouri River Basin Lewis & Clark Center, July 26, 2021; and site visit, Sacajawea Interpretive, Cultural, and Educational Center, July 31, 2021.

13   *to procure bison meat:* Scott Thybony, *The Tipi: Portable Home of the Plains* (Tucson: Western National Parks Association, 2003), 4, 7–9; and site visit, Sacajawea Interpretive, Cultural, and Educational Center, July 31, 2021.

14   *to form tipis:* Thybony, *The Tipi*, 4, 7–9; and site visit, Sacajawea Interpretive, Cultural, and Educational Center, July 31, 2021.

14   *assisting her mother:* Site visit, Sacajawea Interpretive, Cultural, and Educational Center, July 31, 2021; and Summit, *Sacagawea*, 5.

14   *weakened by smallpox outbreaks:* Site visit, Sacajawea Interpretive, Cultural, and Educational Center, July 31, 2021.

14   *could get horses:* Site visit, Sacajawea Interpretive, Cultural, and Educational Center, July 31, 2021.

14   *enslave Shoshone women:* Site visit, Sacajawea Interpretive, Cultural, and Educational Center, July 31, 2021.

14   *thick groves of cottonwood trees:* Meriwether Lewis, July 28, 1805, in *Journals of Lewis and Clark Expedition*, accessed at https://lewisandclarkjournals.unl.edu/item/lc.jrn.1805-07-28#lc.jrn.1805-07-28.02 on July 6, 2021; John Ordway, July 30, 1805, in *Journals of Lewis and Clark Expedition*, accessed at https://lewisandclarkjournals.unl.edu/item/lc.jrn.1805-07-30 on August 13, 2021.

14   *chased them:* Meriwether Lewis, July 28, 1805, in *Journals of Lewis and Clark Expedition*, accessed at https://lewisandclarkjournals.unl.edu/item/lc.jrn.1805-07-28#lc.jrn.1805-07-28.02 on July 6, 2021.

14   *and plunged in:* John Ordway, July 30, 1805, in *Journals of Lewis and Clark Expedition*, accessed at https://lewisandclarkjournals.unl.edu/item/lc.jrn.1805-07-30 on August 13, 2021.

14   *they spared Sacagawea:* Meriwether Lewis, July 28, 1805, in *Journals of Lewis and Clark Expedition*, accessed at https://lewisandclarkjournals.unl.edu/item/lc.jrn.1805-07-28#lc.jrn.1805-07-28.02 on July 6, 2021.

14   *one of five villages:* These were the villages of Mitutanka, Ruptáre, Mahawha, Metaharta, and Menetarra. James P. Ronda, "The Mandan Winter," in *Lewis and Clark Among the Indians* (Lincoln: University of Nebraska Press, 1984), accessed at https://lewisandclarkjournals.unl.edu/item/lc.sup.ronda.01.04 on July 2, 2021; and William Clark, October 24, 1804, in *Journals of Lewis and Clark Expedition*, accessed at https://lewisandclarkjournals.unl.edu/item/lc.jrn.1804-10-24 on August 10, 2021.

14   *more than three thousand:* Site visit, Lewis and Clark Interpretive Center, July 28, 2021.

15   *Sacagawea likely helped:* The information regarding Sacagawea's life among the Hidatsa comes from a site visit I made to the Knife River Indian Villages National Historic Site, July 28, 2021.

15   *Because she was enslaved:* Christina Snyder, *Slavery in Indian Country: The Changing Face of Captivity in Early America* (Cambridge: Harvard University Press, 2010), 128.

15   *sold her into marriage:* William L. Lang, "Toussaint Charbonneau," *Oregon Encyclopedia,* accessed at https://www.oregonencyclopedia.org/articles/charbonneau _toussaint/#.YRPwfohKg2w on August 11, 2021.

15   *Otter Woman:* Scholars believe her name was Otter Woman. "Sacagawea," *American Battlefield Trust,* accessed at https://www.battlefields.org/learn /biographies/sacagawea on July 6, 2021; and Lang, "Toussaint Charbonneau."

15   *abusing Indigenous women:* Susan M. Colby, *Sacagawea's Child: The Life and Times of Jean-Baptiste (Pomp) Charbonneau* (Norman: University of Oklahoma Press, 2009), 34.

15   *lacked the support:* Summit, *Sacagawea,* 5.

15   *promised in marriage:* Meriwether Lewis, August 19, 1805, in *Journals of Lewis and Clark Expedition,* accessed at https://lewisandclarkjournals.unl.edu/item /lc.jrn.1805-08-19 on July 6, 2021.

16   *group of foreign men:* William Clark, October 27, 1804, in *Journals of Lewis and Clark Expedition,* accessed at https://lewisandclarkjournals.unl.edu/item /lc.jrn.1804-10-27 on July 7, 2021.

16   *brought boiled hominy:* William Clark, October 28, 1804, in *Journals of Lewis and Clark Expedition,* accessed at https://lewisandclarkjournals.unl.edu/item /lc.jrn.1804-10-28 on July 7, 2021.

16   *offered goods in return:* John Ordway, October 29, 1804, in *Journals of Lewis and Clark Expedition,* accessed at https://lewisandclarkjournals.unl.edu/item /lc.jrn.1804-10-29 on July 7, 2021.

16   *fired their guns:* William Clark, October 29, 1804, in *Journals of Lewis and Clark Expedition,* accessed at https://lewisandclarkjournals.unl.edu/item/lc .jrn.1804-10-29 on July 7, 2021.

16   *selection of a campsite:* William Clark, November 2, 1804, in *Journals of Lewis and Clark Expedition,* accessed at https://lewisandclarkjournals.unl.edu/item /lc.jrn.1804-11-02 on July 7, 2021.

16   *approached their campsite:* William Clark, November 4, 1804, in *Journals of Lewis and Clark Expedition,* accessed at https://lewisandclarkjournals.unl.edu/item /lc.jrn.1804-11-04 on July 15, 2021. The description in the rest of the paragraph comes from the same source.

16   *Shawnee and French Canadian:* Stephen E. Ambrose, *Undaunted Courage: Meriwether Lewis, Thomas Jefferson, and the Opening of the American West* (New York: Simon & Schuster, 1996), 187.

17   *The forty-five men:* "Corps of Discovery," National Park Service, accessed at https:// www.nps.gov/jeff/learn/historyculture/corps-of-discovery.htm on July 8, 2021.

17   *former Army lieutenant:* In 1803, Meriwether Lewis requested that William Clark receive a promotion to the rank of captain, but this appeal was denied.

Lewis considered Clark to be his co-captain and equal in commanding the expedition. The members of the Corps of Discovery believed that both men were captains. "William Clark," *National Museum of the United States Army*, accessed at https://www.thenmusa.org/biographies/william-clark/ on July 6, 2023 and "William Clark's Commission," National Park Service, accessed at https://www.nps.gov/articles/000/william-clark-s-commission.htm on August 10, 2023.

17 *the Louisiana Purchase:* Having gained independence from Great Britain through the Revolutionary War (1775–83), the United States was finally able to make its own treaties at the turn of the nineteenth century.

17 *most of it remained unceded:* During the two centuries that followed, the United States would pay an additional $418 million to Indian nations in 222 distinct cession transactions, an amount that undervalues the true cost of the land. Currency amounts are valued in US dollars in 1803. Robert Lee, "Accounting for Conquest: The Price of the Louisiana Purchase of Indian Country," *Journal of American History* 103, no. 4 (2017): 921, 923, 931, 933, 939.

17 *5.3 million people:* This number excludes Native Americans who were not taxed. "Return of the Whole Number of Persons within the Several Districts of the United States," US Census of 1800, accessed at https://www2.census.gov/library/publications/decennial/1800/1800-returns.pdf on July 27, 2023.

17 *populated by Native Americans:* Susan Sleeper Smith, "Were They Really Pioneers?" *Middle West Review* 7, no. 1 (2020): 138.

17 *Little was known:* Thomas Jefferson to Meriwether Lewis, November 16, 1803, accessed at https://founders.archives.gov/documents/Jefferson/01-42-02-0005-0001 on August 11, 2021; and Alvin M. Josephy, Jr., *Lewis and Clark Through Indian Eyes: Nine Indian Writers on the Legacy of the Expedition* (New York: Knopf Doubleday Publishing Group, 2008), xiv.

18 *"The object of your mission":* Thomas Jefferson to Meriwether Lewis, June 20, 1803, accessed at https://www.monticello.org/thomas-jefferson/louisiana-lewis-clark/preparing-for-the-expedition/jefferson-s-instructions-to-lewis/ on July 8, 2021.

18 *he sought to secure:* Gary E. Moulton, "Introduction: The Journals of the Lewis and Clark Expedition," *Journals of the Lewis and Clark Expedition*, accessed at https://lewisandclarkjournals.unl.edu/item/lc.jrn.introduction.general on July 8, 2021.

18 *Cantonese merchants:* James R. Gibson, *Otter Skins, Boston Ships, and China Goods: The Maritime Fur Trade of the Northwest Coast, 1785–1841* (Montreal: McGill-Queen's Press, 1992), xi; and Gregory P. Shrine, "North West Company," *Oregon Encyclopedia*, accessed at https://www.oregonencyclopedia.org/articles/north_west_company/ on October 19, 2023.

18 *mapped the portion:* "George Vancouver," *Encyclopaedia Britannica*, accessed at https://www.britannica.com/biography/George-Vancouver on July 9, 2021; Landon Y. Jones, ed., *The Essential Lewis and Clark* (New York: HarperCollins, 2000), xiii; and Jim Mockford, "Before Lewis and Clark, Lt. Broughton's River of Names: The Columbia River Exploration of 1792," *Oregon Historical Quarterly* 106, no. 4 (2005): 543.

18 *penetrated the interior:* "George Vancouver," *Encyclopaedia Britannica*, accessed at https://www.britannica.com/biography/George-Vancouver on July 9, 2021; Jones, ed., *The Essential Lewis and Clark*, xiii; and Mockford, "Before Lewis and Clark, Lt. Broughton's River of Names," 543.

18   *Jefferson also asked:* Thomas Jefferson to Meriwether Lewis, June 20, 1803, accessed at https://www.monticello.org/thomas-jefferson/louisiana-lewis-clark /preparing-for-the-expedition/jefferson-s-instructions-to-lewis/ on July 8, 2021.

18   *he expressed curiosity:* Jefferson to Lewis, June 20, 1803.

19   *"the vital matter":* Ronda, "The Mandan Winter."

19   *"the party destined":* Meriwether Lewis, May 20, 1804, in *Journals of Lewis and Clark Expedition,* accessed at https://lewisandclarkjournals.unl.edu/item/lc. jrn.1804-05-20 on July 8, 2021; William Clark, May 21, 1804, in *Journals of Lewis and Clark Expedition,* accessed at https://lewisandclarkjournals.unl.edu /item/lc.jrn.1804-05-21 on July 8, 2021; and Jones, ed., *The Essential Lewis and Clark,* xix.

19   *soldiers and hunters:* "Members of the Expedition: Privates," US Army Center of Military History, accessed at https://history.army.mil/LC/The%20People /privates.htm on August 11, 2021. Subsequent descriptions of the men in this paragraph come from the same source.

19   *keeled boat's precious cargo:* Site visit, Missouri River Basin Lewis and Clark Center, July 26, 2021.

19   *lugged rifles:* "Lewis's Packing List," accessed at https://www.monticello.org /thomas-jefferson/louisiana-lewis-clark/preparing-for-the-expedition/lewis -s-packing-list/ on August 11, 2021.

20   *Spanish and British merchants:* John A. Alwin, "Pelts, Provisions, and Perceptions: The Hudson's Bay Company Mandan Indian Trade, 1795–1812," *Montana: The Magazine of Western History* 29, no. 3 (1979), in *Journals of the Lewis and Clark Expedition,* accessed at https://lewisandclarkjournals.unl.edu/item /lc.sup.alwin.01 on July 8, 2021; and W. Raymond Wood, "Missouri Company," *Encyclopedia of the Great Plains,* accessed at http://plainshumanities.unl .edu/encyclopedia/doc/egp.ha.029 on August 11, 2021.

20   *The delicate script:* "Lewis and Clark map, with annotations in brown ink by Meriwether Lewis, tracing showing the Mississippi, the Missouri for a short distance above Kansas, Lakes Michigan, Superior, and Winnipeg, and the country onwards to the Pacific," probably 1803, Library of Congress, accessed at https://www.loc.gov/resource/g4126s.ct000071/ on July 12, 2021; and "Mapping and Lewis and Clark," US Department of the Interior, accessed at https://www2.usgs.gov/features/lewisandclark/Mapping2.html on July 12, 2021.

20   *make their own maps:* "William Clark: A Master Cartographer," National Park Service, accessed at https://www.nps.gov/articles/william_clark_cartographer .htm on July 12, 2021.

20   *made their way:* "The Lewis and Clark Expedition: Interactive Map," Gilder Lehrman Center for American History, accessed at https://www.gilderlehr man.org/history-resources/online-exhibitions/lewis-and-clark-expedition -interactive-map on July 12, 2021.

20   *Herds of woolly buffalo:* William Clark, August 23, 1804, in *Journals of Lewis and Clark Expedition,* accessed at https://lewisandclarkjournals.unl.edu/item /lc.jrn.1804-08-23 on July 13, 2021.

20   *the corps met:* William Clark, August 2, 14, and 30, 1804, in *Journals of Lewis and Clark Expedition,* accessed at https://lewisandclarkjournals.unl .edu/item/lc.jrn.1804-08-02; https://lewisandclarkjournals.unl.edu/item/lc.jrn

.1804-08-14 on July 13, 2021; and https://lewisandclarkjournals.unl.edu/item /lc.jrn.1804-08-30 on July 13, 2021.

20   *exchanged goods:* William Clark, August 30, 1804, in *Journals of Lewis and Clark Expedition,* accessed at https://lewisandclarkjournals.unl.edu/item/lc .jrn.1804-08-30 on July 13, 2021.

20   *Clark wrote admiringly:* William Clark, August 30, 1804, in *Journals of Lewis and Clark Expedition,* accessed at https://lewisandclarkjournals.unl.edu/item /lc.jrn.1804-08-30 on March 6, 2023.

20   *also delivered speeches:* Ronda, "The Voyage Begins," and "The Mandan Winter," in *Lewis and Clark Among the Indians.*

21   *a smallpox outbreak:* Footnote for William Clark, August 14, 1804, in *Journals of Lewis and Clark Expedition,* accessed at https://lewisandclarkjournals.unl.edu /item/lc.jrn.1804-08-14 on July 13, 2021.

21   *since the earliest days of European contact:* James C. Riley, "Smallpox and American Indians Revisited," *Journal of the History of Medicine and Allied Sciences* 65, no. 4 (2010): 445, 449.

21   *Up to 95 percent:* Jared Diamond, *Guns, Germs, and Steel: The Fates of Human Societies* (New York: W. W. Norton, 1997), 211.

21   *"their frenzy":* William Clark, August 14, 1804, in *Journals of Lewis and Clark Expedition,* accessed at https://lewisandclarkjournals.unl.edu/item/lc .jrn.1804-08-14 on July 13, 2021.

21   *also weakened the nation:* William Clark, August 14, 1804, in *Journals of Lewis and Clark Expedition,* accessed at https://lewisandclarkjournals.unl.edu/item /lc.jrn.1804-08-14 on July 13, 2021.

21   *The barren fields:* William Clark, John Ordway, and Charles Floyd, August 14, 1804, in *Journals of Lewis and Clark Expedition,* accessed at https://lewisand clarkjournals.unl.edu/item/lc.jrn.1804-08-14 on July 13, 2021.

21   *Floyd suddenly took ill:* William Clark, August 19–20, 1804, in *Journals of Lewis and Clark Expedition,* accessed at https://lewisandclarkjournals.unl .edu/item/lc.jrn.1804-08-19 and https://lewisandclarkjournals.unl.edu/item /lc.jrn.1804-08-20 on July 13, 2021.

21   *believed to be appendicitis:* Joseph Whitehouse, August 20, 1804, in *Journals of Lewis and Clark Expedition,* accessed at https://lewisandclarkjournals.unl .edu/item/lc.jrn.1804-08-20 on July 14, 2021; and David Dary, *Frontier Medicine: From the Atlantic to the Pacific, 1492–1941* (New York: Vintage Books, 2008), 47.

21   *"firmness and determined":* William Clark, August 20, 1804, in in *Journals of Lewis and Clark Expedition,* accessed at https://lewisandclarkjournals.unl.edu /item/lc.jrn.1804-08-20 on July 13, 2021.

21   *the men buried Floyd:* William Clark and John Ordway, August 20, 1804, in *Journals of Lewis and Clark Expedition,* accessed at https://lewisandclarkjournals .unl.edu/item/lc.jrn.1804-08-20 on July 14, 2021.

22   *"beautiful country":* William Clark, October 24, 1804, in *Journals of Lewis and Clark Expedition,* accessed at https://lewisandclarkjournals.unl.edu/item/lc .jrn.1804-10-24 on July 14, 2021.

22   *The expedition had reached:* "Fort Mandan Miscellany," in *Journals of Lewis and Clark Expedition,* accessed at https://lewisandclarkjournals.unl.edu/item/lc

.jrn.1804-1805.winter.introduction on August 21, 2020; and "Lewis and Clark map," probably 1803, Library of Congress.

22 *Bitterly cold winds:* William Clark, October 25, 1804, in *Journals of Lewis and Clark Expedition*, accessed at https://lewisandclarkjournals.unl.edu/item/lc .jrn.1804-10-25 on July 14, 2021.

22 *devastating smallpox epidemic:* Footnote for William Clark, October 27, 1804, in *Journals of Lewis and Clark Expedition*, accessed at https://lewisandclark journals.unl.edu/item/lc.jrn.1804-10-27 on October 20, 2023.

22 *Lewis and Clark hosted:* William Clark, October 26–28, 1804, in *Journals of Lewis and Clark Expedition*, accessed at https://lewisandclarkjournals.unl .edu/item/lc.jrn.1804-10-26, https://lewisandclarkjournals.unl.edu/item/lc.jrn .1804-10-27, and https://lewisandclarkjournals.unl.edu/item/lc.jrn.1804-10-28 on July 14, 2021; and Ronda, "The Mandan Winter."

22 *"I went in the pirogue":* William Clark, October 30, 1804, in *Journals of Lewis and Clark Expedition*, accessed at https://lewisandclarkjournals.unl.edu/item /lc.jrn.1804-10-30 on March 6, 2023.

22 *began on Fort Mandan:* William Clark, November 2, 1804, in *Journals of Lewis and Clark Expedition*, accessed at https://lewisandclarkjournals.unl.edu/item /lc.jrn.1804-11-02 on August 12, 2021.

22 *cottonwood trees:* Patrick Gass, November 16, 1804, in *Journals of Lewis and Clark Expedition*, accessed at https://lewisandclarkjournals.unl.edu/item/lc .jrn.1804-11-16-28#lc.jrn.1804-11-16-28.01 on July 16, 2021.

22 *seven interior dwelling rooms:* Site visit, Fort Mandan State Historic Site, July 28, 2021.

23 *Charbonneau returned:* William Clark, November 11, 1804, in *Journals of Lewis and Clark Expedition*, accessed at https://lewisandclarkjournals.unl.edu/item /lc.jrn.1804-11-11 on July 15, 2021.

23 *warm buffalo robes:* John Ordway, November 11, 1804, in *Journals of Lewis and Clark Expedition*, accessed at https://lewisandclarkjournals.unl.edu/item/lc .jrn.1804-11-11 on July 15, 2021.

23 *move into Fort Mandan:* Harold P. Howard, *Sacajawea* (Norman: University of Oklahoma Press, 1971), 18–19.

23 *likely shared a large room:* Glenn F. Williams, "For Want of an Interpreter," US Army Center for Military History, accessed at https://history.army.mil/lc /The%20People/interpreter.htm on July 23, 2021; and site visit, Fort Mandan State Historic Site, July 28, 2021.

24 *described a ceremony:* William Clark, January 5, 1805, in *Journals of Lewis and Clark Expedition*, accessed at https://lewisandclarkjournals.unl.edu/item /lc.jrn.1805-01-05 on July 15, 2021; and Alice B. Kehoe, "The Function of Ceremonial Sexual Intercourse Among the Northern Plains Indians," *Plains Anthropologist* 15, no. 48 (1970): 99–100. Subsequent descriptions of the ceremony in this paragraph come from the journal of William Clark on January 5, 1805.

24 *one of the men:* William Clark, November 22, 1804, in *Journals of Lewis and Clark Expedition*, accessed at https://lewisandclarkjournals.unl.edu/item/lc .jrn.1804-11-22 on July 15, 2021. Further discussion of this event in this paragraph draws from the same source.

24   *Clark also recorded:* William Clark, January 3, 1805, in *Journals of Lewis and Clark Expedition*, accessed at https://lewisandclarkjournals.unl.edu/item/lc .jrn.1805-01-03 on July 15, 2021.

24   *an event that prompted:* Meriwether Lewis, August 14, 1805, in *Journals of Lewis and Clark Expedition*, accessed at https://lewisandclarkjournals.unl.edu/item /lc.jrn.1805-08-14 on August 1, 2021.

25   *sexually transmitted diseases:* William Clark, January 14, 1805, in *Journals of Lewis and Clark Expedition*, accessed at https://lewisandclarkjournals.unl.edu /item/lc.jrn.1805-01-14#lc.jrn.1805-01-14.01 on July 16, 2021.

25   *Clark noted that:* William Clark, March 30, 1805, in *Journals of Lewis and Clark Expedition*, accessed at https://lewisandclarkjournals.unl.edu/item/lc.jrn.1805 -03-30#lc.jrn.1805-03-30.02 on July 16, 2021.

25   *Lewis brought pewter:* David Lavender, *The Way to the Western Sea: Lewis and Clark Across the Continent* (Lincoln: University of Nebraska Press, 2001), accessed at https://lewisandclarkjournals.unl.edu/item/lc.sup.lavender.01.03 on July 19, 2021; and Dary, *Frontier Medicine*, 45, 47.

25   *numerous men came:* Volney Steele, *Bleed, Blister, and Purge: A History of Medicine on the American Frontier* (Missoula: Mountain Press Publishing Company, 2005), 58.

25   *contracted syphilis:* Steele, *Bleed, Blister, and Purge*, 59.

25   *death by suicide:* Meriwether Lewis most likely died by shooting himself in the head in 1809 after battling hereditary depression for years. Larry E. Morris, *The Fate of the Corps: What Became of the Lewis and Clark Explorers After the Expedition* (New Haven: Yale University Press, 2004), 67-69; and Steele, *Bleed, Blister, and Purge*, 59–60.

25   *delivered their babies:* Steele, *Bleed, Blister, and Purge*, 31.

25   *"labor was tedious":* Meriwether Lewis, February 11, 1805, in *Journals of Lewis and Clark Expedition*, accessed at https://lewisandclarkjournals.unl.edu/item /lc.jrn.1805-02-11 on August 11, 2021.

25   *French Canadian interpreter:* Meriwether Lewis, February 11, 1805, in *Journals of Lewis and Clark Expedition*, accessed at https://lewisandclarkjournals.unl .edu/item/lc.jrn.1805-02-11; and Steele, *Bleed, Blister, and Purge*, 32. Further descriptions of Sacagawea's labor in this paragraph come from Lewis's journal on February 11, 1805.

26   *nickname him Pomp:* Clark also referred to him as "Pompey." Sacagawea," National Women's History Museum, accessed at https://www.womenshistory.org /education-resources/biographies/sacagawea on July 29, 2021.

26   *little time to rest:* Steele, *Bleed, Blister, and Purge*, 32.

26   *Sacagawea had been chosen:* Ella E. Clark and Margot Edmonds, *Sacagawea of the Lewis and Clark Expedition* (Berkeley: University of California Press, 1979), 15.

26   *in a "bier":* Joseph A. Mussulman, "Mosquito Netting," *Discovering Lewis & Clark*, accessed at https://lewis-clark.org/article/1029 on August 14, 2021; and Colby, *Sacagawea's Child*, 56.

26   *the keeled boat returned:* "To Thomas Jefferson from Meriwether Lewis," April 7, 1806; "Ft. Mandan Departure Day 1805," National Park Service, accessed at

https://www.nps.gov/articles/ft-mandan-departure-day-1805.htm on July 26, 2021; and "Fort Mandan Miscellany," in *Journals of Lewis and Clark Expedition*.

27 *"two-thousand miles":* Meriwether Lewis, April 7, 1805, in *Journals of Lewis and Clark Expedition*, accessed at https://lewisandclarkjournals.unl.edu/item/lc.jrn.1805-04-07 on July 24, 2021.

27 *return to her birthplace:* Virginia Scharff, *Twenty Thousand Roads: Women, Movement, and the West* (Berkeley: University of California Press, 2003), 17.

27 *spotting blooming flowers:* Meriwether Lewis and William Clark, April 9, 1805, in *Journals of Lewis and Clark Expedition*, accessed at https://lewisandclarkjournals.unl.edu/item/lc.jrn.1805-04-09#n11040916 on July 25, 2021.

27 *locate wild artichokes:* Meriwether Lewis, April 9, 1805, in *Journals of Lewis and Clark Expedition*, accessed at https://lewisandclarkjournals.unl.edu/item/lc.jrn.1805-04-09#n11040916 on July 25, 2021.

27 *foraged for onions:* Meriwether Lewis, April 11, 1805, and April 12, 1805, in *Journals of Lewis and Clark Expedition*, accessed at https://lewisandclarkjournals.unl.edu/item/lc.jrn.1805-04-11 and https://lewisandclarkjournals.unl.edu/item/lc.jrn.1805-04-12#ln11041202 on July 25, 2021.

27 *nearly flipped on its side:* Meriwether Lewis, April 13, 1805, in *Journals of Lewis and Clark Expedition*, accessed at https://lewisandclarkjournals.unl.edu/item/lc.jrn.1805-04-13 on July 26, 2021. Subsequent descriptions of this event come from the same source.

27 *a squall again upset:* Meriwether Lewis, May 14, 1805, and May 16, 1805, in *Journals of Lewis and Clark Expedition*, accessed at https://lewisandclarkjournals.unl.edu/item/lc.jrn.1805-05-14 and https://lewisandclarkjournals.unl.edu/item/lc.jrn.1805-05-16 on July 26, 2021; and Elin Woodger and Brandon Toropov, *Encyclopedia of the Lewis and Clark Expedition* (New York: Facts on File, Inc., 2014), 203. Further discussion of this event in the following two paragraphs comes from Lewis's diary of May 14, 1805.

28 *Lewis noticed that:* Meriwether Lewis, May 17, 1805, in *Journals of Lewis and Clark Expedition*, accessed at https://lewisandclarkjournals.unl.edu/item/lc.jrn.1805-05-17 on July 26, 2021.

28 *they caught a glimpse:* William Clark, May 26, 1805, in *Journals of Lewis and Clark Expedition*, accessed at https://lewisandclarkjournals.unl.edu/item/lc.jrn.1805-05-26 on July 26, 2021.

29 *Lewis took a look:* Meriwether Lewis, May 26, 1805, in *Journals of Lewis and Clark Expedition*, accessed at https://lewisandclarkjournals.unl.edu/item/lc.jrn.1805-05-26 on July 26, 2021.

29 *"While I viewed these":* Meriwether Lewis, May 26, 1805, in *Journals of Lewis and Clark Expedition*, accessed at https://lewisandclarkjournals.unl.edu/item/lc.jrn.1805-05-26 on July 26, 2021.

29 *forced to stop:* Meriwether Lewis, June 3, 1805, in *Journals of Lewis and Clark Expedition*, accessed at https://lewisandclarkjournals.unl.edu/item/lc.jrn.1805-06-03 on August 13, 2021; and Howard, *Sacajawea*, 34.

29 *but they failed:* Clark and Edmonds, *Sacagawea of the Lewis and Clark Expedition*, 18.

29 *fell seriously ill:* Footnotes for Meriwether Lewis, June 10, 1805, in *Journals of*

Lewis and Clark Expedition, accessed at https://lewisandclarkjournals.unl.edu /item/lc.jrn.1805-06-16 on July 26, 2021.

29 *Clark sought to nurse:* William Clark, June 10, 1805, June 11, 1805, and June 15, 1805, accessed at https://lewisandclarkjournals.unl.edu/item/lc.jrn .1805-06-10, https://lewisandclarkjournals.unl.edu/item/lc.jrn.1805-06-11, and https://lewisandclarkjournals.unl.edu/item/lc.jrn.1805-06-15 on August 13, 2021; and Peter J. Kastor and Conevery Bolton Valencius, "Sacagawea's 'Cold': Pregnancy and the Written Record of the Lewis and Clark Expedition," *Bulletin of the History of Medicine* 82, no. 2 (2008): 282.

29 *in some of these entries:* Entries for Meriwether Lewis and William Clark, June 10, 1805, in *Journals of Lewis and Clark Expedition,* accessed at https:// lewisandclarkjournals.unl.edu/item/lc.jrn.1805-06-10 on March 8, 2023.

29 *Sacagawea's condition:* Entries for Meriwether Lewis and William Clark, June 11, 1805, in *Journals of Lewis and Clark Expedition,* accessed at https:// lewisandclarkjournals.unl.edu/item/lc.jrn.1805-06-11 on July 26, 2021.

29 *Lewis was distressed:* Meriwether Lewis, June 16, 1805, in *Journals of Lewis and Clark Expedition,* accessed at https://lewisandclarkjournals.unl.edu/item /lc.jrn.1805-06-16 on July 26, 2021. Further comments from Lewis in this paragraph are from the same source.

30 *additional poultices:* Meriwether Lewis, June 16, 1805, in *Journals of Lewis and Clark Expedition,* accessed at https://lewisandclarkjournals.unl.edu/item/lc .jrn.1805-06-16 on July 26, 2021.

30 *recorded in his journal:* Meriwether Lewis, June 17, 1805, in *Journals of Lewis and Clark Expedition,* accessed at https://lewisandclarkjournals.unl.edu/item /lc.jrn.1805-06-17 on July 26, 2021.

30 *The following day:* Meriwether Lewis, June 18, 1805, in *Journals of Lewis and Clark Expedition,* accessed at https://lewisandclarkjournals.unl.edu/item/lc .jrn.1805-06-18#ln15061801 on July 26, 2021.

30 *on June 21:* Meriwether Lewis, June 21, 1805, in *Journals of Lewis and Clark Expedition,* accessed at https://lewisandclarkjournals.unl.edu/item/lc.jrn.1805 -06-21 on July 26, 2021.

30 *a difficult portage:* "Great Falls Portage," National Park Service, accessed at https://www.nps.gov/places/great-falls-portage-mt.htm on July 30, 2021; and Kenneth Thomasma, *The Truth About Sacagawea* (Jackson: Grandview Publishing, 1997), 43.

30 *Keenly aware of:* Thomasma, *The Truth About Sacagawea,* 48–49.

30 *mused Lewis:* Meriwether Lewis, July 4, 1805, in *Journals of Lewis and Clark Expedition,* accessed at https://lewisandclarkjournals.unl.edu/item/lc.jrn.1805 -07-04 on August 13, 2021.

30 *Sacagawea recognized:* Meriwether Lewis, July 22, 1805, in *Journals of Lewis and Clark Expedition,* accessed at https://lewisandclarkjournals.unl.edu/item /lc.jrn.1805-07-22 on August 13, 2021. Lewis's subsequent comments about this event come from the same source.

31 *The expedition arrived:* Meriwether Lewis, July 27, 1805, in *Journals of Lewis and Clark Expedition,* accessed at https://lewisandclarkjournals.unl.edu/item /lc.jrn.1805-07-27 on August 13, 2021.

31 *she told the men:* Meriwether Lewis, July 28, 1805, in *Journals of Lewis and*

Clark Expedition, accessed at https://lewisandclarkjournals.unl.edu/item/lc
.jrn.1805-07-28 on July 30, 2021. Sacagawea's account of this event in the paragraph comes from this source.

31 *Lewis began contemplating:* Meriwether Lewis, July 27, 1805, in *Journals of Lewis
and Clark Expedition*, accessed at https://lewisandclarkjournals.unl.edu/item
/lc.jrn.1805-07-27 on August 13, 2021.

32 *Thus, Lewis concluded:* Meriwether Lewis, July 31, 1805, and August 1, 1805,
accessed at https://lewisandclarkjournals.unl.edu/item/lc.jrn.1805-07-31 and
https://lewisandclarkjournals.unl.edu/item/lc.jrn.1805-08-01 on August 13,
2021.

32 *Lewis and a contingent:* Lewis's men included George Drouillard, Toussaint
Charbonneau, and Patrick Gass. Meriwether Lewis, August 1, 1805, in *Journals of Lewis and Clark Expedition*, accessed at https://lewisandclarkjournals
.unl.edu/item/lc.jrn.1805-08-01 on August 13, 2021.

32 *The sun scorched:* Meriwether Lewis, August 1–3, 1805, in *Journals of Lewis and
Clark Expedition*, accessed at https://lewisandclarkjournals.unl.edu/item/lc
.jrn.1805-08-01; https://lewisandclarkjournals.unl.edu/item/lc.jrn.1805-08-02;
and https://lewisandclarkjournals.unl.edu/item/lc.jrn.1805-08-03 on July 31,
2021.

32 *"a track which he":* Meriwether Lewis, August 3, 1805, in *Journals of Lewis and
Clark Expedition*, accessed at https://lewisandclarkjournals.unl.edu/item/lc
.jrn.1805-08-03 on July 31, 2021.

32 *the parties reunited:* Meriwether Lewis, August 6–7, 1805, in *Journals of Lewis
and Clark Expedition*, accessed at https://lewisandclarkjournals.unl.edu/item
/lc.jrn.1805-08-06 and https://lewisandclarkjournals.unl.edu/item/lc.jrn.1805
-08-07 on July 31, 2021.

32 *Sacagawea expressed:* Meriwether Lewis, August 8, 1805, in *Journals of Lewis
and Clark Expedition*, accessed at https://lewisandclarkjournals.unl.edu/item
/lc.jrn.1805-08-08 on July 31, 2021.

32 *a small party of men:* John Ordway, August 9, 1805, in *Journals of Lewis and
Clark Expedition*, accessed at https://lewisandclarkjournals.unl.edu/item/lc
.jrn.1805-08-09 on July 31, 2021.

32 *found the headwaters:* Meriwether Lewis, August 12, 1805, in *Journals of Lewis
and Clark Expedition*, accessed at https://lewisandclarkjournals.unl.edu/item
/lc.jrn.1805-08-12 on July 6, 2023.

33 *Agaidika Shoshone:* Meriwether Lewis, August 13, 1805, in *Journals of Lewis
and Clark Expedition*, accessed at https://lewisandclarkjournals.unl.edu/item
/lc.jrn.1805-08-13#n19081313 on August 1, 2021; and "First Flag Unfurling
Site," National Park Service, accessed at https://www.nps.gov/places/first-flag
-unfurling-site.htm on August 1, 2021. Further descriptions of this meeting
come from Lewis's diary of the same date.

33 *The next day:* Meriwether Lewis, August 14, 1805, in *Journals of Lewis and
Clark Expedition*, accessed at https://lewisandclarkjournals.unl.edu/item/lc
.jrn.1805-08-13#n19081313 on August 1, 2021. Additional material on their
negotiation comes from the same source.

33 *the following day:* Meriwether Lewis, August 15, 1805, in *Journals of Lewis and
Clark Expedition*, accessed at https://lewisandclarkjournals.unl.edu/item/lc
.jrn.1805-08-15 on August 1, 2021.

33   *explaining multiple times:* Meriwether Lewis, August 16, 1805, in *Journals of Lewis and Clark Expedition*, accessed at https://lewisandclarkjournals.unl.edu /item/lc.jrn.1805-08-16 on August 1, 2021.

33   *Sacagawea danced:* William Clark and Meriwether Lewis, August 17, 1805, in *Journals of Lewis and Clark Expedition*, accessed at https://lewisandclarkjournals .unl.edu/item/lc.jrn.1805-08-17#ln20081701 on March 12, 2023. Additional information about her reunion comes from the same sources.

34   *negotiations for horses:* Meriwether Lewis, August 17, 1805, in *Journals of Lewis and Clark Expedition*, accessed at https://lewisandclarkjournals.unl.edu/item /lc.jrn.1805-08-17#ln20081701 on March 12, 2023. Further discussion of this event comes from the same source.

34   *westward overland trek:* William Clark, August 30, 1805, in *Journals of Lewis and Clark Expedition*, accessed at https://lewisandclarkjournals.unl.edu/item /lc.jrn.1805-08-30 on August 2, 2021.

34   *would lead them:* William Clark and John Whitehouse, August 30, 1805, in *Journals of Lewis and Clark Expedition*, accessed at https://lewisandclarkjournals .unl.edu/item/lc.jrn.1805-08-30 on August 2, 2021; and Howard, *Sacajawea*, 66.

34   *A Shoshone man:* Meriwether Lewis, August 19, 1805, in *Journals of Lewis and Clark Expedition*, accessed at https://lewisandclarkjournals.unl.edu/item /lc.jrn.1805-08-19 on March 12, 2023. Further information about this event comes from the same source.

35   *fatigued the members:* Howard, *Sacajawea*, 66–68,

35   *depleted their morale:* William Clark, September 3–4, 16, 21, 1805, in *Journals of the Lewis and Clark Expedition*, accessed at https://lewisandclarkjournals .unl.edu/item/lc.jrn.1805-09-03; https://lewisandclarkjournals.unl.edu/item /lc.jrn.1805-09-04; https://lewisandclarkjournals.unl.edu/item/lc.jrn.1805 -09-16; and https://lewisandclarkjournals.unl.edu/item/lc.jrn.1805-09-21 on August 2, 2021. Subsequent descriptions of conditions come from the same source.

35   *Lewis gloomily recorded:* Meriwether Lewis, September 21, 1805, in *Journals of Lewis and Clark Expedition*, accessed at https://lewisandclarkjournals.unl.edu /item/lc.jrn.1805-09-21 on August 2, 2021.

35   *Weippe Prairie in Idaho:* "September 22, 1805," *Discover Lewis & Clark*, accessed at https://lewis-clark.org/day-by-day/22-sep-1805/ on March 12, 2023.

35   *The very next day:* Meriwether Lewis, September 22, 1805, in *Journals of Lewis and Clark Expedition*, accessed at https://lewisandclarkjournals.unl.edu/item /lc.jrn.1805-09-22#ln22092204 on August 2, 2021. Subsequent remarks from Lewis that day come from the same source.

35   *More than a hundred:* Site visit, Lewis and Clark Interpretive Center at Cape Disappointment, Ilwaco, Washington, on August 2, 2021.

35   *the Clearwater River:* William Clark, September 23, 1805, in *Journals of Lewis and Clark Expedition*, accessed at https://lewisandclarkjournals.unl.edu/item /lc.jrn.1805-09-23 on August 2, 2021; and Howard, *Sacajawea*, 70.

35   *the men set to work:* William Clark, September 27, 1805, in *Journals of Lewis and Clark Expedition*, accessed at https://lewisandclarkjournals.unl.edu/item /lc.jrn.1805-09-27 on August 4, 2021.

35   *hollowed them out:* William Clark, September 27, 1805, in *Journals of Lewis

and Clark Expedition, accessed at https://lewisandclarkjournals.unl.edu/item
/lc.jrn.1805-09-27 on August 4, 2021.

35 *On October 7:* William Clark, October 7, 1805, in *Journals of Lewis and
Clark Expedition,* accessed at https://lewisandclarkjournals.unl.edu/item/lc
.jrn.1805-10-07 on August 4, 2021.

35 *Pacific Northwest nations:* William Clark, October 8, 10, 12, 1805, *Journals of Lewis and Clark Expedition,* accessed at https://lewisandclarkjournals
.unl.edu/item/lc.jrn.1805-10-08; https://lewisandclarkjournals.unl.edu/item/lc
.jrn.1805-10-10; and https://lewisandclarkjournals.unl.edu/item/lc.jrn.1805
-10-12; on August 4, 2021.

35 *Clark wrote on November 7:* William Clark, November 7, 1805, in *Journals of
Lewis and Clark Expedition,* accessed at https://lewisandclarkjournals.unl.edu
/item/lc.jrn.1805-11-07#ln25110710 on August 4, 2021.

36 *could not complete:* William Clark, November 8, 1805, in *Journals of Lewis and
Clark Expedition,* accessed at https://lewisandclarkjournals.unl.edu/item/lc
.jrn.1805-11-08 on August 13, 2021.

36 *the "dismal nitch":* William Clark, November 15, 1805, in *Journals of Lewis and
Clark Expedition,* accessed at https://lewisandclarkjournals.unl.edu/item/lc
.jrn.1805-11-15 on August 13, 2021.

36 *Gass took a moment:* Patrick Gass, November 16, 1805, in *Journals of Lewis and
Clark Expedition,* accessed at https://lewisandclarkjournals.unl.edu/item/lc
.jrn.1805-11-16 on March 12, 2023.

36 *Whitehouse described:* Joseph Whitehouse, November 16, 1805, in *Journals of
Lewis and Clark Expedition,* accessed at https://lewisandclarkjournals.unl.edu
/item/lc.jrn.1805-11-16 on August 13, 2021.

36 *Clark moaned:* William Clark, November 22, 1805, in *Journals of Lewis and
Clark Expedition,* accessed at https://lewisandclarkjournals.unl.edu/item/lc
.jrn.1805-11-22 on March 12, 2023.

37 *decided to hold a vote:* William Clark, November 24, 1805, in *Journals of Lewis
and Clark Expedition,* accessed at https://lewisandclarkjournals.unl.edu/item
/lc.jrn.1805-11-24#ln26112405 on August 13, 2021; and site visit, the Lewis
and Clark Interpretive Center, August 2, 2021. Additional information about
this important event comes from the same sources.

37 *hardly "pacific":* William Clark, December 1, 1805, in *Journals of Lewis and Clark
Expedition,* accessed at https://lewisandclarkjournals.unl.edu/item/lc.jrn.1805
-12-01#lc.jrn.1805-12-01.02 on August 4, 2021.

37 *prevented from visiting the Pacific Ocean:* Clark only permitted men to join him
on the first excursion to shores of the Pacific Ocean. William Clark, November 17, 1805, in *Journals of Lewis and Clark Expedition,* accessed at https://
lewisandclarkjournals.unl.edu/item/lc.jrn.1805-11-17 on July 28, 2023.

37 *a compelling case:* Meriwether Lewis, January 6, 1806, in *Journals of Lewis and
Clark Expedition,* accessed at https://lewisandclarkjournals.unl.edu/item/lc
.jrn.1806-01-06 on August 13, 2021.

37 *they were relieved:* Site visit, Fort Clatsop Visitor Center, August 2, 2021.

37 *they bid farewell:* Meriwether Lewis, March 23, 1806, in *Journals of Lewis and
Clark Expedition,* accessed at https://lewisandclarkjournals.unl.edu/item/lc
.jrn.1806-03-23 on August 13, 2021.

37 *locating a passage:* This passage became known as Bozeman Pass. William Clark, July 13, 1806, in *Journals of Lewis and Clark Expedition*, at https://lewisandclarkjournals.unl.edu/item/lc.jrn.1806-07-13#lc.jrn.1806-07-13.03 on August 13, 2021. Clark's words of appreciation come from the same source.

38 *decided to return:* John Ordway, August 14, 1806, in *Journals of Lewis and Clark Expedition*, accessed at https://lewisandclarkjournals.unl.edu/item/lc.jrn.1806-08-14 on August 13, 2021.

38 *paid Charbonneau:* William Clark, August 17, 1806, in *Journals of Lewis and Clark Expedition*, accessed at https://lewisandclarkjournals.unl.edu/item/lc.jrn.1806-08-17 on August 13, 2021.

38 *writing to Charbonneau:* William Clark to Toussaint Charbonneau, August 20, 1806, quoted in Colby, *Sacagawea's Child*, 64.

38 *an offer made:* William Clark, August 17, 1806, in *Journals of Lewis and Clark Expedition*, accessed at https://lewisandclarkjournals.unl.edu/item/lc.jrn.1806-08-17 on August 5, 2021.

38 *turned out in droves:* William Clark, September 23, 1806, in *Journals of Lewis and Clark Expedition*, accessed at https://lewisandclarkjournals.unl.edu/item/lc.jrn.1806-09-23 on August 5, 2021.

38 *178 new plants:* "Scientific Encounters," National Park Service, accessed at https://www.nps.gov/articles/scientific-encounters.html on August 21, 2020; and site visit, Lewis and Clark Interpretive Trails and Visitor Center, July 26, 2021.

38 *pages of their journals:* "Images and Maps," *Journals of Lewis and Clark Expedition*, accessed at https://lewisandclarkjournals.unl.edu/images on August 13, 2021.

38 *"Clark's Map of 1810":* "Clark's Map of 1810," *Library of Congress*, accessed at https://www.loc.gov/item/2021668419/ on October 23, 2023; "William Clark: A Master Cartographer," National Park Service, accessed at https://www.nps.gov/articles/william_clark_cartographer.htm on October 23, 2023; and site visit, Lewis and Clark Interpretive Trails and Visitor Center, July 26, 2021.

39 *Alvin Josephy believed:* Josephy, *Lewis and Clark through Indian Eyes*, 20.

39 *Sacagawea received nothing:* Colby, *Sacagawea's Child*, 71. Additional information about Sacagawea's life after her time with the Corps of Discovery is from the same source, pages 74–75.

39 *gave birth to a daughter:* Colby, *Sacagawea's Child*, 16.

39 *Sacagawea died:* Colby, *Sacagawea's Child*, 16; and Howard, *Sacajawea*, 160.

39 *the legal guardian:* Morris, *The Fate of the Corps*, 115, 117.

39 *Wind River Indian Reservation:* Members of the Eastern Shoshone and Northern Arapaho tribes live on this reservation today. Thomas P. Slaughter, *Exploring Lewis and Clark: Reflections on Men and Wilderness* (New York: Knopf Doubleday Publishing Group, 2004), 87-92; and Helen Addison Howard, "The Mystery of Sacagawea's Death," *Pacific Northwest Quarterly* 58, no. 1 (1967): 5.

2. JAMES BECKWOURTH: THE MOUNTAIN MAN

41 *James Marshall approached:* Site visit, Marshall Gold Discovery State Historic Park, August 3, 2021; John Marshall's account of his discovery, "Gold!"

in Richard Goldstein, ed., *Mine Eyes Have Seen: A First-Person History of the Events That Shaped America* (New York: Simon & Schuster, 1997), 132; and General John A. Sutter, "The Discovery of Gold in California," *Hutchings' Illustrated California Magazine*, November 1857, accessed at http://www.sf museum.org/photos4/sutpix.html on October 1, 2021. Unless otherwise noted, the information in the rest of the paragraph comes from the same sources.

41   *The local Nisenan:* "Marshall Gold Discovery State Historic Park," California Department of Parks and Recreation, accessed at http://www.parks.ca.gov /?page_id=484 on October 1, 2021.

41   *delved into the bedrock:* Robert F. Heizer, "Archaeological Investigation of Sutter Sawmill Site in 1947," *California Historical Society Quarterly* 26, no. 2 (1947): 144.

42   *"It made my heart":* In this paragraph, Marshall's account of his discovery comes from "Gold!" in Goldstein, ed., *Mine Eyes Have Seen*, 132–33.

42   *Sutter's private offices:* Sutter, "The Discovery of Gold in California."

42   *Sutter, who concurred:* In this paragraph, Sutter's account of Marshall's discovery comes from the preceding source.

42   *child of an enslaved woman:* Jay H. Buckley, "Beckwourth, James," *Encyclopedia of the Great Plains*, University of Nebraska–Lincoln, accessed at http://plains humanities.unl.edu/encyclopedia/doc/egp.afam.008 on October 1, 2021.

43   *possibly named Miss Kill:* Elinor Wilson, *Jim Beckwourth: Black Mountain Man and War Chief of the Crows* (Norman: University of Oklahoma Press, 1972), 14, 20; and Buckley, "Beckwourth, James."

43   *who moved their family:* Wilson, *Jim Beckwourth*, 20–21; and Buckley, "Beckwourth, James."

43   *According to one observer:* John M. Letts, *California Illustrated*, 93–94, quoted in Wilson, *Jim Beckwourth*, 124.

43   *Ina Coolbrith described:* Coolbrith was one of the first immigrants to cross into Sierra Valley over Beckwourth Pass in 1851. Account of Ina Coolbrith's speech at a luncheon in San Francisco, 1927, per *Oroville Mercury Register*, August 13, 1939, quoted in Wilson, *Jim Beckwourth*, 135.

43   *an economic hub:* Site visit, Museum at the Gateway Arch, July 24, 2021. Further descriptions of life in Saint Louis at the turn of the nineteenth century are informed by my visit to this site.

43   *In his memoir:* James Beckwourth, *The Life and Adventures of James P. Beckwourth, Mountaineer, Scout, and Pioneer, and Chief of the Crow Nation of Indians* (New York: Harper and Brothers, 1856), 18.

44   *began an apprenticeship:* Beckwourth, *The Life and Adventures of James P. Beckwourth*, 18; and Wilson, *Jim Beckwourth*, 26. Subsequent information about his apprenticeship comes from the same sources and pages.

44   *Bernard DeVoto:* Bernard Augustine DeVoto, ed., *The Journals of Lewis and Clark* (Boston: Houghton Mifflin, 1953), lii, quoted in Robert V. Hine and John Mack Faragher, *Frontiers: A Short History of the American West* (New Haven: Yale University Press, 2007), 55.

44   *Subsequent Jeffersonian expeditions:* Site visit, Museum at the Gateway Arch, July 24, 2021; "Beyond Lewis and Clark—Timeline 1806–1807," Kansas Historical Society, accessed at https://www.kshs.org/p/beyond-lewis-and-clark

-timeline-1806-1807/10577 on October 2, 2021; "Zebulon Pike," *Encyclopaedia Britannica,* accessed at https://www.britannica.com/biography/Zebulon-Pike on October 2, 2021; John Spurgeon, "Freeman and Custis Red River Expedition," *Encyclopedia of Arkansas,* accessed at https://encyclopediaofarkansas.net /entries/freeman-and-custis-red-river-expedition-3541/ on October 2, 2021; and Trey Berry, "Hunter-Dunbar Expedition," *Encyclopedia of Arkansas,* accessed at https://encyclopediaofarkansas.net/entries/hunter-dunbar-expedition-2205/ on October 2, 2021.

44   *the lucrative fur trade:* Hine and Faragher, *Frontiers,* 56, 58.

44   *In the Pacific Northwest:* Hine and Faragher, *Frontiers,* 57.

44   *Global demand for pelts:* Michael Schaller, Robert Schulzinger, John Bezis-Selfa, Janette Thomas Greenwood, Andrew Kirk, Sarah J. Purcell, and Aaron Sheehan-Dean, *American Horizons: US History in a Global Context, Volume 1, To 1877* (New York: Oxford University Press, 2016), 295.

45   *Hawaii, where some gentlemen wore:* "New Goods," *Polynesian,* June 15, 1844, page 16.

45   *disrupted the fur trade:* "Trade and Commerce," National Park Service, accessed at https://www.nps.gov/stsp/learn/historyculture/trade-commerce.htm on March 13, 2023.

45   *to migrate further westward:* Schaller, Schulzinger, Bezis-Selfa, et al., *American Horizons,* 300.

45   *Settlers to the region:* Barbara J. Steinson, "Rural Life in Indiana, 1800–1950," *Indiana Magazine of History* 90, no. 3 (1994): 204, 207, 208.

45   *Mississippi River Valley:* Charles Lowery, "The Great Migration to the Mississippi Territory, 1798–1819," *Mississippi History Now,* Mississippi Historical Society, accessed at http://www.mshistorynow.mdah.ms.gov/articles/169/the -great-migration-to-the-mississippi-territory-1798-1819 on August 24, 2021.

45   *the traditional practice:* Barton Barbour, "Fur Trade in Oregon Country," *Oregon Encyclopedia,* accessed at https://www.oregonencyclopedia.org/articles /fur_trade_in_oregon_country/ on October 2, 2021; and Dana Dick, "Arrows, Guns, and Buffalo," National Park Service, accessed at https://www.nps.gov /fous/learn/historyculture/arrows-guns-and-buffalo.htm on October 2, 2021.

45   *Smallpox and measles:* Schaller, Schulzinger, Bezis-Selfa, et al., *American Horizons,* 295.

45   *disrupted the environment:* David Rich Lewis, "Native Americans and the Environment: A Survey of Twentieth-Century Issues," *American Indian Quarterly* 19, no. 3 (1995): 423.

46   *legally emancipated him:* Jennings Beckwith affirmed in court in Saint Louis in 1824, 1825, and 1826 that he had executed a deed of emancipation for his son, James. Wilson, *Jim Beckwourth,* 19, 200.

46   *biographer Elinor Wilson:* Wilson, *Jim Beckwourth,* 29.

46   *Beckwourth journeyed:* Wilson, *Jim Beckwourth,* 26–27, 30–31; and "Johnson, James," *Biographical Directory of the United States Congress,* accessed at https:// bioguide.congress.gov/search/bio/J000143 on October 2, 2021.

46   *hoped to make a fortune:* "Lead Mining in Southwestern Wisconsin," Wisconsin Historical Society, accessed at https://wisconsinhistory.org/turningpoints /tp-026/?action=more_essay on October 2, 2021.

46 *a three-year lease:* Bethel Saler, *The Settlers' Empire: Colonialism and State For-mation in America's Old Northwest* (Philadelphia: University of Pennsylvania, 2015), 148–50.

46 *A stern armed group:* Beckwourth, *The Life and Adventures of James P. Beck-wourth*, 20.

46 *pressure from US soldiers:* Saler, *The Settlers' Empire*, 148–50.

46 *The next year, Beckwourth:* Wilson, *Jim Beckwourth*, 29.

46 *dug large holes:* "Lead Mining in Southwestern Wisconsin," Wisconsin Histor-ical Society.

46 *seeking guidance from:* Beckwourth, *The Life and Adventures of James P. Beck-wourth*, 22.

47 *Beckwourth wrote:* Beckwourth, *The Life and Adventures of James P. Beck-wourth*, 23.

47 *assembled an expedition:* In her biography of Beckwourth, Elinor Wilson as-serts that there is some confusion regarding the date of Beckwourth's first foray into the Rocky Mountains as a trapper. It could have occurred earlier, in 1819, before Beckwourth traveled to Wisconsin to mine for lead. Wilson, *Jim Beck-wourth*, 30–31; Beckwourth, *The Life and Adventures of James P. Beckwourth*, 23; and "William Ashley," National Park Service, accessed at https://www.nps.gov/bica/learn/historyculture/william-ashley.htm on October 2, 2021.

47 *In autumn 1824:* Wilson, *Jim Beckwourth*, 30.

47 *honed his hunting skills:* Beckwourth, *The Life and Adventures of James P. Beck-wourth*, 35.

47 *gained confidence in:* Beckwourth, *The Life and Adventures of James P. Beck-wourth*, 45.

47 *a long-lost member:* Beckwourth, *The Life and Adventures of James P. Beckwourth*, 139–40, 145; and Wilson, *Jim Beckwourth*, 46–50.

47 *became successfully integrated:* "Jim Beckwourth," *Encyclopaedia Britannica*, ac-cessed at https://www.britannica.com/biography/Jim-Beckwourth on Oc-tober 25, 2023; and "James Beckwourth: African Americans in History and the West," History Colorado, accessed at https://www.historycolorado.org/story/2020/02/19/james-beckwourth-african-americans-history-and-west on October 2, 2021.

47 *to trade goods:* Beckwourth, *The Life and Adventures of James P. Beckwourth*, 272; and Buckley, "Beckwourth, James."

47 *one of whom he physically assaulted:* Beckwourth tells a disturbing story in his autobiography in which he hits one of his Indigenous wives in the head with the side of a battle-ax. He did so out of anger when she participated in a dance against his will, celebrating the scalping of three white men. Beckwourth con-tends that her father defended his action as reflecting their peoples' customs, saying: "She was the wife of the trader, I gave her to him. When your wives dis-obey your commands, you kill them; that is your right." *The Life and Adventures of James P. Beckwourth* includes an illustration depicting Beckwourth standing over her body, captioned: "Beckwourth punishes his disobedient wife." After leaving her on the ground and taking a new wife, Beckwourth writes that his first wife returned to him that night after recovering from the blow. Alonzo Delano, a satirist whose descriptions of California during the Gold Rush and

the western frontier included sensational material, claimed he heard a story in which the woman never recovered—Beckwourth's blow killed her. Beckwourth, *The Life and Adventures of James P. Beckwourth*, 112–20, 170, 247; and Wilson, *Jim Beckwourth*, 42.

47   *the level of chief:* Beckwourth, *The Life and Adventures of James P. Beckwourth*, 267.

48   *his stated advantage:* Beckwourth, *The Life and Adventures of James P. Beckwourth*, 268.

48   *passed for a native Crow:* Beckwourth, *The Life and Adventures of James P. Beckwourth*, 177.

48   *he reminisced in:* Beckwourth, *The Life and Adventures of James P. Beckwourth*, 370–71. All subsequent quotations in this paragraph come from the same pages within this source.

48   *to the Yellowstone River:* Beckwourth, *The Life and Adventures of James P. Beckwourth*, 377; and Wilson, *Jim Beckwourth*, 78.

48   *abandoned them:* Wilson, *Jim Beckwourth*, 78; and Beckwourth, *The Life and Adventures of James P. Beckwourth*, 376–78.

48   *He soon left Saint Louis:* Beckwourth, *The Life and Adventures of James P. Beckwourth*, 407; and Wilson, *Jim Beckwourth*, 84–85.

48   *The Santa Fe trail:* "Santa Fe Trail," *Encyclopaedia Britannica*, accessed at https://www.britannica.com/topic/Santa-Fe-Trail on October 2, 2021; and "Fort Vasquez," *Colorado Encyclopedia*, accessed at https://coloradoencyclopedia .org/article/fort-vasquez on October 2, 2021.

48   *a route for suppliers:* "Santa Fe National Historic Trail: History and Culture," National Park Service, accessed at https://www.nps.gov/safe/learn/history culture/index.htm on October 2, 2021; and Susan M. Colby, *Sacagawea's Child*, 142.

48   *some twelve hundred miles:* Karen Berger, *America's National Historic Trails: Walking the Trails of History* (New York: Rizzoli, 2020), 156; and "Fort Union and the Santa Fe Trail," National Park Service, accessed at https://www.nps .gov/foun/learn/historyculture/santa-fe-trail.htm on September 2, 2021.

49   *Upon reaching Santa Fe:* Berger, *America's National Historic Trails*, 156.

49   *decimated the beaver fur trade:* Colby, *Sacagawea's Child*, 139; "The Fur Trade in Colorado," *Colorado Encyclopedia*, accessed at https://coloradoencyclopedia.org /article/fur-trade-colorado on October 2, 2021; and "Beaver," *Colorado Encyclopedia*, accessed at https://coloradoencyclopedia.org/article/beaver on October 2, 2021.

49   *silk top hats:* "Hats," *The Country Gentleman*, February 17, 1853, volume 1 (7), page 108; and Emilie Hambleton, "Our Grandmothers," *Page Monthly*, July 1, 1858, page 1.

49   *much of the landscape:* James A. Little, *What I Saw on the Old Santa Fe Trail: A Condensed Story of Frontier Life Half a Century Ago* (Plainsfield: The Friends Press, 1904), 31–32. The quotation that follows is from the same pages in the same source.

49   *Inclement weather appeared:* Berger, *America's National Historic Trails*, 156.

49   *the prospect of attack:* Little, *What I Saw on the Old Santa Fe Trail*, 34.

49   *one newspaper recorded:* "Santa Fe, July 9," *Herald of the Times*, August 9, 1849, page 3.

49   *two men with:* Beckwourth, *The Life and Adventures of James P. Beckwourth*, 422; "Fort Vasquez," *Colorado Encyclopedia*, accessed at https://coloradoencyclopedia .org/article/fort-vasquez on October 2, 2021; and Buckley, "Beckwourth, James."

50   *a dusty trading post:* Beckwourth, *The Life and Adventures of James P. Beckwourth*, 422, 444, 450; "James Pierson Beckwourth," *Missouri Encyclopedia*, accessed at https://missouriencyclopedia.org/people/beckwourth-beckwith-james-pierson on October 2, 2021; and L.R. Hafen, "The Early Fur Trade Posts on the South Platte," *Mississippi Valley Historical Review* 12, no. 3 (1925): 337.

50   *Beckwourth also traded:* Beckwourth, *The Life and Adventures of James P. Beckwourth*, 453, 456–57, 464; and Wilson, *Jim Beckwourth*, 101, 103.

50   *He likely befriended:* Colby, *Sacagawea's Child*, 124–25, 146.

50   *land of remarkable variety:* Kevin Starr, *California: A History* (New York: Modern Library, 2005), 7–13.

50   *these diverse tribes:* Starr, *California: A History*, xv, 13.

51   *Diseases like smallpox:* "California Indian History," State of California Native American Heritage Commission, accessed at http://nahc.ca.gov/resources /california-indian-history/ on October 2, 2021; and Margaret A. Field, "Genocide and the Indians of California, 1769–1873," master's thesis, University of Massachusetts at Boston, 1993, page 15, accessed at https://scholarworks.umb .edu/cgi/viewcontent.cgi?article=1142&context=masters_theses on October 2, 2021.

51   *150,000 Indigenous people:* Sucheng Chan, "A People of Exceptional Character: Ethnic Diversity, Nativism, and Racism in the California Gold Rush," *California History* 79, no. 2 (2000): 50.

51   *cut down to 30,000:* Field, "Genocide and the Indians of California, 1769–1873," 1.

51   *Hispanic settlers:* Malcolm J. Rohrbough, *Days of Gold: The California Gold Rush and the American Nation* (Berkeley: University of California Press, 1997), 7–8; and "Early California Exploration and Settlement," *Calisphere: University of California*, accessed at https://calisphere.org/exhibitions/8/early-california -exploration-and-settlement/ on October 2, 2021.

51   *non-Indigenous people:* Rohrbough, *Days of Gold*, 8.

51   *Beckwourth first traveled:* Beckwourth, *The Life and Adventures of James P. Beckwourth*, 465.

51   *The Old Spanish Trail:* "Old Spanish National Historic Trail," National Park Service, accessed at https://www.nps.gov/nr/travel/american_latino_heritage /Old_Spanish_National_Historic_Trail.html on October 2, 2021.

51   *found eager buyers:* Beckwourth, *The Life and Adventures of James P. Beckwourth*, 465.

51   *arriving in Los Angeles:* Beckwourth, *The Life and Adventures of James P. Beckwourth*, 465; and William Loren Katz, *The Black West: A Documentary and Pictorial History of the African American Role in the Westward Expansion of the United States* (New York: Touchstone, 1996), 117.

51   *Beckwourth hastily returned:* Beckwourth, *The Life and Adventures of James P. Beckwourth*, 475.

51   *wartime mail courier:* Beckwourth, *The Life and Adventures of James P. Beck-wourth*, 476–77.

51   *88 percent of whom:* Abstract for Vincent J. Cirillo, "'More Fatal Than Pow-der and Shot': Dysentery in the US Army During the Mexican War, 1846–48," *Perspectives in Biology and Medicine* 52, no. 3 (2009): 400–13, accessed at https://pubmed.ncbi.nlm.nih.gov/19684375/ on September 7, 2021.

52   *as well as additional territory:* "The Impact of the Mexican-American War on American Society and Politics," *American Battlefield Trust*, accessed at https://www.battlefields.org/learn/articles/impact-mexican-american-war -american-society-and-politics on March 14, 2023; and "Treaty of Guadalupe Hidalgo," National Archives, accessed at https://www.archives.gov/milestone -documents/treaty-of-guadalupe-hidalgo on October 26, 2023.

52   *President James Polk:* James K. Polk, "Special Message," July 6, 1848, *The Amer-ican Presidency Project,* accessed at https://www.presidency.ucsb.edu/documents /special-message-4369 on October 2, 2021. The subsequent quotation is also from the same source.

52   *a subsequent message:* James K. Polk, "December 5, 1848: Fourth Annual Mes-sage to Congress," Miller Center at the University of Virginia, accessed at https://millercenter.org/the-presidency/presidential-speeches/december-5 -1848-fourth-annual-message-congress on October 2, 2021. The quotation that follows this one is also from the same source.

52   *remain in California:* Beckwourth, *The Life and Adventures of James P. Beck-wourth*, 499, 503.

52   *Malcolm Rohrbough:* Rohrbough, *Days of Gold*, 2. The subsequent quotation is from the same source and page.

53   *100,000 people:* Katz, *The Black West*, 120.

53   *The new inhabitants:* H. W. Brands, *The Age of Gold: The California Gold Rush and the New American Dream* (New York: Doubleday, 2002), 24.

53   *One miner who reached:* J. S. Holiday, *The World Rushed In: The California Gold Rush Experience* (New York: Simon & Schuster, 1981), 312.

53   *in his autobiography:* Beckwourth, *The Life and Adventures of James P. Beck-wourth*, 506.

53   *One merchant noted:* "The California Mines," in *Weekly Miners' Express*, March 27, 1849, page 4.

53   *Duels were commonplace:* Roger D. McGrath, "A Violent Birth: Disorder, Crime, and Law Enforcement, 1849–1890," *California History* 81, no. 3/4, (2003): 39.

54   *held up stagecoaches:* McGrath, "A Violent Birth," 27–28.

54   *Beckwourth remembered:* Beckwourth, *The Life and Adventures of James P. Beck-wourth*, 506.

54   *seized purported criminals:* Beckwourth, *The Life and Adventures of James P. Beck-wourth*, 506.

54   *a gruesome letter:* "Lynch Law in California," *Anti-Slavery Bugle,* Decem-ber 8, 1849, page 2. Both quotations in this paragraph are from the same source.

54   *Approximately a thousand:* Chan, "A People of Exceptional Character," 68.

54   *Fugitive Slave Law:* Tyree Boyd-Pates and Taylor Bythewood-Porter, "California Bound: Slavery on the New Frontier, 1848–1865," California African American Museum, accessed at https://caamuseum.org/exhibitions/2018 /california-bound-slavery-on-the-new-frontier-18481865 on November 18, 2022.

54   *De facto segregation:* Chan, "A People of Exceptional Character," 68–69. All of the information in this paragraph, including the quotation, comes from the same source and pages.

55   *for the local commissariat:* Beckwourth, *The Life and Adventures of James P. Beckwourth,* 503.

55   *the first US steamship:* Beckwourth, *The Life and Adventures of James P. Beckwourth,* 507.

55   *desperate gold-seekers:* Raymond A. Rydell, "The Cape Horn Route to California, 1849," *Pacific Historical Review* 17, no. 2 (1948): 155–56.

55   *The* California *stopped:* Rydell, "The Cape Horn Route to California, 1849," 156; and "Rising Tide," Smithsonian National Postal Museum, accessed at https://postalmuseum.si.edu/research-articles/making-way/rising-tide on October 2, 2021.

55   *Beckwourth boarded:* Beckwourth, *The Life and Adventures of James P. Beckwourth,* 507; and "Rising Tide," Smithsonian National Postal Museum.

55   *one newspaper reported:* "Three Days Later from California," *Liberator,* May 4, 1849, 18–19. The subsequent quotation comes from the same source.

55   *prospectors stocked up:* Chan, "A People of Exceptional Character," 68, 69; Rudolph M. Lapp, *Blacks in Gold Rush California* (New Haven: Yale University Press, 1977), 116; and site visit, Marshall Gold Discovery State Historic Park, August 3, 2021.

55   *small mining town:* "A Short History," City of Sonora, accessed at https://www .sonoraca.com/downtown-sonora/sonora-california-history/short-history/ on October 3, 2021.

55   *sold his valuable goods:* Beckwourth, *The Life and Adventures of James P. Beckwourth,* 507.

56   *Beckwourth stayed in Sonora:* Beckwourth, *The Life and Adventures of James P. Beckwourth,* 507.

56   *most populous group:* Site visit, Marshall Gold Discovery State Historic Park, August 3, 2021.

56   *settlers who often cheated:* Clifford E. Trafzer and Joel R. Hyer, *Exterminate Them: Written Accounts of the Murder, Rape, and Enslavement of Native Americans During the California Gold Rush* (East Lansing: Michigan State University Press, 1999), 16.

56   *even cruelly urged: Emigrant's Guide to California,* 1849, excerpted in "Early Accounts of Indians in the California Gold Rush," *SHEC: Resources for Teachers,* accessed at https://shec.ashp.cuny.edu/items/show/1731 on October 3, 2021.

56   *the Daily Alta California:* "Our Indian Difficulties," *Daily Alta California,* January 15, 1851, in Trafzer and Hyer, *Exterminate Them,* 37; and Beckwourth, *The Life and Adventures of James P. Beckwourth,* 507.

56   *Beckwourth eventually found that:* Beckwourth, *The Life and Adventures of James P. Beckwourth,* 509.

56   *temporary business partnership:* Beckwourth, *The Life and Adventures of James P. Beckwourth*, 509.

56   *the men ran an inn:* Beckwourth, *The Life and Adventures of James P. Beckwourth*, 509; and Wilson, *Jim Beckwourth*, 127.

57   *industrial prospecting operation:* "Placer Diggings," *Mining Magazine*, January 1, 1854, page 63. The following quotation is from the same source.

57   *mythical city of gold:* Willie Drye, "El Dorado," *National Geographic*, accessed at https://www.nationalgeographic.com/history/article/el-dorado on October 3, 2021.

57   *the Gold Rush era:* Karen Clay and Randall Jones, "Migrating to Riches? Evidence from the California Gold Rush," *Journal of Economic History* 68, no. 4 (2008): 997.

57   *Beckwourth glimpsed:* Beckwourth, *The Life and Adventures of James P. Beckwourth*, 515.

57   *But Jim had other plans:* Beckwourth, *The Life and Adventures of James P. Beckwourth*, 515.

58   *Beckwourth described as:* Beckwourth, *The Life and Adventures of James P. Beckwourth*, 515. Additional descriptions of Beckwourth's discovery in the American Valley come from the same source, pages 515–16.

58   *California's Maidu people:* "Maidu People," *Encyclopaedia Britannica*, accessed at https://www.britannica.com/topic/Maidu on October 3, 2021.

58   *massive grizzly bears:* Beckwourth, *The Life and Adventures of James P. Beckwourth*, 510.

58   *spanning two hundred miles:* Angel Morgan and Karen Mitchell, "The Beckwourth Emigrant Trail: Using Historical Accounts to Guide Archeological Fieldwork in the Plumas National Forest," *SCA Proceedings* 25 (2011): 1, accessed at chrome-extension://efaidnbmnnnibpcajpcglclefindmkaj/https://scahome.org/publications/proceedings/Proceedings.25Morgan.pdf on March 15, 2023.

58   *a local investor named Mr. Turner:* Beckwourth, *The Life and Adventures of James P. Beckwourth*, 516. The information that follows in this paragraph and the next comes from the same source, page 516–17.

59   *traveling on the California Trail:* "Truckee Trail Historical Markers," *The Historical Marker Database*, accessed at https://www.hmdb.org/results.asp?Search=Series&SeriesID=520 on July 2, 2023.

59   *with a strange rash:* Beckwourth, *The Life and Adventures of James P. Beckwourth*, 517–18; and "Erysipelas," *Rare Disease Database*, accessed at https://rarediseases.org/rare-diseases/erysipelas/ on October 2, 2021. Subsequent detail about Beckwourth's illness comes from the same two sources and pages.

60   *they climbed down:* Beckwourth, *The Life and Adventures of James P. Beckwourth*, 518. Further descriptions of his encounter with these travelers come from the same source and page.

60   *They may have tried:* Stephen Berry and Tracy L. Barnett, "The Graveyard of Old Diseases," *CSI: Dixie,* the University of Georgia, accessed at https://csidixie.org/numbers/mortality-census/graveyard-old-diseases on July 2, 2023.

60   *5,221 feet:* "Emigrant Trail: Beckwourth Pass," historical marker, Beckwourth Pass, California; site visit on August 3, 2021.

60   *miles of extra travel:* Morgan and Mitchell, "The Beckwourth Emigrant Trail," 1.

60   *Beckwourth recalled feeling:* Beckwourth, *The Life and Adventures of James P. Beckwourth*, 518; and Wilson, *Jim Beckwourth*, 136.

60   *good on their promises:* Wilson, *Jim Beckwourth*, 136; and "Beckwourth Cabin/Trading Post," National Park Service, accessed at https://www.nps.gov/park history/online_books/5views/5views2h10.htm on October 3, 2021.

60   *had spent $1,600:* Beckwourth, *The Life and Adventures of James P. Beckwourth*, 518–19.

60   *bitterness he felt:* Beckwourth, *The Life and Adventures of James P. Beckwourth*, 518.

60   *Beckwourth remarked sourly:* Beckwourth, *The Life and Adventures of James P. Beckwourth*, 519. Beckwourth's subsequent comments in this paragraph come from the same source and page.

61   *Beckwourth Valley:* "Beckwourth Cabin/Trading Post," National Park Service.

61   *A verdant place:* Beckwourth, *The Life and Adventures of James P. Beckwourth*, 519, 526–28. Additional descriptions of Beckwourth Valley in this paragraph come from the same source, pages 526–28.

61   *described to visitors:* Beckwourth, *The Life and Adventures of James P. Beckwourth*, 526–27; Obed G. Wilson, *My Adventures in the Sierras*, 195–96, quoted in Wilson, *Jim Beckwourth*, 159.

61   *Many of them stopped:* "Beckwourth Cabin/Trading Post," National Park Service.

61   *Beckwourth recalled fondly:* Beckwourth, *The Life and Adventures of James P. Beckwourth*, 519.

61   *his nineteenth-century readers:* Beckwourth, *The Life and Adventures of James P. Beckwourth*, 519. Beckwourth's following comments in this paragraph come from the same source, page 526.

62   *agreed to dictate:* Wilson, *Jim Beckwourth*, 152–53; and Beckwourth, *The Life and Adventures of James P. Beckwourth*, front cover.

62   *the* National Era: Review of "The Life and Adventures of James P. Beckwourth," *National Era* 10, no. 503, August 21, 1856. The quotation that follows is also from the same source.

62   *returned to Missouri:* Wilson, *Jim Beckwourth*, 160.

62   *After leaving California:* Wilson, *Jim Beckwourth*, 162, 164, 169, 171, 175–77.

62   *among the Crow people:* Beckwourth likely died somewhere in present-day Colorado, Montana, or Wyoming. Wilson, *Jim Beckwourth*, 184; "Jim Beckwourth," *Encyclopaedia Britannica*, accessed at https://www.britannica.com/biography /Jim-Beckwourth; Buckley, "Beckwourth, James"; and "Homelands: Crow Nation," Smithsonian National Museum of the American Indian, accessed at https://americanindian.si.edu/nk360/plains-belonging-homelands/crow-nation on July 2, 2023.

62   *the* Gold Hill Daily News: "Death of Jim Beckwourth," *Gold Hill Daily News*, February 6, 1867, page 2.

63   *the nation's largest human migration:* "California National Historic Trail," National Park Service, accessed at https://www.nps.gov/nr/travel/american _latino_heritage/California_National_Historic_Trail.html on October 1, 2021.

63 *reshaped the environment:* "Resource 6–1a: California Population by Ethnic Groups, 1790–1880," accessed at http://explore.museumca.org/goldrush /curriculum/1stcalifornians/resourcesix.htm on October 1, 2021.

3. LAURA INGALLS WILDER: THE HOMESTEADER

64 *Eleven million acres:* Judith Royster, "Of Surplus Lands and Landfills: The Case of the Yankton Sioux," *South Dakota Law Review* 43 (1998): 285; Thomas Constantine Maroukis, *Peyote and the Yankton Sioux: The Life and Times of Sam Necklace* (Norman: University of Oklahoma Press, 2004), 39; and Edward Elliott Collins, "A History of Union County, South Dakota, to 1880," master's thesis, University of South Dakota, 1937, accessed at http://files.usgwarchives .net/sd/union/history/collins1937.txt on October 7, 2021.

64 *Thirteen Yankton men:* One of the thirteen men, Charles Picotte, also signed on behalf of Little White Swan, Pretty Boy, and White Medicine Cow that Stands, bringing the total number of signatures to sixteen. Patrick Coleman, "A Rare Find: The Treaty of Washington, 1858," *Minnesota History* 59 (Spring 2005): 197; and Caroline Fraser, *Prairie Fires: The American Dreams of Laura Ingalls Wilder* (New York: Metropolitan Books, 2017), 10.

64 *Their names evoked:* "Yankton Sioux Treaty Monument," National Park Service, accessed at https://www.nps.gov/mnrr/learn/historyculture/yankton-sioux -treaty-monument.htm on October 6, 2021.

64 *The damp streets:* "Georgetown," *Evening Star,* April 19, 1858, page 3.

64 *cede their territory:* Royster, "Of Surplus Lands and Landfills," 285; Maroukis, *Peyote and the Yankton Sioux,* 39; and Collins, "A History of Union County, South Dakota, to 1880."

64 *As compensation:* Maroukis, *Peyote and the Yankton Sioux,* 39.

64 *Newspapers hailed:* Letter from J. R. Brown of Washington, DC, "Indian Treaties," *Weekly Pioneer and Democrat,* May 13, 1858, page 4.

65 *One white traveler:* "From the Upper Country," *Sioux City Register,* July 21, 1859, page 2.

65 *disputed the terms:* "From the Upper Country," page 2.

65 *the Yankton relocated:* "Treaty with the Yankton Sioux, 1858," Oklahoma State Library, accessed at https://dc.library.okstate.edu/digital/collection/kapplers /id/26617/ on October 7, 2021.

65 *bright, pleasant morning:* Laura Ingalls Wilder, *Pioneer Girl: The Annotated Autobiography,* ed. Pamela Smith Hill (Pierre: South Dakota Historical Society Press, 2014), 153.

65 *one of the boundaries:* "Treaty with the Yankton Sioux, 1858," Oklahoma State Library.

65 *"a big meadow":* Wilder, *Pioneer Girl,* 153.

65 *faintly visible path:* Wilder, *Pioneer Girl,* 153–55.

65 *appeared uninhabited:* Wilder, *Pioneer Girl,* 155.

66 *evidence of bison:* Wilder, *Pioneer Girl,* 155.

66 *the impressive mound:* Wilder, *Pioneer Girl,* 160.

66 *a peripatetic life:* Paula M. Nelson, "Women's Place: Family, Home, and Farm,"

in Nancy Tystad Koupal, ed., *Pioneer Girl Perspectives: Exploring Laura Ingalls Wilder* (Pierre: South Dakota Historical Society Press, 2017), 179; and "Laura Ingalls Wilder," *Kansapedia*, Kansas Historical Society, accessed at https:// www.kshs.org/kansapedia/laura-ingalls-wilder/15563 on October 9, 2021.

66   *the Free-Soil Party:* Fraser, *Prairie Fires*, 12.

66   *produced a platform:* "Republican Party Platform of 1856," June 18, 1856, *The American Presidency Project,* accessed at https://www.presidency.ucsb.edu /documents/republican-party-platform-1856 on October 18, 2021; and Hannah L. Anderson, "That Settles It: The Debate and Consequences of the Homestead Act of 1862," *The History Teacher* 45, no. 1 (2011): 119.

67   *Ryan Hall:* Ryan Hall, "Chaos and Conquest: The Civil War and Indigenous Crisis on the Upper Missouri, 1861–1865," *Journal of the Civil War Era* 12, no. 2 (2022): 147. This paragraph's information about the upheaval experienced by Indigenous people during the Civil War comes from the same source and page.

67   *also able to participate:* Richard Edwards, Jacob K. Friefeld, and Rebecca S. Wingo, *Homesteading the Plains: Toward a New History* (Lincoln: University of Nebraska Press, 2017), 2.

67   *All applicants had:* "Transcript of Homestead Act (1862)," National Archives, accessed at https://www.ourdocuments.gov/doc.php?flash=false&doc =31&page=transcript on October 11, 2021; Edwards, Friefeld, and Wingo, *Homesteading the Plains,* 10; and "150th Anniversary of the Homestead Act," US Fish and Wildlife Service, accessed at https://www.fws.gov/refuges /news/150thAnniversaryHomesteadAct_04272012.html on October 11, 2021.

67   *what some scholars:* Edwards, Friefeld, and Wingo, *Homesteading the Plains,* 7.

67   *270 million acres:* Todd Arrington, "Homesteading by the Numbers," National Park Service, April 24, 2007, accessed at https://www.nps.gov/home/learn /historyculture/bynumbers.htm on October 11, 2021.

67   *filed a claim:* Nancy Cleaveland and Penny Linsenmayer, "Charles Ingalls and the U.S. Public Land Laws" (SeventhWinter Press, 2001), 8–9, accessed at http://pioneergirl.com/cpi_land_laws.pdf on October 18, 2021.

68   *Charles had found work:* Pamela Smith Hill, *Laura Ingalls Wilder: A Writer's Life* (Pierre: South Dakota State Historical Society Press, 2007), 36; and Fraser, *Prairie Fires*, 100.

68   *advertised tickets:* "Home Seekers' Excursions to the Northwest, West, and Southwest, and Colonist Low Rates West," *Aberdeen Democrat*, January 15, 1904, page 1.

68   *some 80 percent:* John E. Miller, *Laura Ingalls Wilder's Little Town: Where History and Literature Meet* (Lawrence: University Press of Kansas, 1994), 22; and Marta McDowell, *The World of Laura Ingalls Wilder: The Frontier Landscapes That Inspired the Little House Books* (Portland: Timber Press, 2017), 139.

68   *American prospectors trespassing:* Fraser, *Prairie Fires,* 93; Terry Mort, *Thieves' Road: The Black Hills Betrayal and Custer's Path to Little Bighorn* (Amherst: Prometheus Books, 2015), 290.

68   *Standing Bear, remembered:* Account of the Battle of Little Bighorn, Standing Bear, translated by Veleda Goulden, with contributions by Renate Maasen, early 1930s, located at the Pine Ridge Reservation in South Dakota, accessed at https://peelarchivesblog.com/2022/06/08/your-friend-standing-bear-coming -together-through-repatriation/#Standing-Bear on November 18, 2022.

68   *government implicitly admitted:* "Annual Report of the Secretary of the Interior" (Washington: US Government Printing Office, 1875), 660, accessed at https://www.google.com/books/edition/Abridgment_Containing_the_Annual_Message/-91XAAAAYAAJ?hl=en&gbpv=0.

69   *Battle of Little Bighorn:* Site visit, Little Bighorn Battlefield National Monument, July 29, 2021; and "Context and Story of the Battle," National Park Service, accessed at https://www.nps.gov/libi/learn/historyculture/battle-story.htm on October 18, 2021.

69   *authorize a document:* "United States v. Sioux Nation of Indians, 448 US 371 (1980)," *Justia: US Supreme Court,* accessed at https://supreme.justia.com/cases/federal/us/448/371/ on October 18, 2021.

69   *settling 41 percent:* Edwards, Friefeld, and Wingo, *Homesteading the Plains,* 10.

69   *he later lamented:* Account of the Battle of Little Bighorn, Standing Bear, translated by Veleda Goulden, with contributions by Renate Maasen, early 1930s, located at the Pine Ridge Reservation in South Dakota.

69   *for Dakota Territory:* Charles Ingalls arrived in Dakota Territory in the spring of 1879; his family followed in September of that year. They rode the train from Walnut Grove, Minnesota, to Tracy, Minnesota, then proceeded to Dakota Territory by wagon. Wilder, *Pioneer Girl,* 145.

69   *South Dakota's population:* "Population of South Dakota by Counties and Minor Civil Divisions," *Census Bulletin* (Washington, DC, February 1, 1901), 1, accessed at https://www2.census.gov/library/publications/decennial/1900/bulletins/demographic/47-population-sd.pdf on October 19, 2021. All population data in this paragraph comes from the same source.

69   *South Dakota's terrain:* Fraser, *Prairie Fires,* 94.

69   *Surveyor General George Hill:* Kenneth M. Hammer, "Come to God's Country: Promotional Efforts in Dakota Territory, 1861–1889," *South Dakota History* 10, no. 4 (1980): 291–96; and Fraser, *Prairie Fires,* 94–95.

70   *Laura recorded:* Wilder, *Pioneer Girl,* 160. Other quotations from Laura in this paragraph are from the same source and page.

70   *a single tree:* Wilder, *Pioneer Girl,* 158.

70   *often been ignored:* Margaret Walsh, "State of the Art: Women's Place on the American Frontier," *Journal of American Studies* 29, no. 2 (1995): 241.

70   *"invisible helpmate":* Walsh, "State of the Art," 241.

70   *While some farmed:* Edwards, Friefeld, and Wingo, *Homesteading the Plains,* 132.

70   *single women and widows:* Edwards, Friefeld, and Wingo, *Homesteading the Plains,* 143.

71   *Laura described as:* Laura Ingalls Wilder to Rose Wilder Lane, January 25, 1938, Box 13, File 194, Lane Papers, quoted in Wilder, *Pioneer Girl,* 158.

71   *there was little privacy:* Wilder, *Pioneer Girl,* 158, 160, 162.

71   *the grueling nature:* Frederick W. Kaul, L. A. Rollins, and Mr. John Grosvenor, "Interview with Mr. John Grosvenor," Nebraska, November 1938, Library of Congress, accessed at https://www.loc.gov/item/wpalh000923/ on October 25, 2021. All John Grosvenor's comments in this paragraph come from the same source.

71   *saw a brawl develop:* Wilder, *Pioneer Girl,* 162–64. Other details in this paragraph about the lawlessness of the camp come from the same source and pages.

72   *soon after her arrival:* Wilder, *Pioneer Girl*, 165.

72   *an outbreak of violence:* Wilder, *Pioneer Girl*, 165–67. All details and quotations in the following two paragraphs comes from the same source and pages.

72   *December 1879 brought:* Wilder, *Pioneer Girl*, 174.

72   *Laura marveled at:* Wilder, *Pioneer Girl*, 178.

73   *out of the crude shanty:* Wilder, *Pioneer Girl*, 176.

73   *the house felt:* Wilder, *Pioneer Girl*, 178.

73   *procuring additional provisions:* Wilder, *Pioneer Girl*, 174. Further information about these activities come from the same source, pages 174–78.

73   *tracked the wolves:* Wilder, *Pioneer Girl*, 183.

73   *an attractive plot:* Wilder, *Pioneer Girl*, 178.

73   *the town of Brookings:* Fraser, *Prairie Fires*, 105.

73   *terms of the application:* "Sample Homestead File (for Charles Ingalls)," National Archives, accessed at https://www.archives.gov/research/land/ingalls on October 27, 2021.

74   *what Laura described:* Wilder, *Pioneer Girl*, 183. Laura's additional comments about winter on this page come from the same source, pages 181–83.

74   *wolves were reported:* "Southern Dakota," *Daily Press and Dakotaian*, December 20, 1879, page 3.

74   *One newspaper reported:* "Dakota News," *Canton Advocate*, May 6, 1880, page 3.

75   *Eliza Jane Wilder:* Eliza Jane Wilder, *A Wilder in the West: The Story of Eliza Jane Wilder*, ed. William Anderson (Anderson Publications, 1985), 10–11.

75   *first structure in:* Wilder, *Pioneer Girl*, 189.

75   *moved the family:* Wilder, *Pioneer Girl*, 189.

75   *late April 1880:* Wilder, *Pioneer Girl*, 192.

75   *By late spring:* Wilder, *Pioneer Girl*, 192; and "Population of De Smet, SD," accessed at https://population.us/sd/de-smet/ on October 29, 2021.

75   *a claim jumper:* Wilder, *Pioneer Girl*, 194.

75   *was not uncommon:* "Charged with Perjury," *Canton Advocate*, October 14, 1880, page 3.

75   *Laura remembered fondly:* Wilder, *Pioneer Girl*, 194.

75   *The claim was situated:* William Anderson, *Laura Ingalls Wilder Country: The People and Places in Laura Ingalls Wilder's Life and Books* (New York: Harper, 1990), 53–55.

75   *lodging was modest:* Wilder, *Pioneer Girl*, 194–95.

75   *The family wasted no time:* Wilder, *Pioneer Girl*, 196–98.

76   *Laura and her mother:* Wilder, *Pioneer Girl*, 198.

76   *Homesteaders came from:* "David Imes," Homestead National Historical Park, accessed at https://www.nps.gov/people/david-imes.htm on October 31, 2021; "Robert Ball Anderson," Homestead National Historical Park, accessed at https://www.nps.gov/people/robert-ball-anderson.htm on October 31, 2021; and "Caleb Benson," Homestead National Historical Park, accessed at https://

www.nps.gov/people/caleb-benson.htm on October 30, 2023; and "Moses and Susan Cropps-Kirke Speese," Homestead National Historical Park, accessed at https://www.nps.gov/people/moses-and-susan-cropps-kirke-speese.htm on October 30, 2023.

76 *Civil War battlefields:* "Abraham Hall," Homestead National Historical Park, accessed at https://www.nps.gov/people/abraham-hall.htm on October 31, 2021; "Daniel Freeman," Homestead National Historical Park, accessed at https://www.nps.gov/people/daniel-freeman.htm on August 5, 2023; "The Homestead Act of 1862," National Archives, accessed at https://www.archives.gov/education/lessons/homestead-act on November 20, 2021; and "Robert Ball Anderson," Homestead National Historical Park.

76 *to the Midwest:* "Maggie Walz," Homestead National Historical Park, accessed at https://www.nps.gov/people/maggie-walz.htm on October 31, 2021; and "Rachel Bella Calof," Homestead National Historical Park, accessed at https://www.nps.gov/people/rachel-bella-calof.htm on October 30, 2023.

76 *Female homesteaders:* "Amanda Neal," Homestead National Historical Park, accessed at https://www.nps.gov/people/amanda-neal.htm on October 31, 201; and "Henry and Julia Gordon," Homestead National Historical Park, accessed at https://www.nps.gov/people/henry-and-julia-gordon.htm on October 31, 2021.

76 *a blizzard assailing:* Wilder, *Pioneer Girl*, 201–2; and Fraser, *Prairie Fires*, 107.

76 *fifteen degrees Fahrenheit:* Wilder, *Pioneer Girl*, 202; and Constance Potter, "De Smet, Dakota Territory, Little Town in the National Archives," *Prologue Magazine* 35, no. 4 (2003), accessed at https://www.archives.gov/publications/prologue/2003/winter/little-town-in-nara on March 18, 2023.

77 *its icy blast killing:* Wilder, *Pioneer Girl*, 203; and Eliza Jane Wilder, *A Wilder in the West*, 14.

77 *"storm without parallel":* "From the Borean Belt," *Daily Press and Dakotaian*, October 16, 1880, page 4. This paragraph contains further quotations from the same source.

77 *an elderly Indigenous man:* Wilder, *Pioneer Girl*, 203.

77 *The Ingalls family moved:* Wilder, *Pioneer Girl*, 203.

77 *three-day blizzard:* Wilder, *Pioneer Girl*, 203–7.

77 *the wintry weeks:* Wilder, *Pioneer Girl*, 209; and Cindy Wilson, *The Beautiful Snow: The Ingalls Family, the Railroads, and the Hard Winter of 1880–81* (Saint Paul: Beavers Pond Press, 2020), 24.

77 *Eliza Jane Wilder recalled:* Eliza Jane Wilder, *A Wilder in the West*, 15.

78 *woefully unprepared:* Wilder, *Pioneer Girl*, 210; and Wilson, *The Beautiful Snow*, 12.

78 *that cruel winter:* Wilder, *Pioneer Girl*, 210. The following quotation is also from the same source and page.

78 *thirty-two degrees below:* Potter, "De Smet, Dakota Territory, Little Town in the National Archives."

78 *could not safely hunt:* Wilder, *Pioneer Girl*, 210–12.

78 *almost no milk:* Wilder, *Pioneer Girl*, 212–13.

78 *In town, Laura:* Wilder, *Pioneer Girl*, 213. Further details in this paragraph about life in town during the winter come from the same source, pages 213–15.

79   *also running low:* Wilder, *Pioneer Girl*, 219–20.

79   *Eliza Jane recorded:* The members of the Ingalls family were forced to eat seed wheat, ground in a coffee mill, when their flour ran out. Eliza Jane Wilder, *A Wilder in the West*, 15; and Wilder, *Pioneer Girl*, 219.

79   *Word began circulating:* Wilder, *Pioneer Girl*, 220.

79   *Laura remembered:* Wilder, *Pioneer Girl*, 220.

79   *the two young men:* Eliza Jane Wilder, *A Wilder in the West*, 5; Fraser, *Prairie Fires*, 113; and "Margaret Garland Family/Cap Garland," *Laura Ingalls Wilder A–Z*, accessed at http://www.pioneergirl.com/blog/archives/5115 on November 20, 2021.

79   *successfully made the trek:* Wilder, *Pioneer Girl*, 220–21. The quotations in the following paragraph are from the same source, pages 219–20.

79   *the boys returned:* Wilder, *Pioneer Girl*, 221. Information in the rest of the paragraph regarding this event comes from the same source and page.

80   *first signs of spring:* Wilder, *Pioneer Girl*, 223. Wilder's descriptions in the rest of this paragraph come from the same source, page 225.

80   *correspondent in Cheyenne:* "The Wyoming Range," *Daily Press and Dakotaian*, June 8, 1881, page 1. The quotation that follows is also from the same source and page.

81   *"a beautiful green":* Wilder, *Pioneer Girl*, 225.

81   *The population exploded:* Wilder, *Pioneer Girl*, 229.

81   *Laura alternately welcomed:* Wilder, *Pioneer Girl*, 189.

81   *their permanent disappearance:* Wilder, *Pioneer Girl*, 183.

81   *distinctly regretful feelings:* Wilder, *Pioneer Girl*, 231.

81   *She wrote admiringly:* Wilder, *Pioneer Girl*, 231–34.

81   *express her thoughts:* Site visit, De Smet, South Dakota, July 27, 2021.

81   *what she described as:* Wilder, *Pioneer Girl*, 260–61.

81   *began a courtship:* Wilder, *Pioneer Girl*, 264, 322.

82   *Almanzo's tree claim:* Wilder, *Pioneer Girl*, 322, 324; and Rebecca Brammer and Phil Greetham, "The Homestead Act of 1862," *Laura Ingalls Wilder: Frontier Girl*, accessed at http://liwfrontiergirl.com/homestead.html on March 18, 2023.

82   *They would endure:* Wilder, *Pioneer Girl*, 287.

82   *proposed a compromise:* Laura Ingalls Wilder, *The First Four Years* (New York: HarperCollins, 1971), 4–7.

82   *wildness of the prairie:* Wilder, *The First Four Years*, 6.

82   *spent her days:* Wilder, *The First Four Years*, 22.

82   *These hardships:* Laura Ingalls Wilder to Congressman Clarence E. Kilburn, 1945, in *The Selected Letters of Laura Ingalls Wilder*, ed. William Anderson (New York: Harper, 2016), 269; James I. Stewart, "The Economics of American Farm Unrest, 1865–1900," EH.Net Encyclopedia, edited by Robert Whaples, 2008, accessed at https://eh.net/encyclopedia/the-economics-of-american -farm-unrest-1865-1900/ on March 18, 2023; and Brian Solender, "'Farm-

ing Don't Pay': The Anatomy of the 19th-Century Western Farm Mortgage Industry," senior thesis, Department of History, Columbia University, 2017, accessed at chrome-extension://efaidnbmnnnibpcajpcglclefindmkaj/https://history.columbia.edu/wp-content/uploads/sites/20/2016/06/Solender-Thesis.pdf on March 18, 2023.

82   *a hailstorm destroyed:* Wilder, *The First Four Years,* 55–56.

83   *Laura gave birth:* The Wilders moved to their homestead claim on August 25, 1886. Rose was born on December 5, 1886. Wilder, *The First Four Years,* 71.

83   *Diphtheria struck:* Wilder, *The First Four Years,* 87; and "Claim," *Laura Ingalls Wilder A–Z,* accessed at http://www.pioneergirl.com/blog/archives/10912 on November 20, 2021.

83   *expenses proved unmanageable:* Wilder, *The First Four Years,* 89–90; and "Claim," *Laura Ingalls Wilder A–Z.*

83   *for a fourth year:* Wilder, *The First Four Years,* 100-01.

83   *Overwhelmed, Laura began:* Wilder, *The First Four Years,* 119.

83   *eight inches of rain:* Fraser, *Prairie Fires,* 148.

83   *ruining their harvest:* Wilder, *The First Four Years,* 121.

83   *a pre-emption claim:* Wilder, *The First Four Years,* 122; and Fraser, *Prairie Fires,* 150.

83   *a baby boy:* Fraser, *Prairie Fires,* 150.

83   *His death shattered:* Wilder, *The First Four Years,* 127. The quotations that precede and follow are from the same source and page.

83   *overwhelming grief:* Wilder, *The First Four Years,* 128-29; and Fraser, *Prairie Fires,* 151.

84   *returned to the kitchen:* Wilder, *The First Four Years,* 128.

84   *all the Wilders' possessions:* Wilder, *The First Four Years,* 128–30.

84   *plan to sell their land:* Fraser, *Prairie Fires,* 153.

84   *flee Dakota Territory:* Fraser, *Prairie Fires,* 153–54.

84   *Caroline Fraser:* Fraser, *Prairie Fires,* 153.

84   *migrants like the Ingallses:* Fraser, *Prairie Fires,* 154.

84   *departed De Smet:* Fraser, *Prairie Fires,* 157, 163.

84   *permanently moved into town:* Fraser, *Prairie Fires,* 166.

84   *their final move:* Fraser, *Prairie Fires,* 168, 174.

85   *its advantageous climate:* Fraser, *Prairie Fires,* 174.

85   *the family's arrangements:* Fraser, *Prairie Fires,* 175; and Laura Ingalls Wilder, *On the Way Home: The Diary of a Trip from South Dakota to Mansfield, Missouri, in 1894* (New York: HarperCollins, 1962), 10.

85   *Bidding farewell:* Wilder, *On the Way Home,* 11–12, 15.

85   *Laura made notes:* Wilder, *On the Way Home,* 15.

85   *wondered at the beauty:* Wilder, *On the Way Home,* 23–24.

85   *the Native Americans:* Wilder, *On the Way Home,* 24.

85   *letter to a friend:* Laura Ingalls Wilder to Ida Carson, March 25, 1946, in Anderson, *The Selected Letters of Laura Ingalls Wilder,* 275–76.

86   *sixty million copies:* Fraser, *Prairie Fires,* 2.

4. JOHN MUIR: THE PRESERVATIONIST

87   *opened their petals:* John Muir, "Yosemite Valley: Beauties of the Landscape in Early Summer," *San Francisco Daily Evening Bulletin,* June 21, 1889, in William F. Kimes and Maymie B. Kimes, *John Muir: A Reading Bibliography by Kimes* (Fresno: Panorama West Books, 1986), 190, accessed at https://scholarly commons.pacific.edu/jmb/190 on December 30, 2021.

87   *sounds of birdsong:* Muir, "Yosemite Valley: Beauties of the Landscape in Early Summer."

87   *the words of John Muir:* Muir, "Yosemite Valley." Further descriptions of Yosemite Valley that summer and comments from Muir on this page are from the same source.

88   *Born in 1838:* John Muir, "A Boyhood in Scotland," in *The Story of My Boyhood and Youth* (Boston: Houghton Mifflin Company, 1913), accessed at https://vault.sierraclub.org/john_muir_exhibit/writings/the_story_of_my_boyhood_and_youth/chapter_1.aspx on November 26, 2021; Donald Worster, *A Passion for Nature: The Life of John Muir* (New York: Oxford University Press, 2008), 13; and Stephen R. Fox, *John Muir and His Legacy: The American Conservation Movement* (Boston: Little, Brown, and Company, 1981), 28.

88   *His hometown overlooked:* Site visit, home of John Muir in Dunbar, Scotland, October 11, 2022.

88   *Called Johnnie:* Muir, "A Boyhood in Scotland."

88   *family's three-story house:* Site visit, home of John Muir, October 11, 2022; "Number 1.04–John Muir's Childhood Home," John Muir's Birthplace, accessed at https://www.jmbt.org.uk/wp-content/uploads/2017/04/1_04.pdf on November 21, 2022; and Muir, "A Boyhood in Scotland."

88   *Johnnie and his playmates:* Muir, "A Boyhood in Scotland."

89   *a surprising announcement:* John Muir, "A New World," in *The Story of My Boyhood and Youth,* accessed at https://vault.sierraclub.org/john_muir_exhibit/writings/the_story_of_my_boyhood_and_youth/chapter_2.aspx on November 26, 2021; and Fox, *John Muir and His Legacy,* 30.

89   *Campbellite Disciples of Christ:* Fox, *John Muir and His Legacy,* 30.

89   *recalled his desire:* Muir, "A New World."

89   *train bound for Glasgow:* Muir, "A New World."

89   *As Muir looked back:* Muir, "A New World." The quotation expressing Muir's feeling at this moment, which follows in this paragraph, comes from the same source.

90   *The sea crossing:* Muir, "A New World"; and Worster, *A Passion for Nature,* 44. Information in this paragraph about the voyage comes from the same sources.

90   *completed Erie Canal:* The Erie Canal was fully opened for public use in 1825. John Muir, *The Wilderness World of John Muir,* ed. Edwin Way Teal (Boston: Houghton Mifflin, 1954), 27.

90   *160 acres of land:* Muir, "A New World"; Fox, *John Muir and His Legacy,* 30; and Worster, *A Passion for Nature,* 47.

90   *"wandering in the fields":* Muir, "A New World."

90   *their new life:* Fox, *John Muir and His Legacy,* 30.

90   *Scottish forests:* "History of Scotland's Woodlands," NatureScot, accessed at https://www.nature.scot/professional-advice/land-and-sea-management /managing-land/forests-and-woodlands/history-scotlands-woodlands on March 18, 2023; "Dunbar in the 1840s," John Muir's Birthplace, accessed at chrome-extension://efaidnbmnnnibpcajpcglclefindmkaj/https://www.jmbt.org.uk/wp -content/uploads/2017/04/3_08.pdf on March 18, 2023; and "Map of Scotland, England and Wales, 1840s–1880s," National Library of Scotland, accessed at https://maps.nls.uk/geo/explore on March 18, 2023.

91   *"This sudden splash":* Muir, "A New World.

91   *state's population increased:* "Population of Wisconsin, 1820–1990," Wisconsin Historical Society, accessed at https://www.wisconsinhistory.org/Records /Article/CS1816 on November 30, 2021.

91   *Immigrants came from:* David Long and Dan Veroff, "A Brief History of Immigration in Wisconsin," December 13, 2010, accessed at https://cdn.apl.wisc .edu/publications/WI_Immigration_History.pdf on November 30, 2021.

91   *Meskwaki, Ho-Chunk, Sauk:* "American Indians in Wisconsin: History," Wisconsin Department of Health Services, accessed at https://www.dhs .wisconsin.gov/minority-health/population/amind-pophistory.htm on December 1, 2021.

91   *millions of acres:* "The Territorial Era: 1787–1848," Wisconsin Historical Society, accessed https://www.wisconsinhistory.org/Records/Article/CS3586 on December 1, 2021; and "American Indians in Wisconsin: History," Wisconsin Department of Health Services.

91   *tens of thousands:* Jeanne Kay, "The Fur Trade and Native American Population Growth," *Ethnohistory* 31, no. 4 (1984): 265, 284.

91   *on their farm:* John Muir, "Life on a Wisconsin Farm," in *The Story of My Boyhood and Youth,* accessed at https://vault.sierraclub.org/john_muir_exhibit /writings/the_story_of_my_boyhood_and_youth/chapter_3.aspx on November 30, 2021.

91   *an Indian mound:* Muir, "Life on a Wisconsin Farm"; and Worster, *A Passion for Nature,* 49.

91   *Muir marveled at:* John Muir, "Young Hunters," in *The Story of My Boyhood and Youth,* accessed at https://vault.sierraclub.org/john_muir_exhibit/writings /the_story_of_my_boyhood_and_youth/chapter_5.aspx on December 2, 2021. The two quotations that follow in this paragraph are from the same source.

92   *by performing hard labor:* Fox, *John Muir and His Legacy,* 32.

92   *Muir "repudiated":* Worster, *A Passion for Nature,* 57.

92   *luck as an inventor:* Worster, *A Passion for Nature,* 66.

92   *time-keeping invention:* John Muir, "The World and the University," in *The Story of My Boyhood and Youth,* accessed at https://vault.sierraclub.org/john _muir_exhibit/writings/the_story_of_my_boyhood_and_youth/chapter_8 .aspx on December 3, 2021.

92   *University of Wisconsin for:* Muir, "The World and the University"; and Worster, *A Passion for Nature*, 90.

92   *a new adventure:* Muir, "The World and the University"; and "Chronology (Timeline) of the Life and Legacy of John Muir," Sierra Club, accessed at https://vault.sierraclub.org/john_muir_exhibit/life/chronology.aspx on December 30, 2021. The quotation that immediately follows is from the same source.

93   *called Yosemite "Ahwahnee":* Mark Spence, "Dispossessing the Wilderness: Yosemite Indians and the National Park Ideal, 1864–1930," *Pacific Historical Review* 65, no. 1 (1996): 31; and "Their Lifeways," National Park Service, accessed at https://www.nps.gov/yose/learn/historyculture/their-lifeways.htm on December 8, 2021.

93   *hunted for rabbits:* Stephen Powers, *Tribes of California* (Washington, DC: US Government Printing Office, 1877), accessed at http://www.yosemite.ca.us /library/powers/23.html on December 8, 2021; and "Tools and Trade," National Park Service, accessed at https://www.nps.gov/yose/learn/historyculture /tools-and-trade.htm on December 8, 2021.

93   *Brush houses sheltered:* Powers, *Tribes of California*.

93   *served as currency:* Powers, *Tribes of California*: and "Tools and Trade," National Park Service.

93   *the Indigenous population:* Spence, "Dispossessing the Wilderness," 29.

93   *Act for the Government:* "An Act for the Government and Protection of Indians" (1850), Library of Congress, accessed at https://www.loc.gov/resource /cph.3c16681/ on December 8, 2021; and Kimberly Johnston-Dodds, "Early California Laws and Policies Related to California Indians," prepared at the request of Senator John L. Burton (Sacramento: California Research Bureau, 2002), 1, 4, accessed at https://digitalcommons.csumb.edu/cgi/viewcontent .cgi?article=1033&context=hornbeck_usa_3_d on December 30, 2021.

93   *The Mariposa Battalion:* Alfred Runte, *Yosemite: The Embattled Wilderness* (Lincoln: University of Nebraska Press, 1990), accessed at https://www.nps.gov /parkhistory/online_books/runte2/chap1.htm on December 30, 2021; and "Destruction and Disruption," National Park Service, accessed at https://www .nps.gov/yose/learn/historyculture/destruction-and-disruption.htm on December 8, 2021.

93   *militia set fire:* "Destruction and Disruption," National Park Service.

94   *an 1853 attack:* Worster, *A Passion for Nature*, 167.

94   *published an article:* Lafayette Houghton Bunnell, "How the Yo-Semite Valley Was Discovered and Named," *Hutchings' Illustrated California Magazine* 35, May (1859): 498–504.

94   *tourists began trickling:* Spence, "Dispossessing the Wilderness," 30.

94   *to sign an act:* "An Act Authorizing a Grant to the State of California of the 'Yo-Semite Valley,' and of the Land Embracing the 'Mariposa Big Tree Grove,'" approved June 30, 1864 (13 Stat. 325), accessed at https://www.nps.gov/park history/online_books/anps/anps_1a.htm on December 8, 2021.

94   *period of national rebuilding:* Spence, "Dispossessing the Wilderness," 30.

94   *Between 1863 and 1867:* Worster, *A Passion for Nature*, 91, 97–98, 103; and "Chronology (Timeline) of the Life and Legacy of John Muir," Sierra Club.

94 *Muir noticed one day:* John Muir, "Tribute to Catharine Merrill," published in *The Man Shakespeare and Other Essays by Catharine Merrill with Impressions and Reminiscences of the Author by Melville B. Anderson, and with Some Words of Appreciation from John Muir* (Indianapolis: The Bowen-Merrill Company, 1902), accessed at https://vault.sierraclub.org/john_muir_exhibit/writings/people/catharine_merrill_tribute.aspx on December 27, 2021.

94 *As Muir handled:* John Muir to Sarah Muir, May 1867, in William Frederic Badè, *The Life and Letters of John Muir* (Boston: Houghton Mifflin Company, 1924), accessed at https://vault.sierraclub.org/john_muir_exhibit/life/life_and_letters/chapter_5.aspx on December 10, 2021.

95 *"perfect darkness":* John Muir to Sarah Muir, May 1867, in Badè, *The Life and Letters of John Muir.*

95 *sympathetic ophthalmia:* Kara C. LaMattina, "Sympathetic Opthalmia," Merck Manual, 2022, accessed at https://www.merckmanuals.com/home/eye-disorders/uveitis-and-related-disorders/sympathetic-ophthalmia on March 19, 2023.

95 *Catharine Merrill:* Catharine Merrill became the second female professor in the United States, teaching at North Western Christian University in the English Department from 1869 to 1885. "Catharine Merrill," Sierra Club, accessed at https://vault.sierraclub.org/john_muir_exhibit/people/catharine_merrill.aspx on December 27, 2021.

95 *Muir later wrote:* Muir, "Tribute to Catharine Merrill."

95 *Muir promised himself:* John Muir to Sarah Muir, May 1867, in Badè, *The Life and Letters of John Muir.*

95 *Muir would subsequently:* John Muir to Sarah Muir, May 1867, in Badè, *The Life and Letters of John Muir.*

95 *thousand-mile journey:* John Muir, *A Thousand-Mile Walk to the Gulf,* edited by William Frederic Badè (Boston: Houghton, Mifflin, and Company, 1916), accessed at https://vault.sierraclub.org/john_muir_exhibit/writings/a_thousand_mile_walk_to_the_gulf/ on December 10, 2021; and "Cumberland Gap," National Park Service, accessed at http://npshistory.com/publications/cuga/index.htm on December 10, 2021. Muir's comments in the following two paragraphs about this journey and his decision to travel to California come from the same source.

96 *Muir later described:* John Muir, "The Treasures of the Yosemite," *Century Magazine* 40, no. 4, (1890), accessed at https://vault.sierraclub.org/john_muir_exhibit/writings/the_treasures_of_the_yosemite/ on December 23, 2021. The quotation that follows is from the same source.

97 *he looked for work:* Worster, *A Passion for Nature*, 153. The subsequent information regarding Muir's first job as a shepherd is from the same source, pages 153–59.

97 *spent his days climbing:* John Muir, *My First Summer in the Sierra* (Boston: Houghton, Mifflin, and Company, 1911), 3, accessed at https://www.gutenberg.org/files/32540/32540-h/32540-h.htm on September 3, 2020. The quotations in this paragraph come from the same source, pages 13–15.

97 *documented his observations:* Muir, *My First Summer in the Sierra*, 13, 36, 39.

98 *"How interesting":* Muir, *My First Summer in the Sierra*, 240–41.

98 *wanted nothing more:* Muir, *My First Summer in the Sierra*, 16–17.

98   *vowed to return:* Muir, *My First Summer in the Sierra*, 241.

98   *fragile seedlings:* Muir, *My First Summer in the Sierra*, 96–97, 195. The quotation that follows is from the same source, page 97.

98   *"The harm they do":* Muir, *My First Summer in the Sierra*, 195.

99   *begun to articulate:* Roderick Frazier Nash, *Wilderness and the American Mind* (New Haven: Yale University Press, 2014), 122.

99   *became enamored of:* John Muir, "In the Yo-semite: Holidays Among the Rocks," *New York Tribune*, January 1, 1872, accessed at https://www.yosemite.ca.us /john_muir_writings/yosemite_in_winter.html on December 30, 2021.

99   *Louisa Strentzel:* "John Muir," National Park Service, accessed at https://www .nps.gov/yose/learn/historyculture/muir.htm on December 29, 2021.

99   *camped without so much:* C. Hart Merriam, "To the Memory of John Muir," *Sierra Club Bulletin* 10, no. 2 (January 1917): 148, accessed at C. Hart Merriam Papers, BANC MS 83/129, Box 1555: 112, Bancroft Library, University of California.

99   *"glad cascade songs":* Muir, *My First Summer in the Sierra*, 106, 147.

99   *he began writing:* Worster, *A Passion for Nature*, 197–98.

99   *the relentless building:* Muir, "In the Yo-semite: Holidays Among the Rocks."

100   *recent census data:* Robert Porter, quoted in Gerald D. Nash, "The Census of 1890 and the Closing of the Frontier," *Pacific Northwest Quarterly* 71, no. 3 (1980): 98.

100   *famous 1893 essay:* Frederick Jackson Turner, "The Frontier in American History," delivered in 1893 at the Annual Meeting of the American Historical Association and published by Henry Holt and Company, New York, 1921, accessed at http://www.gutenberg.org/files/22994/22994-h/22994-h.htm on February 14, 2020.

100   *approximately ten million:* "The West," *American Yawp*, accessed at https://www .americanyawp.com/text/17-conquering-the-west/ on December 15, 2021; and Michael Schaller, Janette Thomas Greenwood, Andrew Kirk, Sarah J. Purcell, Aaron Sheehan-Dean, and Christina Snyder, *American Horizons: US History in a Global Context, Volume 2,* 3rd ed. (New York: Oxford University Press, 2018), 566.

100   *George Brewerton described:* George D. Brewerton, "In the Buffalo Country," *Harper's New Monthly Magazine*, September 1862, pages 448, 460.

101   *"scenes of butchery":* George Bird Grinnell, "The Last of the Buffalo," *Scribner's Magazine* 12, no. 3, September 1892, 282, 286. The quotation that follows is from the same source and pages.

101   *In the words of Grinnell:* Grinnell, "The Last of the Buffalo," 278.

101   *loss of their territory:* "The West," *American Yawp.*

101   *Zitkala-Ša:* Zitkala-Ša, *American Indian Stories* (Washington, DC: Hayworth Publishing House, 1921), accessed at https://digital.library.upenn.edu/women /zitkala-sa/stories/stories.html#impressions on December 21, 2021. All quotations from Zitkala-Ša are from the same source.

102   *her cultural inheritance:* "Zitkala-Ša (Red Bird / Gertrude Simmons Bonnin)," National Park Service, accessed at https://www.nps.gov/people/zitkala-sa.htm on December 22, 2021.

102   *divided reservation land:* Schaller, Greenwood, Kirk, et al., *American Horizons*, 568.

102   *80 percent of the land:* "Wounded Knee Massacre," *Encyclopaedia Britannica*, accessed at https://www.britannica.com/event/Wounded-Knee-Massacre on December 21, 2021. Information about the Wounded Knee Massacre in this paragraph is from the same source.

103   *zinc and copper:* Schaller, Greenwood, Kirk, et al., *American Horizons*, 580.

103   *to mine coal:* Schaller, Greenwood, Kirk, et al., *American Horizons*, 579.

103   *a rise in GDP:* Hugh Rockoff, "Great Fortunes of the Gilded Age," National Bureau of Economic Research Working Paper (2008), page 6, accessed at https://www.nber.org/system/files/working_papers/w14555/w14555.pdf on December 20, 2021.

103   *Most American gold:* Randall Rohe, "Man and the Land: Mining's Impact in the Far West," *Arizona and the West* 28, no. 4 (1986): 300–1. Information about placer mining in this paragraph is from the same source, page 309.

103   *construct sluice boxes:* "What Is Placer Mining?" National Park Service, accessed at https://www.nps.gov/yuch/learn/historyculture/placer-mining.htm on July 31, 2023.

103   *chopped down trees:* Rohe, "Man and the Land," 302–5.

103   *a shocking rate:* "St. Louis Lumber Trade for 1873," *Lumberman's Gazette* 4, no. 1 (1874): 11.

103   *a federal report:* Donald J. Pisani, "Forests and Conservation, 1865–1890," *Journal of American History* 72, no. 2 (1985): 343.

103   *the* Lumberman's Gazette: "St. Louis Lumber Trade for 1873," *Lumberman's Gazette*. The quotation that follows is from the same source and page.

104   *In Ohio:* R. Harvey Reed, "Climatic Changes in Ohio, from the Destruction of the Forests and the Drainage of the Land, and Its Effects on the Health of the People," *Sanitarian* 186, May 1, 1885, page 409.

104   *what one man called:* G. W. Putnam, *Harper's Magazine*, reprinted in "Destruction of Our Forests," *Lumberman's Gazette* 15, no. 3 (1879): 7.

104   *total urban population:* US Department of Commerce, *Historical Statistics of the United States: 1789–1945*, Series B 145–159—Population—Urban Size-Groups and Rural Territory: 1790 to 1940, page 29.

104   *"No dogma taught":* John Muir, "Wild Wool," *Overland Monthly*, April 1875, accessed at https://vault.sierraclub.org/john_muir_exhibit/writings/wild_wool.aspx on December 28, 2021.

104   *To Muir, who possessed:* A. J. Willis, "The Ecosystem: An Evolving Concept Viewed Historically," *Functional Ecology* 11, no. 2 (1997): 268.

105   *"Plants, animals":* Muir, "Wild Wool."

105   *whom he increasingly viewed:* Worster, *A Passion for Nature*, 374–75.

105   *wrote in his journal:* John Muir, *Alaska Fragments*, 1890, accessed at https://vault.sierraclub.org/john_muir_exhibit/writings/favorite_quotations.aspx on December 24, 2021. The quotation that follows is from the same source.

105   *Roderick Nash:* Nash, *Wilderness and the American Mind*, 122.

105   *Muir brought readers:* John Muir, "The Hetch Hetchy Valley," *Weekly Transcript*,

March 25, 1873, accessed at https://vault.sierraclub.org/john_muir_exhibit/
writings/muir_hh_boston_25mar1873.aspx on December 28, 2021.

106   *Willard Glazier:* Willard Glazier, *Peculiarities of American Cities* (Philadelphia:
Hubbard Brothers, Publishers, 1886), 128. The quotation that follows is from
the same source and page.

106   *"Pollution, defilement":* Muir, quoted in *John of the Mountains: The Unpublished
Journals of John Muir*, ed. Linnie Marsh Wolfe (Boston: Houghton Mifflin
and Company, 1979), 222, accessed at https://vault.sierraclub.org/john_muir
_exhibit/writings/favorite_quotations.aspx on December 30, 2021.

106   *spilled far more ink:* Rebecca Solnit, "John Muir in Native America," *Sierra*,
March 2, 2021, accessed at https://www.sierraclub.org/sierra/2021-2-march
-april/feature/john-muir-native-america on December 28, 2021.

106   *using insensitive language:* For instance, Solnit observes in her article, Muir described an Indigenous woman in *My First Summer in the Sierra* as follows: "Her
dress was calico rags, far from clean. In every way she seemed sadly unlike
Nature's neat well-dressed animals, though living like them on the bounty of
the wilderness." Muir, *My First Summer in the Sierra*, accessed at https://www
.gutenberg.org/files/32540/32540-h/32540-h. on December 28, 2021.

107   *His assignment was:* Robert Underwood Johnson, *Remembered Yesterdays* (Boston: Little, Brown, and Company, 1923), 278, 280.

107   *found himself overcome:* Johnson, *Remembered Yesterdays*, 280. The quotations
that follow come from the same source, pages 281 and 283, respectively.

107   *Johnson proposed:* Johnson, *Remembered Yesterdays*, 287. Additional information in this paragraph regarding the plan comes from the same source, pages
287–88.

107   *In text that accompanied:* John Muir, "Features of the Proposed Yosemite National Park," *Century Magazine* 40, no. 5 (September 1890), accessed at https://
vault.sierraclub.org/john_muir_exhibit/writings/features_of_the_proposed
_yosemite_national_park/ on December 30, 2021. Muir's additional comments
in this paragraph come from the same source.

108   *a powerful presentation:* Johnson, *Remembered Yesterdays*, 288; and Runte, " National Park," chapter 4 in *Yosemite*.

108   *Johnson confessed that:* Johnson, *Remembered Yesterdays*, 288.

108   *select business interests:* Runte, "National Park," chapter 4 in *Yosemite*.

108   *Yosemite National Park:* "Act of October 1, 1890 (26 STAT., 650)," accessed at
https://vault.sierraclub.org/john_muir_exhibit/writings/the_yosemite/appendix
_a.aspx on December 30, 2021.

109   *preservation rather than conservation:* Gifford Pinchot was the most prominent
advocate of environmental conservation during the late nineteenth and early
twentieth centuries. Robert Hudson Westover, "Conservation vs. Preservation?" Forest Service: US Department of Agriculture, accessed at https://www
.fs.usda.gov/features/conservation-versus-preservation on December 29, 2021;
and "History: Sierra Club Timeline," Sierra Club, accessed at http://vault.sierra
club.org/history/timeline.aspx on December 30, 2021.

109   *a growing understanding:* The scholar Richard Fleck argues that Muir "would
evolve and change from his somewhat ambivalent stance toward various Indian
cultures to a positive admiration." Cited in Justin Nobel, "The Miseducation of

John Muir," *Atlas Obscura*, July 26, 2016, accessed at https://www.atlasobscura .com/articles/the-miseducation-of-john-muir on December 30, 2021.

109   *Tlingit, Haida, and Inupiat people:* Worster, *A Passion for Nature*, 247.

109   *establish national parks:* "Legacy," John Muir National Historic Site, accessed at https://www.nps.gov/museum/exhibits/jomu/legacy.html on December 29, 2021.

109   *Muir succeeded in:* John Muir, *Our National Parks* (Boston: Houghton, Mifflin, and Company, 1901), accessed at https://vault.sierraclub.org/john_muir _exhibit/writings/our_national_parks/chapter_1.aspx on December 29, 2021.

110   *he created images:* Muir, *Our National Parks*, accessed at https://vault.sierraclub .org/john_muir_exhibit/writings/our_national_parks/ on November 7, 2023.

5. FLORENCE MERRIAM BAILEY: THE CONSERVATIONIST

111   *The* New York Sun: "A Country Without Song Birds," *New York Sun*, November 22, 1885, page 8. Subsequent quotations in this paragraph from the *Sun* are from the same page.

112   *estimated that 50 percent:* Harriet Kofalk, *No Woman Tenderfoot: Florence Merriam Bailey, Pioneer Naturalist* (College Station: Texas A&M University Press, 1989), 79.

112   *less than 1 percent:* Thomas D. Snyder, ed., "120 Years of American Education: A Statistical Portrait," National Center for Education Statistics, US Department of Education, 1993, accessed at https://nces.ed.gov/pubs93/93442.pdf on January 6, 2022.

112   *one contemporary described:* "The Influence of the A. O. U. on Bird Protection," *American Museum Journal* 18, 1918, page 481.

113   *an avid astronomer:* Kofalk, *No Woman Tenderfoot*, 4.

113   *friend of John Muir:* Jonathan Wolfe, "Overlooked No More: Florence Merriam Bailey, Who Defined Modern Bird-Watching," *New York Times,* July 17, 2019, accessed at https://www.nytimes.com/2019/07/17/obituaries/florence -merriam-bailey-overlooked.html on February 25, 2022.

113   *The two men first:* Kofalk, *No Woman Tenderfoot*, 9. The sentence that follows is from the same source and page.

113   *they hiked up:* C. Hart Merriam, "To the Memory of John Muir," *Sierra Club Bulletin* 10, no. 2 (January 1917): 147, accessed at C. Hart Merriam Papers, BANC MS 83/129, Box 1555: 112, Bancroft Library, University of California.

114   *Muir would later:* Muir went camping with Hart, Florence, and Vernon Bailey in 1901. C. Hart Merriam, "To the Memory of John Muir," 147; Kolfak, *No Woman Tenderfoot*, 102; and Worster, *A Passion for Nature*, 195.

114   *regularly strolled:* Kofalk, *No Woman Tenderfoot*, 3, 7, 91.

114   *pleasure in excursions:* Diary of Florence Merriam Bailey, March 21, 1874, Smithsonian Institution Archives, Record Unit 7417, Florence Merriam Bailey Papers, Box 1, accessed at https://transcription.si.edu/transcribe/7429/SIA -SIA2014-01946 on January 12, 2022.

114   *On one occasion:* Diary of Florence Merriam Bailey, January 3, 1874, Smithsonian Institution Archives, Record Unit 7417, Florence Merriam Bailey Papers,

Box 1, accessed at https://collections.si.edu/search/slideshow_embedded?xml =http://siarchives.si.edu/sites/default/files/viewers/csc/viewer_fa_007417 _B01_F01.xml on January 6, 2022.

114 *describing the bird as:* Diary of Florence Merriam Bailey, February 6, 1874, Smithsonian Institution Archives, Record Unit 7417, Florence Merriam Bailey Papers, Box 1, accessed at https://transcription.si.edu/transcribe/7429/SIA -SIA2014-01924 on January 6, 2022.

114 *she marveled at:* Diary of Florence Merriam Bailey, January 8, 1874, and March 5, 1874, Smithsonian Institution Archives, Record Unit 7417, Florence Merriam Bailey Papers, Box 1, accessed at https://transcription.si.edu /transcribe/7429/SIA-SIA2014-01910, and https://transcription.si.edu /transcribe/7429/SIA-SIA2014-01938 on January 6, 2022.

114 *a budding naturalist:* Kofalk, *No Woman Tenderfoot*, 9; Wolfe, "Overlooked No More"; and "Florence Merriam Bailey," *Journal of the Sierra College Natural History Museum*, accessed at https://www.sierracollege.edu/ejournals/jscnhm /v6n1/bailey.html on February 25, 2022.

114 *While they ate:* Kofalk, *No Woman Tenderfoot*, 4.

114 *private collection:* "Clinton Hart Merriam," University of California–Davis Department of Anthropology Museum, accessed at http://anthromuseum.ucdavis .edu/c-hart-merriam-biography.html on February 25, 2022; Madelyn Holmes, "Florence Merriam Bailey," in *American Women Conservationists: Twelve Profiles* (Jefferson: McFarland and Company, 2004), 39; and "Florence Merriam Bailey," *Journal of the Sierra College Natural History Museum*.

114 *Lewis described how:* Meriwether Lewis, December 18, 1805, in *Journals of Lewis and Clark Expedition*, accessed at https://lewisandclarkjournals.unl.edu /item/lc.jrn.1805-12-18#lc.jrn.1805-12-18.03 on January 11, 2022. The quotation that follows is from the same source.

115 *John James Audubon:* "John James Audubon," Audubon Society, accessed at https://www.audubon.org/content/john-james-audubon on January 11, 2022.

115 *his artistic process:* John James Audubon, "Myself," reprinted from *Scribner's Magazine*, March 1893, page 267, accessed at https://www.gutenberg.org /files/39975/39975-h/39975-h.htm#Footnote_3 on January 11, 2022.

115 *advocated taxidermy:* "Taxidermy," *Turf, Field, and Farm* 19, no. 26, December 25, 1874, page 454.

115 *memorable window display:* "Foreign Game Birds: Splendid Specimens of Taxidermy," *Turf, Field, and Farm* 23, no. 8, August 25, 1876, page 113. The descriptions and quotations that follow in this paragraph are from the same source.

115 *attracting Americans:* Daniel Justin Herman, "The Hunter's Aim: The Cultural Politics of American Sport Hunters, 1880–1910," *Journal of Leisure Research* 35, no. 4 (2003): 455.

115 *located on Long Island:* Mike Wallace, *Greater Gotham: A History of New York City from 1898 to 1919* (New York: Oxford University Press, 2017), 299.

116 *thirty-nine journals:* Herman, "The Hunter's Aim," 455.

116 *Audubon recalled how:* John James Audubon, quoted in Richard Rhodes, *John James Audubon: The Making of an American* (New York: Alfred K. Knopf, 2005), 111–12.

116   Cultivator and Country Gentleman*:* N., "Ornithology: Some Common Birds—V," *Cultivator and Country Gentleman* 42, no. 1279, August 2, 1877, page 495.

116   *proponent of pigeon hunting:* "Wild Pigeons and Trap Shooting," *Turf, Field, and Farm* 26, no. 25, June 21, 1878, 391. The quotations and descriptions that follow in this paragraph are from the same source and page.

116   *destruction of natural habitats:* Ann E. Chapman, "Nineteenth Century Trends in American Conservation," National Park Service, accessed at https://www .nps.gov/articles/000/nineteenth-century-trends-in-american-conservation .htm on January 13, 2022.

116   *Martha died:* Barry Yeoman, "Why the Passenger Pigeon Went Extinct," *Audubon Magazine,* May–June 2014, accessed at https://www.audubon.org /magazine/may-june-2014/why-passenger-pigeon-went-extinct on November 8, 2023.

116   *herons and egrets:* Chapman, "Nineteenth Century Trends in American Conservation."

116   *the 1870s and 1880s:* Carolyn Merchant, *Spare the Birds! George Bird Grinnell and the First Audubon Society* (New Haven: Yale University Press, 2016), 22; and "Lady's Bird-of-Paradise Hat and Fur Coats: The Hidden Costs of Beauty," NYC Gilded Ages Exhibit, Barnard College, accessed at https://bt.barnard .edu/nycgildedages/exhibit/project/a-primi/ on November 17, 2020.

117   *"walking dresses":* *Godey's Lady's Book and Magazine* 89 (1874): 100. Further quotations and descriptions in this paragraph come from the same source and page.

117   *rose dramatically:* Robert Henry Welker, *Birds and Men* (Cambridge: Belknap Press of Harvard University Press, 1955), 196.

117   *five million birds:* Amelia Birdsall, "A Woman's Nature: Attitudes and Identities of the Bird Hat Debate at the Turn of the 20th Century," senior thesis, Haverford College Department of History, 2002, 29, accessed at https:// scholarship.tricolib.brynmawr.edu/bitstream/handle/10066/675/2002Birds allA.pdf?sequence=7&isAllowed=y on January 16, 2022.

117   *Mary Thatcher:* Mary Thatcher, "The Slaughter of the Innocents," *Harper's Bazaar,* May 22, 1875, 338, cited in Birdsall, "A Woman's Nature." The quotations that follow in this paragraph are from the same source.

117   *Thatcher may have been:* Birdsall, "A Woman's Nature," 29.

118   *Chapman later remembered:* Frank Michler Chapman, *Autobiography of a Bird-Lover* (New York: D. Appleton-Century Company, 1933), 38–39. Further details in this paragraph from Chapman's study come from the same source and pages.

118   *After he published his data:* Chapman's findings appeared in *Forest and Stream* and *Bulletin No. I* of the American Ornithologists' Union.

119   *Outside the classroom:* Kofalk, *No Woman Tenderfoot,* 25, 27, 29.

119   *befriend John Muir:* C. Hart Merriam, "To the Memory of John Muir," 145.

119   *first female associate member:* Kofalk, *No Woman Tenderfoot,* 29.

119   *Back on campus:* Kofalk, *No Woman Tenderfoot,* 30–31.

119   *prepared an article:* Kofalk, *No Woman Tenderfoot,* 29–31.

119   *Brockway revealed:* Florence Merriam to Beman Brockway, *Watertown Daily Times,* November 11, 1885, page 3.

119 *As she put it:* Merriam to Brockway, *Watertown Daily Times*. The quotation that follows in this paragraph is from the same source and page.

120 *three hundred undergraduates:* Florence Merriam, "Our Smith College Audubon Society," *Audubon Magazine* 1, no. 1 (1887), 176.

120 *established a formal club:* Kofalk, *No Woman Tenderfoot*, 35.

120 *Dr. George Bird Grinnell:* Carolyn Merchant, "George Bird Grinnell's Audubon Society: Bridging the Gender Divide in Conservation," *Environmental History* 15, no. 1 (2010): 11.

120 *described her strategy:* Merriam, "Our Smith College Audubon Society," 176. The quotation that follows is from the same source, page 175.

120 *the famous naturalist John Burroughs:* Merriam, "Our Smith College Audubon Society," 177; and Kofalk, *No Woman Tenderfoot*, 32.

120 *Burroughs would later call:* Diary of John Burroughs, "A Visit from John Muir," June 26, 1896, Sierra Club, accessed at https://vault.sierraclub.org/john_muir _exhibit/life/a-visit-from-john-muir-by-john-burroughs-1896.aspx on March 24, 2023.

120 *wanted to inform women:* John Burroughs, "Bird Enemies," in *Century Illustrated Monthly Magazine* 31, no. 2, December 1885.

121 *Burroughs inspired them:* Merriam, "Our Smith College Audubon Society," 177.

121 *Merriam prepared for:* Merriam, "Our Smith College Audubon Society," 176; and Kofalk, *No Woman Tenderfoot*, 34.

121 *Merriam remembered proudly:* Merriam, "Our Smith College Audubon Society," 176. Additional information in this paragraph is from the same source, pages 176–77; and Mark Barrow, *A Passion for Birds: American Ornithology after Audubon* (Princeton: Princeton University Press, 2000), 119.

121 *the winter of 1887:* Kofalk, *No Woman Tenderfoot*, 39, 40, 43.

121 *a widowed woman:* Jacob A. Riis, *How the Other Half Lives: Studies Among the Tenements of New York* (New York: Charles Scribner's Sons, 1890), 240. The quotations that follow are from the same source, pages 240–42.

122 *Merriam began volunteering:* Kofalk, *No Woman Tenderfoot*, 43–44.

122 *whom Dodge described:* "Miss Dodge and Her Work," *Advance* 24, no. 1358 (1891): 858.

122 *Merriam published a few:* Florence Merriam, *Birds Through an Opera Glass* (Cambridge: The Riverside Press, 1889), vii; Kolfak, *No Woman Tenderfoot*, 49.

122 *accessible, appealing writing:* Kofalk, *No Woman Tenderfoot*, 49.

122 *her father had made his fortune:* "Merriam, Clinton Levi," United States House of Representatives, accessed at https://history.house.gov/People/Listing/M /MERRIAM,-Clinton-Levi-(M000653)/ on January 28, 2022. The sentence that follows is from the same source.

123 *Gilded Age mores:* Cecelia Tichi, *What Would Mrs. Astor Do? The Essential Guide to the Manners and Mores of the Gilded Age* (New York: New York University Press, 2018), 60–65.

123 *Horseback riding:* Tichi, *What Would Mrs. Astor Do?*, 167–69.

123 *Crowninshield warned:* Francis W. Crowninshield, *Manners for the Metropolis:*

An Entrance Key to the Fantastic Life of the 400 (New York: D. Appleton and Company, 1909), 116.

123   *During the daytime:* Diary of Florence Merriam Bailey, February 12, 1887, and March 12, 1887, accessed at https://edan.si.edu/transcription/pdf_files/7430 .pdf on February 2, 2022.

123   *and once enjoyed:* Diary of Florence Merriam Bailey, February 13, 1887, accessed at https://edan.si.edu/transcription/pdf_files/7430.pdf on February 2, 2022.

123   *Frederick Law Olmsted:* Diary of Florence Merriam Bailey, February 7, 1887, accessed at https://edan.si.edu/transcription/pdf_files/7430.pdf on February 2, 2022.

123   *A lively dinner party:* Diary of Florence Merriam Bailey, January 23, 1887, accessed at https://edan.si.edu/transcription/pdf_files/7430.pdf on February 2, 2022.

123   *Much of Merriam's time:* Diary of Florence Merriam Bailey, February 23, 1887; March 15, 1887; and March 16, 1887, accessed at https://edan.si.edu/transcription /pdf_files/7430.pdf on February 2, 2022.

124   *felt deep displeasure:* Kofalk, *No Woman Tenderfoot*, 44.

124   *signaled their social status:* Esther Crain, *The Gilded Age in New York, 1870–1910* (New York: Black Dog and Leventhal Publishers, 2016), 60.

124   *McCabe even opined:* James D. McCabe, *New York by Sunlight and Gaslight: A Work Descriptive of the Great American Metropolis* (Philadelphia: Hubbard Brothers, Publishers, 1881), 203. The quotation that follows is from the same source.

124   *suffered from tuberculosis:* Kofalk, *No Woman Tenderfoot*, 46.

124   *at least one occasion:* Diary of Florence Merriam Bailey, March 19, 1887, accessed at https://edan.si.edu/transcription/pdf_files/7430.pdf on February 2, 2022.

124   *an overwhelming experience:* Diary of Florence Merriam Bailey, May 21, 1887, accessed at https://edan.si.edu/transcription/pdf_files/7430.pdf on February 2, 2022.

124   *she still cared about:* Diary of Florence Merriam Bailey, May 20, 1887, accessed at https://edan.si.edu/transcription/pdf_files/7430.pdf on February 2, 2022.

124   *watched gray squirrels:* Diary of Florence Merriam Bailey, April 1, 1887, accessed at https://edan.si.edu/transcription/pdf_files/7430.pdf on February 2, 2022.

124   *Florence went on walks:* Diary of Florence Merriam Bailey, May 9, 1887, and May 15, 1887, accessed at https://edan.si.edu/transcription/pdf_files/7430.pdf on February 2, 2022.

125   *a grand adventure:* Kofalk, *No Woman Tenderfoot*, 48. Further information about Florence's mother's tuberculosis comes from the same source, page 46.

125   *the rugged peaks:* Kofalk, *No Woman Tenderfoot*, 66–67.

125   *returning to New York:* Kofalk, *No Woman Tenderfoot*, 48.

125   *A groundbreaking tome:* Merriam, *Birds Through an Opera Glass*, v and vi; and Kofalk, *No Woman Tenderfoot*, 49.

125   *Merriam urged her readers:* Merriam, *Birds Through an Opera Glass*, 3.

125   *"It is not merely":* Merriam, *Birds Through an Opera Glass*, vi–vii; cited in Kofalk, *No Woman Tenderfoot*, 49.

126   *the humble robin:* Merriam, *Birds Through an Opera Glass*, 4.

126   *the unappreciated crow:* Merriam, *Birds Through an Opera Glass*, 10. The quotations that follow in the next paragraph are from the same source, pages 210 and 3.

126   *Reviews of the book:* Kofalk, *No Woman Tenderfoot*, 51.

126   *The* Atlantic Monthly: "Outdoors and Indoors," *Atlantic Monthly*, November 1889, page 704.

127   *to divide her time:* Kofalk, *No Woman Tenderfoot*, 54.

127   *her mother's health:* Kofalk, *No Woman Tenderfoot*, 54. The information that follows in this paragraph pertaining to her mother's death comes from the same source, pages 54–55.

127   *Merriam remembered how:* Florence Merriam Bailey, *My Summer in a Mormon Village* (Cambridge: The Riverside Press, 1894), 20–21.

127   *Harriet Mann Miller:* Bailey, *My Summer in a Mormon Village*, 1; and "Olive Thorne Miller," *Condor* 21, no. 2 (1919): 69–70.

127   *Utah Territory:* Bailey, *My Summer in a Mormon Village*, 2.

127   *in a boardinghouse:* Bailey, *My Summer in a Mormon Village*, 4, 8.

127   *her newfound independence:* Bailey, *My Summer in a Mormon Village*, 21.

128   *She remembered how:* Bailey, *My Summer in a Mormon Village*, 26–27.

128   *she purchased a horse:* Bailey, *My Summer in a Mormon Village*, 33.

128   *She and Jumbo spent:* Bailey, *My Summer in a Mormon Village*, 42.

128   *Merriam reveled in:* Bailey, *My Summer in a Mormon Village*, 21.

128   *an abandoned mine:* Bailey, *My Summer in a Mormon Village*, 43–44.

128   *surrounded by birds:* Bailey, *My Summer in a Mormon Village*, 22, 24, 48, 50, 164.

128   *more than doubled:* "Population of Utah by Counties and Minor Civil Divisions," *Twelfth Census of the United States*, February 4, 1901, accessed at https://www2.census.gov/library/publications/decennial/1900/bulletins/demographic/50-population-ut.pdf on February 10, 2022; and Richard L. Jensen, "Immigration to Utah," *Utah History Encyclopedia*, accessed at https://www.uen.org/utah_history_encyclopedia/i/IMMIGRATION.shtml on February 10, 2022.

128   *Merriam marveled at:* Bailey, *My Summer in a Mormon Village*, 66.

129   *"The silver question":* Bailey, *My Summer in a Mormon Village*, 70, 72.

129   *train's piercing whistle:* Bailey, *My Summer in a Mormon Village*, 72.

129   *An excursion to Alta:* Bailey, *My Summer in a Mormon Village*, 151.

129   *bragged to Merriam:* Bailey, *My Summer in a Mormon Village*, 163.

129   *"It was a rude shock":* Bailey, *My Summer in a Mormon Village*, 163–64.

129   *back to Twin Oaks:* Kofalk, *No Woman Tenderfoot*, 63.

129   *bird-watching excursions:* Florence Merriam Bailey, *A-Birding on a Bronco* (Cambridge: The Riverside Press, 1896), iii–iv; and Kofalk, *No Woman Tenderfoot*, 63.

129 *refused to carry a gun:* Bailey, *A-Birding on a Bronco*, 1. The descriptions and quotations that follow in the next paragraph are from the same source, pages 1–2, 4–5, 6–7, 9, 15, 18, 66, 70.

130 *her chief purposes:* Kofalk, *No Woman Tenderfoot*, 70.

130 *her home base:* Bailey, *A-Birding on a Bronco*, 2.

130 *She acknowledged:* Bailey, *A-Birding on a Bronco*, 140. The quotation that follows is from the same source, page 141.

131 *One reviewer:* "Books of the Hour," *Saint Paul Globe*, March 21, 1897, page 12. The quotation that follows is from the same source.

131 *To another reviewer:* "Birds of Village and Field," *Los Angeles Herald*, April 24, 1898, page 22.

131 *continued to wear feathers:* Kofalk, *No Woman Tenderfoot*, 75.

131 *as much as 50 percent:* Witmer Stone, "Report of the Committee on the Protection of North American Birds for the Year 1900," *Auk* 18, no. 1 (1901): 68–69.

131 *Harriet Kofalk calls:* Kofalk, *No Woman Tenderfoot*, 79.

131 *Theodore Cart:* Theodore Whaley Cart, "The Lacey Act: America's First Wildlife Statute," *Forest History Newsletter* 17, no. 3 (1973): 6.

132 *sought to inform:* Cart, "The Lacey Act," 6, 10.

132 *AOU members reported:* Stone, "Report on the Protection of North American Birds for the Year 1900," 71.

132 *passed the Lacey Act:* 18 USC 42–43 16 USC 3371–3378 Lacey Act, accessed at https://www.fws.gov/le/pdffiles/Lacey.pdf on February 28, 2022.

132 *the AOU reported:* Stone, "Report of the Committee on the Protection of North American Birds," 71–72.

132 Pioneer Press: "Big Game Seizure," *Pioneer Press*, May 2, 1901, page 7.

6. WILLIAM SHEPPARD: THE MISSIONARY

137 *rocky transatlantic crossing:* William Henry Sheppard, *Presbyterian Pioneers in Congo* (Richmond: Presbyterian Committee of Publication, 1917), 20; Samuel Lapsley to Sarah and James Lapsley, February 26, 1890, in *Life and Letters of Samuel Norvell Lapsley: Missionary to the Congo Valley, West Africa, 1866–1892* (Richmond: Whittet & Shepperson, 1893), 25; and "The White Star Line steamship 'Adriatic' Leaving Liverpool," Royal Museums Greenwich, accessed at https://www.rmg.co.uk/collections/objects/rmgc-object-14646 on April 8, 2022.

137 *a small tugboat:* Samuel Lapsley to Sarah and James Lapsley, February 26, 1890, in *Life and Letters of Samuel Norvell Lapsley*, 26; and Sheppard, *Presbyterian Pioneers in Congo*, 19.

137 *The well-wishers:* "Transatlantic Steamship Routes," Supplement to the Pilot Chart of the North Atlantic Ocean, January 1893, University of North Texas, accessed at https://texashistory.unt.edu/ark:/67531/metapth190428/m1/1/zoom/?resolution=6&lat=6745.81816923196&lon=5333.257976854928 on April 8, 2022.

138 *he later remembered:* Sheppard, *Presbyterian Pioneers in Congo*, 20.

138   *white Protestant missionaries:* Adam Hochschild, *King Leopold's Ghost: A Story of Greed, Terror, and Heroism in Colonial Africa* (Boston: Houghton Mifflin, 1998), 154; Michael Parker, "William Sheppard: Missionary to the Belgian Congo," *Mission Crossroads* (Spring 2017), accessed at https://www.presbyterianmission.org/story/april-3-2017/ on April 8, 2022; and Marvin D. Markowitz, "The Missions and Political Development in the Congo," *Africa: Journal of the International Africa Institute* 40, no. 3 (1970): 234.

138   *"It is sad to leave":* Sheppard, *Presbyterian Pioneers in Congo,* 19.

138   *His father, who served:* Sheppard, *Presbyterian Pioneers in Congo,* 15.

138   *When night fell:* Sheppard, *Presbyterian Pioneers in Congo,* 15.

139   *wheat, corn, rye:* "Augusta County: Slaves as a Percentage of Total Population, 1820–1860," US Census Office, Eighth Census, Population of the United States in 1860, Compiled from the Original Returns of the Eighth Census, Washington, D.C., 1864, accessed at https://valley.lib.virginia.edu/VoS/tablesandstats/augusta/slaverypop.html on April 8, 2022; and "Franklin and Augusta: 1860 Agricultural Production, by Percentages," accessed at https://valley.lib.virginia.edu/VoS/tablesandstats/comparison/1860agprod.html on April 8, 2022.

139   *numbered 586 people:* Ellen Eslinger, "Free Black Residency in Two Antebellum Virginia Counties: How the Laws Functioned," *Journal of Southern History* 79, no. 2 (2013): 266; and Nancy DuPree and Robert DuPree, "William Henry Sheppard," *Encyclopedia of Alabama,* accessed at http://encyclopediaofalabama.org/article/h-3785 on April 8, 2022.

139   *some of whom had fled:* "The Abolition of Slavery in Virginia," *Encyclopedia of Virginia,* accessed at https://encyclopediavirginia.org/entries/the-abolition-of-slavery-in-virginia/ on April 8, 2022.

139   *During his childhood:* DuPree and DuPree, "William Henry Sheppard."

139   *Young William remembered:* Sheppard, *Presbyterian Pioneers in Congo,* 16. The information in the rest of the paragraph comes from the same source.

139   *a kind woman:* Sheppard, *Presbyterian Pioneers in Congo,* 15–16. The description and quotations that follow in this paragraph are from the same source and pages.

139   *When he heard about:* Sheppard, *Presbyterian Pioneers in Congo,* 16.

140   *Booker T. Washington:* Booker T. Washington, *Up from Slavery: An Autobiography* (Garden City: Doubleday and Company, 1900), 53.

140   *Born into slavery:* Washington, *Up from Slavery,* 46. The quotations that follow in this paragraph are from the same source, pages 73–74.

140   *Sheppard enrolled:* Pagan Kennedy, *Black Livingstone: A True Tale of Adventure in the Nineteenth-Century Congo* (New York: Viking, 2022), 10.

140   *before attending classes:* Sheppard, *Presbyterian Pioneers in Congo,* 17. The information and quotation that follow in this paragraph are from the same source and page.

140   *"Curiosity Room":* James T. Campbell, *Middle Passages: African American Journeys to Africa, 1787–2005* (New York: Penguin Press, 2006), 148.

141   *Tuscaloosa Theological Institute:* William Sheppard, "Reminiscences of My Early Days at Stillman Institute," 1925, RG 457, Box 1, Folder 7, William H. Sheppard Papers, Presbyterian Historical Society, accessed at https://digital.history.pcusa.org/islandora/object/islandora:15928#page/1/mode/1up on April 8,

2022; DuPree and DuPree, "William Henry Sheppard"; and Ira Dworkin, *Congo Love Song: African American Culture and the Crisis of the Colonial State* (Chapel Hill: University of North Carolina Press, 2017), 49. The school would later be known as Stillman College.

141   *service in Africa:* Campbell, *Middle Passages,* 148.

141   *begin pastoral training:* Sheppard, "Reminiscences of My Early Days at Stillman Institute."

141   *gained practical skills:* Sheppard, *Presbyterian Pioneers in Congo,* 18.

141   *During his examination:* Sheppard, *Presbyterian Pioneers in Congo,* 18. Both quotations about his examination are from the same source and page.

141   *he first served:* Sheppard, *Presbyterian Pioneers in Congo,* 18.

141   *a position abroad:* Kennedy, *Black Livingstone,* 15.

141   *to consolidate white power:* Hochschild, *King Leopold's Ghost,* 152–53; and Kennedy, *Black Livingstone,* 14.

141   *lay the groundwork:* Hochschild, *King Leopold's Ghost,* 152–53.

142   *lead a mission to Africa:* Kennedy, *Black Livingstone,* 15; Hochschild, *King Leopold's Ghost,* 153; and John G. Turner, "A 'Black-White' Missionary on the Imperial Stage: William H. Sheppard and Middle-Class Black Manhood," *Journal of Southern Religion* 9 (2006), accessed at https://jsr.fsu.edu/Volume9/Turner.htm on April 8, 2022.

142   *forced to wait:* Hochschild, *King Leopold's Ghost,* 153.

142   *Zion Presbyterian Church:* Ramona Austin, "An Extraordinary Generation: The Legacy of William Henry Sheppard, the 'Black Livingstone' of Africa," *Afrique & Histoire* 4, no. 2 (2005), accessed at https://www.cairn.info/revue -afrique-et-histoire-2005-2-page-73.htm on March 27, 2023.

142   *he recalled:* Sheppard, *Presbyterian Pioneers in Congo,* 18.

142   *Senator Morgan:* Hochschild, *King Leopold's Ghost,* 153.

142   *the arduous voyage:* Sheppard, *Presbyterian Pioneers in Congo,* 18; and Hochschild, *King Leopold's Ghost,* 153.

142   *been a slaveholder:* Turner, "A 'Black-White' Missionary on the Imperial Stage"; and Hochschild, *King Leopold's Ghost,* 153.

143   *12.5 million Africans:* "Trans-Atlantic Slave Trade—Estimates," *Slave Voyages,* accessed at https://www.slavevoyages.org/assessment/estimates on April 8, 2022.

143   *gradual economic shifts:* John Parker and Richard Rathbone, *African History: A Very Short Introduction* (New York: Oxford University Press, 2007), 88.

143   *persisted in some:* Parker and Rathbone, *African History,* 88.

143   *"Scramble for Africa":* Parker and Rathbone, *African History,* 93.

143   *Bantu-speaking peoples:* "The Scramble for Africa," Saint John's College, University of Cambridge, accessed at https://www.joh.cam.ac.uk/library/library _exhibitions/schoolresources/exploration/scramble_for_africa on April 8, 2022; and Hochschild, *King Leopold's Ghost,* 62.

143   Through the Dark Continent: "The Scramble for Africa"; "Henry Morton Stanley," *Encyclopaedia Britannica,* accessed at https://www.britannica.com

/biography/Henry-Morton-Stanley on April 8, 2022; and "Congo Free State," *Encyclopaedia Britannica*, accessed at https://www.britannica.com/place/Congo -Free-State on April 8, 2022.

143   *Welsh-born American immigrant:* Henry Morgan Stanley, *Through the Dark Continent* (New York: Harper and Brothers, 1878), 119–20; and "Henry Morton Stanley," *Encyclopaedia Britannica*.

143   *assured his readers:* Stanley, *Through the Dark Continent*, 401–2.

144   *penned a letter:* Henry Morgan Stanley, writing for the *Daily Telegraph*, November 12, 1877, referenced in Henry Morton Stanley, *The Congo and the Founding of Its Free State* (New York: Harper and Brothers, 1885), vi.

144   *he argued:* Stanley, writing for the *Daily Telegraph*, November 12, 1877, referenced in Stanley, *The Congo and the Founding of Its Free State*, vi.

144   *start building infrastructure:* Hochschild, *King Leopold's Ghost*, 62–63; and "Congo Free State," *Encyclopaedia Britannica*.

144   *million square miles:* Thomas Pakenham, *The Scramble for Africa: The White Man's Conquest of the Dark Continent from 1876 to 1912* (New York: Random House, 1991), 254.

144   *gathered in Berlin:* Hochschild, *King Leopold's Ghost*, 84–86; and Pakenham, *The Scramble for Africa*, 254.

144   *No Africans were invited:* Hochschild, *King Leopold's Ghost*, 84; and G. N. Uzoigwe, "Reflections on the Berlin West Africa Conference," *Journal of the Historical Society of Nigeria* 12, no. 3/4 (1985): 10. Further information in this paragraph is from Hochschild, *King Leopold's Ghost*, 84–87.

144   *struggled to answer:* Sheppard, *Presbyterian Pioneers in Congo*, 18–19.

145   *he later recorded:* Sheppard, *Presbyterian Pioneers in Congo*, 25.

145   *"in full knowledge":* Sheppard, *Presbyterian Pioneers in Congo*, 25–26.

145   *Sheppard felt:* Sheppard, *Presbyterian Pioneers in Congo*, 18–19.

145   *delivered a lecture:* "The Son of a Staunton Colored Barber Goes as a Missionary to the Congo," *Staunton Vindicator*, February 14, 1890, page 3. The quotation that follows in this paragraph is from the same source and page.

145   *eleven-day journey:* Sheppard, *Presbyterian Pioneers in Congo*, 20.

145   *Born in Selma: Life and Letters of Samuel Norvell Lapsley*, 11.

145   *Presbyterian ministers: Life and Letters of Samuel Norvell Lapsley*, 11.

145   *what his brother: Life and Letters of Samuel Norvell Lapsley*, 13.

145   *his brother remembered: Life and Letters of Samuel Norvell Lapsley*, 20–21.

145   *bouts of seasickness:* Samuel Lapsley to Sarah and James Lapsley, February 26, 1890, and Samuel Lapsley to Tillie and Isabel Lapsley, March 1, 1890, in *Life and Letters of Samuel Norvell Lapsley*, 26.

146   *"very modest":* Samuel Lapsley to Tillie Lapsley, March 8, 1890, in *Life and Letters of Samuel Norvell Lapsley*, 34.

146   *Liverpool to London:* Sheppard, *Presbyterian Pioneers in Congo*, 20.

146   *well-connected men:* Kennedy, *Black Livingstone*, 20–24.

146   *traveled to Brussels:* Kennedy, *Black Livingstone*, 23–24.

146   *wrote to his mother:* Samuel Lapsley to Sarah Lapsley, March 23, 1890, in *Life and Letters of Samuel Norvell Lapsley*, 44–45.

146   *Pagan Kennedy argues:* Kennedy, *Black Livingstone*, 26.

146   *the younger man's:* Samuel Lapsley to Sarah Lapsley, March 23, 1890, in *Life and Letters of Samuel Norvell Lapsley*, 44.

146   *Dutch merchant vessel:* Sheppard, *Presbyterian Pioneers in Congo*, 21; and Samuel Lapsley to Sarah Lapsley, April 20, 1890, in *Life and Letters of Samuel Norvell Lapsley*, 47–51.

146   *violent weather:* Sheppard, *Presbyterian Pioneers in Congo*, 21.

147   *his first impressions:* Sheppard, *Presbyterian Pioneers in Congo*, 21.

147   *mangrove trees:* "First Days on the Congo River," in *Life and Letters of Samuel Norvell Lapsley*, 53.

147   *banana trees:* Sheppard, *Presbyterian Pioneers in Congo*, 21–22.

147   *Numerous people:* Sheppard, *Presbyterian Pioneers in Congo*, 21.

147   *Lapsley admired:* "First Days on the Congo River," in *Life and Letters of Samuel Norvell Lapsley*, 53.

147   *seventy white residents:* Sheppard, *Presbyterian Pioneers in Congo*, 22.

147   *began their expedition:* "First Days on the Congo River," in *Life and Letters of Samuel Norvell Lapsley*, 55.

147   *no evangelist had trod:* DuPree and DuPree, "William Henry Sheppard."

147   *malaria cut down:* Sheppard, *Presbyterian Pioneers in Congo*, 22, 24. Additional information and quotations about malaria in this paragraph come from the same source, page 24.

147   *Malaria would claim:* William E. Phipps, *William Sheppard: Congo's African American Livingstone* (Louisville: Geneva Press, 2002), 31.

147   *march deeper inland:* Phipps, *William Sheppard*, 33–34.

148   *Congolese men:* Sheppard, *Presbyterian Pioneers in Congo*, 26.

148   *woke at 5 a.m.:* Sheppard, *Presbyterian Pioneers in Congo*, 27–28.

148   *passage of explorers:* Sheppard, *Presbyterian Pioneers in Congo*, 28.

148   *"We found a road":* Sheppard, *Presbyterian Pioneers in Congo*, 28. The quotations that follow in this paragraph come from the same source, pages 28–29.

148   *reached Leopoldville:* Lapsley, *Life and Letters of Samuel Norvell Lapsley*, 71–72.

148   *trading posts:* Lapsley, *Life and Letters of Samuel Norvell Lapsley*, 71–72; and Phipps, *William Sheppard*, 38.

148   *Lapsley noticed that:* Lapsley, *Life and Letters of Samuel Norvell Lapsley*, 72.

148   *rubber became essential:* Hochschild, *King Leopold's Ghost*, 159; and "Development of the Natural Rubber Industry," *Encyclopaedia Britannica*, accessed at https://www.britannica.com/science/rubber-chemical-compound/Development-of-the-natural-rubber-industry on April 8, 2022.

149   *Teke people:* Sheppard, *Presbyterian Pioneers in Congo*, 33–34.

149   *ivory and wood:* Sheppard, *Presbyterian Pioneers in Congo*, 34.

149 *wrote to his aunt:* Samuel Lapsley to Elsie Lapsley, August 19, 1891, in *Life and Letters of Samuel Norvell Lapsley*, 83; and Phipps, *William Sheppard*, 39.

149 *Lapsley mused:* Samuel Lapsley to Elsie Lapsley, August 19, 1891, in Lapsley, *Life and Letters of Samuel Norvell Lapsley*, 83.

149 *Lapsley left Sheppard:* Sheppard, *Presbyterian Pioneers in Congo*, 35–36. The information and quotations that follow in this paragraph are from the same source and pages.

149 *By late autumn:* Sheppard, *Presbyterian Pioneers in Congo*, 42.

150 *Sheppard recorded:* Phipps, *William Sheppard*, 41; and Sheppard, *Presbyterian Pioneers in Congo*, 43, 47–48.

150 *aboard the* Florida: Sheppard, *Presbyterian Pioneers in Congo*, 51.

150 *"the red waters":* Sheppard, *Presbyterian Pioneers in Congo*, 52.

150 *As they proceeded:* Sheppard, *Presbyterian Pioneers in Congo*, 54. The information and quotations in this paragraph come from the same source and page.

150 *reached a junction:* Sheppard, *Presbyterian Pioneers in Congo*, 61. The information and quotations in this paragraph come from the same source and page.

151 *He would begin:* William Sheppard to Sarah Lapsley, May 26, 1892, cited in Sheppard, *Presbyterian Pioneers in Congo*, 85–86; and DuPree and DuPree, "William Henry Sheppard."

151 *works of art:* Sheppard, *Presbyterian Pioneers in Congo*, 138; Harold G. Cureau, "William H. Sheppard: Missionary to the Congo and Collector of African Art," *Journal of Negro History* 67, no. 4 (1982): 343, 345; and "Dr. William Henry Sheppard," Hampton University Archives, accessed at https://hampton archives.org/content/dr-william-henry-sheppard on April 8, 2022.

151 *American Presbyterian Congo:* DuPree and DuPree, "William Henry Sheppard"; and Diary of Samuel Lapsley, April 23, 1891, in *Life and Letters of Samuel Norvell Lapsley*, 164.

151 *the village of Luebo:* Diary of Samuel Lapsley, April 23, 1891; and Sheppard, *Presbyterian Pioneers in Congo*, 66.

151 *Sheppard remembered how:* Sheppard, *Presbyterian Pioneers in Congo*, 67. The quotation that follows is from the same source and page.

151 *Sheppard wrote positively:* Sheppard, *Presbyterian Pioneers in Congo*, 65.

151 *Sheppard explained:* Sheppard, *Presbyterian Pioneers in Congo*, 75–76. The quotation that follows is from the same source and page.

151 *In a letter:* Samuel Lapsley to the *Missionary*, October 2, 1890, cited in Dworkin, *Congo Love Song*, 52.

152 *the Bantu language:* Sheppard, *Presbyterian Pioneers in Congo*, 63–64. The information that follows in this paragraph is from the same source, pages 63–67.

152 *Sheppard awoke:* Sheppard, *Presbyterian Pioneers in Congo*, 68. The information that follows in this paragraph and the next is from the same source and page.

152 *The two missionaries:* Sheppard, *Presbyterian Pioneers in Congo*, 65–66.

153 *Lapsley made his way:* Sheppard, *Presbyterian Pioneers in Congo*, 79.

153 *Lapsley's sacrifices:* Sheppard, *Presbyterian Pioneers in Congo*, 80.

153   *There was little time:* Sheppard, *Presbyterian Pioneers in Congo*, 82. The quotation that follows is from the same source and page.

153   *low-grade temperature:* Campbell, *Middle Passages*, 150; and W. Holman Bentley to Robert Whyte, May 3, 1892, in *Life and Letters of Samuel Norvell Lapsley*, 229–30.

153   *fell gravely ill:* W. Holman Bentley to Robert Whyte, May 3, 1892, in *Life and Letters of Samuel Norvell Lapsley*, 229–30.

153   *When Sheppard learned:* Sheppard, *Presbyterian Pioneers in Congo*, 84. The quotations that follow in this paragraph are from the same source, pages 84–85.

154   *the capital city:* Campbell, *Middle Passages*, 160; and Phipps, *William Sheppard*, 68.

154   *famously hostile:* Phipps, *William Sheppard*, 68–69.

154   *sent the warriors back:* Sheppard, *Presbyterian Pioneers in Congo*, 99–101.

154   *"a Makuba":* Sheppard, *Presbyterian Pioneers in Congo*, 101.

154   *finally entered Mushenge:* Sheppard, *Presbyterian Pioneers in Congo*, 106, 114.

154   *Kot aMweeky:* Sheppard, *Presbyterian Pioneers in Congo*, 106; and Campbell, *Middle Passages*, 160.

154   *said simply "Wyni":* Sheppard, *Presbyterian Pioneers in Congo*, 107.

154   *During the next four months:* Campbell, *Middle Passages*, 161.

154   *"the only information":* Phipps, *William Sheppard*, 76.

155   *He also brought:* Sheppard, *Presbyterian Pioneers in Congo*, 138.

155   *Sheppard would later:* Cureau, "William H. Sheppard," 349.

155   *Sheppard gained recognition:* Phipps, *William Sheppard*, 93; and Campbell, *Middle Passages*, 163-64.

155   *married his fiancée:* Phipps, *William Sheppard*, 95–96, 98.

155   *ruthlessly pursued the collection of rubber:* Campbell, *Middle Passages*, 173; and "The Free State of the Congo: A Hidden History of Genocide," Ramon Marull Philathelic Collection, accessed at https://ajuntament.barcelona.cat/gabinet postal/lestat-lliure-del-congo-un-genocidi-a-lombra/?lang=en on July 20, 2023. Information that follows in this paragraph comes from the same sources.

155   *other forms of violence:* Phipps, *William Sheppard*, 135.

156   *an open letter:* George Washington Williams, "An Open Letter to His Serene Majesty Leopold II, King of the Belgians and Sovereign of the Independent State of Congo by Colonel, The Honorable Geo. W. Williams, of the United States of America," 1890, accessed at https://www.blackpast.org /global-african-history/primary-documents-global-african-history/george -washington-williams-open-letter-king-leopold-congo-1890/ on April 10, 2022.

156   *briefly gained traction:* Kennedy, *Black Livingstone*, 34–35.

156   *the first whistleblower:* Hochschild, *King Leopold's Ghost*, 102.

156   *village of Ibanche:* Phipps, *William Sheppard*, 140.

156   *the Zappo Zaps:* Phipps, *William Sheppard*, 137–39.

156   *agents of the Congo Free State:* Campbell, *Middle Passages*, 156.

156   *as cannibalistic:* Sheppard, *Presbyterian Pioneers in Congo*, 57; and Phipps, *William Sheppard*, 137.

156   *received orders:* Phipps, *William Sheppard*, 140–41.

156   *the investigation:* Typed transcription of William H. Sheppard's diary, Sept. 13–14, 1899, RG 457, Box 1, Folder 6, William H. Sheppard Papers, Presbyterian Historical Society, Philadelphia, PA, accessed at https:// digital.history.pcusa.org/islandora/object/islandora%3A15919?solr_nav %5Bid%5D=3b91658ffa58a98e4226&solr_nav%5Bpage%5D=0&solr_nav %5Boffset%5D=1#page/1/mode/1up on March 30, 2022; Phipps, *William Sheppard*, 141; and Dworkin, *Congo Love Song*, 53.

156   *days of marching:* Typed transcription of William H. Sheppard's diary, Sept. 13–14, 1899; Phipps, *William Sheppard*, 141; and Dworkin, *Congo Love Song*, 53.

156   *Sheppard stepped forward:* Sheppard, "An African's Work for Africa," *Missionary Review of the World*, (October 1906): 772.

156   *Chembamba remembered him:* Sheppard, "An African's Work for Africa," 772.

157   *M'lumba N'kusa:* Dworkin, *Congo Love Song*, 54.

157   *had murdered members:* Sheppard, "An African's Work for Africa," 771.

157   *evidence of cannibalism:* E. D. Morel, *Red Rubber: The Story of the Rubber Slave Trade Flourishing on the Congo in the Year of Grace 1906* (London: T. Fisher Unwin, 1906), 71.

157   *A skull rested:* Sheppard, "An African's Work for Africa," 772.

157   *when Sheppard inquired:* Typed transcription of William H. Sheppard's diary, Sept. 13–14, 1899.

157   *James Campbell argues:* Campbell, *Middle Passages*, 176.

157   *More than sixty women:* Sheppard, "An African's Work for Africa," 773. Missionaries subsequently requested that Belgian soldiers intervene to rescue the enslaved women. They did so and also arrested M'lumba N'kusa.

157   *went to work:* "Eaten by Zappo Zaps: The Bena Kamba Country Raided by African Cannibals," *Indianapolis Journal*, January 6, 1900, page 5; and Phipps, *William Sheppard*, 142–43.

157   *corroborating testimony:* "Eaten by Zappo Zaps," page 5.

157   *alerting their readers:* "Eaten by Zappo Zaps," page 5.

157   Evening Times: "A Slaughter by Cannibals: Congo Free State Natives Murdered for Not Paying Taxes," *Evening Times*, January 5, 1900, page 3.

157   *credited Sheppard:* "Eaten by Zappo Zaps," page 5; and "A Slaughter by Cannibals," page 3.

157   *The following month:* "The Alleged Atrocities in the Congo Free State," *Times*, February 22, 1900, page 9.

158   *The journalist wrote:* "The Alleged Atrocities in the Congo Free State," *Times*, page 9.

158   *the Belgian state:* "Anglais et Belges," *Le Peuple*, February 27, 1900, page 1.

158   *"Negro Shot":* "Negro Shot to Death," *Indianapolis Journal*, January 6, 1900, page 5.

158   *raise awareness:* Robert Benedetto, *Presbyterian Reformers in Central Africa: A*

Documentary Account of the American Presbyterian Congo Mission and the Human Rights Struggle in the Congo, 1890–1918 (Leiden: E. J. Brill, 1996), 126, n6.

158   *the international public:* Phipps, *William Sheppard*, 146, 150–51.

158   *multiple audiences:* Phipps, *William Sheppard*, 155–56, 161.

158   *back to Luebo:* Phipps, *William Sheppard*, 161.

159   *a condemnatory essay:* Phipps, *William Sheppard*, 162; Kennedy, *Black Livingstone*, 170–71, and Leah Gass, "Rubber Crimes: Sheppard and Morrison Versus the Kasai Rubber Company," Presbyterian Historical Society, accessed at https://www.history.pcusa.org/blog/rubber-crimes-sheppard-and-morrison -versus-kasai-rubber-company on April 5, 2022.

159   *intimidate their accuser:* Jeannette Eileen Jones, *In Search of Brightest Africa: Reimaging the Dark Continent in American Culture, 1884–1936* (Athens: University of Georgia Press, 2011), 71; Phipps, *William Sheppard*, 165–69; and Gass, "Rubber Crimes."

159   *he was acquitted:* Gass, "Rubber Crimes."

159   *seizure of control:* DuPree and DuPree, "William Henry Sheppard."

159   *But little changed:* "Belgian Congo," *Encyclopaedia Britannica*, accessed at https://www.britannica.com/place/Belgian-Congo on March 28, 2023.

159   *permission to resign:* Phipps, *William Sheppard*, 176.

159   *members of the APCM:* "Manuscript Copy of American Presbyterian Congo Mission Acceptance of Sheppard's Resignation, 1909," RG 457, Box 1, Folder 3, William H. Sheppard Papers, Presbyterian Historical Society, accessed at https://digital.history.pcusa.org/islandora/object/islandora%3A15889#page/1 /mode/1up on April 7, 2022.

159   *They felt that:* "Manuscript copy of American Presbyterian Congo Mission Acceptance."

160   *never to return:* DuPree and DuPree, "William Henry Sheppard."

160   *Grace Presbyterian Church:* Parker, "William Sheppard."

7. HARRIET CHALMERS ADAMS: THE GEOGRAPHER

161   *the* Alexandria Gazette: "Independence Proclaimed," *Alexandria Gazette*, November 4, 1903, page 2. The quotation that follows on the next page is from the same source.

161   *expand its profits:* Michael Schaller, Janette Thomas Greenwood, Andrew Kirk, Sarah J. Purcell, Aaron Sheehan-Dean, and Christina Snyder, *American Horizons: US History in a Global Context, Volume 2*, 4th ed. (New York: Oxford University Press, 2021), 671.

162   *failed to finish:* Schaller, Greenwood, Kirk, et al., *American Horizons*, 671.

162   *fifty-mile-wide isthmus:* John Haskell Kemble, "The Gold Rush by Panama, 1848–1851," *Pacific Historical Review* 18, no. 1, (1949): 52; and Augustus Campbell and Colin D. Campbell, "Crossing the Isthmus of Panama, 1849: The Letters of Dr. Augustus Campbell," *California History* 78, no. 4 (1999/2000): 226.

162   *"oppressive" heat:* Augustus Campbell to Sarah Elderkin Campbell, April 27, 1849, in Campbell and Campbell, "Crossing the Isthmus of Panama," 235; and Kemble, "The Gold Rush by Panama," 55.

162   *one newspaper claimed:* "Panama Shelled," *Evening Star*, November 4, 1903, page 1; and "The Hay-Herran Treaty," Theodore Roosevelt Center, accessed at https://www.theodorerooseveltcenter.org/Learn-About-TR/TR -Encyclopedia/Foreign-Affairs/Hay-Herran-Treaty on April 19, 2022.

162   Evening Star: "Anxiety in Panama," *Evening Star*, November 5, 1903, page 7. The quotation that follows is from the same source and page.

162   *Hay–Bunau-Varilla:* "Convention for the Construction of a Ship Canal (Hay-Bunau-Varilla Treaty), November 18, 1903," Lillian Goldman Law Library, Yale University, accessed at https://avalon.law.yale.edu/20th_century/pan001 .asp on April 19, 2022.

163   *pay Panama:* "Hay–Bunau-Varilla Treaty," *Encyclopaedia Britannica*, accessed at https://www.britannica.com/event/Hay-Bunau-Varilla-Treaty on April 19, 2022.

163   *arrived in Panama:* Harriet Chalmers Adam's journal from her 1904-06 South America trip, February 19, 1904, transcribed by Kate Davis.

163   *"There is great":* "Woman Explorer Is Planning Difficult Trip from Mediterranean to India," interview with the *Evening Star*, September 12, 1915, Part 4, page 2. The quotation that follows is from the same source.

163   *Adams's chief hope:* Kathryn Davis, "Harriet Chalmers Adams: Remembering an American Geographer," *California Geographer* 49 (2009): 58–59.

163   *Alexander Chalmers:* Susan Crosby, "Harriet Chalmers Adams," senior thesis, Department of History, San Joaquin Delta College, May 1983, 1; and Durlynn Anema, *Harriet Chalmers Adams: Adventurer and Explorer* (Parker, CO: National Writers Press, 1997), 13.

163   *He made his way:* Crosby, "Harriet Chalmers Adams," 1.

163   *Frances Wilkins:* Crosby, "Harriet Chalmers Adams," 1; "Woman Ahead of Roosevelt," *Boston Globe*, May 24, 1914, page 65; and Davis, "Harriet Chalmers Adams: Remembering an American Geographer," 54.

164   *Stockton served prospectors:* James A. Barr, "Stockton, California," *Journal of Education* 65, no. 24 (1907): 661.

164   *a port of trade:* Barr, "Stockton, California," 661.

164   *a dusty dirt road:* "Stockton—Streets—circa 1880s: Main St. looking west toward California St. Matteson and Williamson," circa 1885, Holt-Atherton Special Collections, University of the Pacific Library, accessed at https:// scholarlycommons.pacific.edu/hsp/4536/ on April 26, 2022; and Anema, *Harriet Chalmers Adams*, 14.

164   *tall glass windows:* Van Covert Martin, "General Stores—Stockton: Alex Chalmers Dry Goods store, El Dorado St. between Main and Weber St.," likely early twentieth century, Holt-Atherton Special Collections, University of the Pacific Library, accessed at https://scholarlycommons.pacific.edu /hsp/7679/ on April 26, 2022.

164   *a modest house:* Anema, *Harriet Chalmers Adams*, 14–15.

164   *Much of her childhood:* Crosby, "Harriet Chalmers Adams," 1–2.

164   *"made me over":* "Woman Ahead of Roosevelt," *Boston Globe*; and Anema, *Harriet Chalmers Adams*, 11.

164   *a public feat:* "Dashes Here and There," *Mail*, August 14, 1886, page 3; and Anema, *Harriet Chalmers Adams*, 12.

165   *impressed a journalist:* "Shifting Sands," *Santa Cruz Surf,* August 26, 1886, page 3.

165   *"drownings [were] terribly":* "Little Boy Drowned," *Stockton Record,* September 20, 1890, page 3.

165   *Formal schooling:* Kathryn Davis, "The Forgotten Life of Harriet Chalmers Adams: Geographer, Explorer, Feminist," master's thesis, Department of History, San Francisco State University, 1995, 16.

165   *charm was undeniable:* "Far Away Places Call Again and Noted Washington Woman Explorer Sets Sail," *Washington Post,* September 4, 1933, page 7.

165   *Highly sociable:* "Miss Bessie Reid Entertains," *Evening Mail,* January 4, 1896, page 2; and "A Bundle Party," *Stockton Daily Record,* August 9, 1896, page 3.

165   *An accomplished musician:* "The Euterpean Circle," *Evening Mail,* December 22, 1893, page 1; and "Hugo Mansfield's Piano Recital," *Evening Mail,* May 2, 1892, page 5.

165   *"the urge to seek":* "Career of Woman Explorer Praised," *Evening Star,* February 20, 1938, page 34.

165   *Henry the Navigator:* "Far Away Places Call Again," *Washington Post.* The quotation that follows is from the same source and page.

165   *studying both Spanish and French:* Davis, "Harriet Chalmers Adams: Remembering an American Geographer," 55. The information and quotation in the paragraph that follows are from the same source and page.

166   *She and her father:* Anema, *Harriet Chalmers Adams,* 19.

166   *"Range of Light":* John Muir, *The Yosemite* (New York: The Century Company, 1912), accessed at https://vault.sierraclub.org/john_muir_exhibit/writings/the _yosemite/chapter_1.aspx on April 26, 2022.

166   *"the big, broad, unconventional":* "An Interview with Harriet Chalmers Adams," *Evening Mail,* October 28, 1907, page 8.

166   *acquired her father's:* "Woman Ahead of Roosevelt," *Boston Globe.*

166   *"learn[ed] the secrets":* John Oliver La Gorce, "In Memory of Harriet Chalmers Adams," *Society of Woman Geographers* (Washington, DC: 1938), 12.

166   *"a sweet musical":* "An Interview with Harriet Chalmers Adams," *Evening Mail.*

167   *a local man:* "Adams, Franklin Pierce," *Who's Who in the Nation's Capital* (Consolidated Publishing Company, 1921), 3.

167   *of California settlers:* Anema, *Harriet Chalmers Adams,* 21.

167   *They were connected:* Davis, "The Forgotten Life of Harriet Chalmers Adams: Geographer, Explorer, Feminist," 16.

167   *Stockton Gas and Electric:* "Adams-Chalmers: A Notable Society Wedding Which Is to Take Place This Evening," *Evening Mail,* October 5, 1899, page 8; and Van Covert Martin, "Gas Companies—Stockton: Gas and Electric Co. exterior, Hazelton and Center St.," Holt-Atherton Special Collections, University of the Pacific Library, accessed at https://scholarlycommons.pacific.edu /hsp/7638/ on April 27, 2022.

167   *"Ward McAllister":* "Adams-Chalmers Wedding," *Evening Mail,* October 6, 1899, page 8.

167   *the wedding:* "Social Notes," *Daily Evening Record,* September 18, 1899, page 4.

167   *a joyful occasion:* "Adams-Chalmers Wedding," *Evening Mail.* Further details in this paragraph about the wedding are from the same source and page.

167   *Harriet's biographer Durlynn Anema:* Anema, *Harriet Chalmers Adams,* 22.

168   *"Washington's aspirations":* Robert E. Hannigan, *The New World Power: American Foreign Policy, 1898–1917* (Philadelphia: University of Pennsylvania Press, 2002), 51. The information that follows in this paragraph is from the same source and page.

168   *local historic sites:* Anema, *Harriet Chalmers Adams,* 25.

168   *additional opportunities:* Anema, *Harriet Chalmers Adams,* 25.

168   *US companies:* Kendall W. Brown, *A History of Mining in Latin America: From the Colonial Era to the Present* (Albuquerque: University of New Mexico Press, 2012), 92.

168   *Inca Mining and Rubber:* Adolfo de Clairmont, *A Guide to Modern Peru: Its Great Advantages and Vast Opportunities* (Toledo: A. de. Clairmont, 1908), 42; and Davis, "Harriet Chalmers Adams: Remembering an American Geographer," 56.

168   New York Tribune: "US Gets Peru Trade: Inter-Country Business Increasing—American Investments Large," *New York Tribune,* May 13, 1906, page 51. The information that follows in this paragraph is from the same source and page.

169   *from San Francisco:* Harriet Chalmers Adam's journal, January 9–10, 1904, and "Weather Bureau Report," *San Francisco Chronicle,* January 8, 1904, page 6.

169   *"perfect day":* Harriet Chalmers Adam's journal, January 9–10, 1904. The quotations in the following paragraph are from the same source.

170   *Harriet was thrilled:* Harriet Chalmers Adam's journal, January 19, 1904. The information that follows in this paragraph is from the same source.

170   *President Theodore Roosevelt:* Theodore Roosevelt to Kermit Roosevelt, November 20, 1906, in *Theodore Roosevelt's Letters to His Children,* ed. Joseph Bishop (New York: Charles Scribner's Sons, 1923), 179–80. The quotations that follow in this paragraph are from the same source, pages 183–84.

170   *When Harriet reached:* Harriet Chalmer Adam's journal, February 18, 1904. The quotation that follows is from the same source on February 19, 1904.

171   *the Culebra Cut:* Likely Max T. Vargas and Martin Chambi, "From the North to the South: Some Sights I Have Seen in South America," Photographic Album, August 1904–May 1905, Charles E. Young Research Library, University of California, Los Angeles; and Theodore Roosevelt to Kermit Roosevelt, November 20, 1906, *Theodore Roosevelt's Letters to His Children,* 182.

171   *"There the huge":* Theodore Roosevelt to Kermit Roosevelt, November 20, 1906, *Theodore Roosevelt's Letters to His Children,* 182–83. The information and quotations that follow in this paragraph are from the same source, page 183.

171   *she photographed:* "Panama's First President," *New York Sun,* February 21, 1904, page 2; and Harriet Chalmers Adam's journal, February 20, 1904.

171   New York Sun *reported:* "Panama's First President," *New York Sun,* February 21, 1904, page 2. The information and quotation that follow in this paragraph are from the same source, page 183.

171   *Harriet and Frank departed:* Harriet Chalmers Adam's journal, February 23, 24, 1904.

171   *mountains of Peru:* Harriet Chalmers Adam's journal, March 1, 1904.

172   *"all the way back":* Harriet Chalmers Adam's journal, March 2, 4, 1904.

172   *She was pleased:* Harriet Chalmers Adam's journal, March 5, 19, 26, 1904.

172   *made their way inland:* Harriet Chalmers Adam's journal, April 5, 1904. The quotations that follow in this paragraph are from the same source, April 5–6, 1904.

172   *La Paz:* Harriet Chalmers Adams, "Kaleidoscopic La Paz: The City of the Clouds," *National Geographic* 20, no. 2 (February, 1909): 119.

172   *"city of the clouds":* Adams, "Kaleidoscopic La Paz," 119.

172   *First inhabited by:* "Bolivia: Early Period," *Encyclopaedia Britannica,* accessed at https://www.britannica.com/place/Bolivia/Early-period on May 10, 2022.

172   *admired La Paz's historicity:* Adams, "Kaleidoscopic La Paz," 119.

172   *During her stay:* Adams, "Kaleidoscopic La Paz," 130. The information and quotations in the paragraph that follows come from the same source, pages 120–21.

173   *"In many of the":* "Far Away Places Call Again," *Washington Post.*

173   *"forced labor":* Adams, "Kaleidoscopic La Paz," 121.

173   *Harriet regretted departing:* Adams, "Kaleidoscopic La Paz," 134.

173   *rode a mule:* Adams, "Kaleidoscopic La Paz," 131. The quotation that follows is from the same source and page.

174   *"the most wonderful":* Harriet Chalmers Adam's journal, April 9, 1904.

174   *gold mines:* Likely Vargas and Chambi, "From the North to the South: Some Sights I Have Seen in South America."

174   *"restless and anxious":* Harriet Chalmers Adam's journal, April 17, 1904.

174   *her gaunt horse:* Harriet Chalmers Adam's journal, April 19, 1904.

174   *the next four days:* Harriet Chalmers Adam's journal, April 19, 20, 1904. The information and quotations that follow in this paragraph come from the same source, April 19–22, 1904.

174   *felt well enough:* Harriet Chalmers Adam's journal, April 30, 1904. The information and quotations that follow in this paragraph come from the same source.

175   *Vampire bats:* "Woman Explorer's Hazardous Trip in South America," *New York Times,* August 18, 1912, page 35.

175   *bat swooped down:* Harriet Chalmers Adam's journal, April 30, 1904.

175   *spotted wild turkeys:* Harriet Chalmers Adam's journal, May 1, 2, 1904.

175   *the weary travelers:* Harriet Chalmers Adam's journal, May 4, 6, 1904.

175   *Harriet even tried:* Harriet Chalmers Adam's journal, May 5, 1904.

175   *a laborious process:* Harriet Chalmers Adam's journal, May 6, 1904.

175   *"The trees are tapped":* William Eleroy Curtis, *Between the Andes and the Ocean: An Account of an Interesting Journey Down the West Coast of South America from the Isthmus of Panama to the Straits of Magellan* (Chicago: Herbert S. Stone and Company, 1900), 345–46.

176 *Potosí, Bolivia:* Adams refers to this region as "Poto." Harriet Chalmers Adam's journal, May 18, 1904.

176 *altitude sickness:* Harriet Chalmers Adam's journal, May 19, 1904.

176 *out of her saddle:* Harriet Chalmers Adam's journal, May 20, 1904.

176 *helped provide warmth:* Harriet Chalmers Adam's journal, May 20, 1904.

176 *"Too ill to go":* Harriet Chalmers Adam's journal, May 20, 1904.

176 *hunkered down:* Harriet Chalmers Adam's journal, May 20, 1904.

176 *she lay awake:* Harriet Chalmers Adam's journal, May 20, 1904.

176 *"Tears came":* "Woman Explorer's Hazardous Trip in South America," *New York Times.*

176 *When Harriet emerged:* Harriet Chalmers Adam's journal, May 21, 1904. The information that follows in this paragraph and the next comes from the same source.

177 *the mining operations:* Likely Vargas and Chambi, "From the North to the South: Some Sights I Have Seen in South America." The information that follows in this paragraph comes from the same source.

177 *She reached Cuzco:* Harriet Chalmers Adam's journal, June 3, 1904; likely Vargas and Chambi, "From the North to the South: Some sights I have seen in South America"; and "Cuzco," *Encyclopaedia Britannica,* accessed at https://www.britannica.com/place/Cuzco on November 23, 2022.

177 *architectural wonders:* Harriet Chalmers Adam's journal, June 4, 1904; and likely Vargas and Chambi, "From the North to the South: Some Sights I Have Seen in South America."

177 *"See many marks":* Harriet Chalmers Adam's journal, June 5, 1904.

177 *"Saw ancient aqueducts":* Harriet Chalmers Adam's journal, June 12, 1904.

177 *the megalithic fortress:* Likely Vargas and Chambi, "From the North to the South: Some Sights I Have Seen in South America."

177 *Harriet was conscious:* Harriet Chalmers Adam's journal, May 23, 1904.

177 *"a great curiosity":* Harriet Chalmers Adam's journal, May 23, 1904.

178 *"These women before me":* "Woman Explorer Is Planning Difficult Trip from Mediterranean to India."

178 *a sacred place:* Dagmara M. Socha, Johan Reinhard, and Ruddy Chávez Perea, "Inca Human Sacrifices on Misti Volcano," *Latin American Antiquity* 32, no. 1 (2021): 138–53.

178 *Harriet began preparing:* Harriet Chalmers Adam's journal, July 9, 1904.

178 *did not reveal it:* "Here's a Woman Who Edged South American Continent," *Sacramento Bee,* February 18, 1910, page 16.

178 *on July 10:* Harriet Chalmers Adam's journal, July 10, 1904.

178 *Flashes of lightning:* Harriet Chalmers Adam's journal, July 10, 1904.

178 *"a mountain of gold":* Harriet Chalmers Adam's journal, July 11, 1904.

178 *"trail in lava":* Harriet Chalmers Adam's journal, July 11, 1904; and "Wilds of South America Described by Mrs. Adams to Geographic Society," *Washington Times,* December 14, 1907, page 3.

179   *terribly taxing ascent:* Harriet Chalmers Adam's journal, July 11, 1904.

179   *"we had the most":* Harriet Chalmers Adam's journal, July 11, 1904.

179   *visiting Chile:* Harriet Chalmers Adam's journal, November 2, 1904.

179   *The next year:* Anema, *Harriet Chalmers Adams*, 55–57.

179   *by way of Panama:* Harriet Chalmers Adam's journal, May 9, 1906.

179   *"A Red Letter Day":* Harriet Chalmers Adam's journal, May 16, 1906.

179   *"intimately acquainted":* "Woman to Lecture on Her Travels in Peru," *San Francisco Call,* October 27, 1907, page 40; and "Woman Has Traveled Far," *Lamar Register,* November 13, 1907, page 3.

179   *praised as "advanced":* Harriet Chalmers Adams, "Some Wonderful Sights in the Andean Highlands," *National Geographic* 19, no. 9 (1908): 597; and Davis, "Harriet Chalmers Adams: Remembering an American Geographer," 59.

180   National Geographic: Introduction to *National Geographic* 1, no. 1 (1888), accessed at https://www.gutenberg.org/cache/epub/49711/pg49711.txt on November 18, 2023.

180   *President Gardiner Hubbard:* Gardiner Hubbard, "Introductory Address," *National Geographic* 1, no. 1 (1888): 7. The quotations that follow in this paragraph come from pages 8–11.

180   *She was thrilled:* Alexa Keefe, "How a Woman Photographer Paved the Way for Adventure in Latin America," *National Geographic,* March 7, 2016, accessed at https://www.nationalgeographic.com/photography/proof/2016/03/07/a-woman-of-passion-courage-and-style-explores-the-world/ on February 20, 2020.

180   *the first American:* Davis, "The Forgotten Life of Harriet Chalmers Adams: Geographer, Explorer, Feminist," 30.

180   *Her first article:* Harriet Chalmers Adams, "Picturesque Paramaribo," *National Geographic* 18, June 1907, 365–73.

180   *twenty additional essays:* Keefe, "How a Woman Photographer Paved the Way."

180   *wrote of her subjects:* Harriet Chalmers Adams, "The East Indians in the New World," *National Geographic* 18, July 1907, 487.

181   *failed to fund:* Davis, "Harriet Chalmers Adams: Remembering an American Geographer," 59.

181   *For instance:* Davis, "Harriet Chalmers Adams: Remembering an American Geographer," 59.

181   *Frank Chapman:* Frank M. Chapman, *The Distribution of Bird Life in the Urubamba Valley of Peru* (Washington: US Government Printing Office, 1921), 11.

181   *Pan American Union:* Davis, "Harriet Chalmers Adams: Remembering an American Geographer," 59.

181   *a war correspondent:* Davis, "Harriet Chalmers Adams: Remembering an American Geographer," 61.

181   *she became a fellow:* Davis, "The Forgotten Life of Harriet Chalmers Adams: Geographer, Explorer, Feminist," 51.

181   *Society of Woman Geographers:* Davis, "Harriet Chalmers Adams: Remembering an American Geographer," 51–52.

181   *"a woman can":* "Daring Women in Thrilling Hunts," *New Britain Herald*, February 6, 1928, page 14.

181   *Mary Vaux Wallcott:* "In Memory of Harriet Chalmers Adams," *Society of Woman Geographers*, February 1938, page 10.

182   *Amelia Earhart:* "Harriet C. Adams, Explorer, Is Dead," *New York Times*, Sunday, July 18, 1937, page 7.

182   *sustaining serious injuries:* Davis, "The Forgotten Life of Harriet Chalmers Adams: Geographer, Explorer, Feminist," 28.

182   *"wondered why men":* Keefe, "How a Woman Photographer Paved the Way."

8. MATTHEW HENSON: THE ARCTIC EXPLORER

183   *the fierce wind:* Matthew Henson, *A Negro Explorer at the North Pole* (New York: Frederick A. Stokes Company Publishers, 1912), 52–54, 77.

183   *413-mile trek:* Henson, *A Negro Explorer at the North Pole*, 70, 155.

183   *The past week:* Henson, *A Negro Explorer at the North Pole*, 71–72, 76. Further information in this paragraph comes from the same source and pages.

184   *Peary's command:* Henson, *A Negro Explorer at the North Pole*, 77.

184   *Henson knew that:* Henson, *A Negro Explorer at the North Pole*, 76.

184   *strove to discover:* Edward J. Larson, *To the Edges of the Earth: 1909, the Race for the Three Poles, and the Climax of the Age of Exploration* (New York: HarperCollins, 2018), 3–4.

184   *William Edward Parry:* Sir William Edward Parry, *Memoirs of Rear-Admiral Sir W. Edward Parry* (London: Longman, Green, Longman, Roberts, and Green, 1863), 56–58.

184   *subsequent generations:* Larson, *To the Edges of the Earth*, 4; and "North Pole," National Geographic Resource Library, accessed at https://www.national geographic.org/encyclopedia/north-pole/ on June 8, 2021.

184   *Edward Larson:* Larson, *To the Edges of the Earth*, 4.

184   *Robert Peary:* Henson, *A Negro Explorer at the North Pole*, vii.

184   *Caroline and Lemuel Henson:* "Matthew Alexander Henson," Archives of Maryland (Biographical Series), MSA SC 3520–12524, accessed at https://msa.maryland.gov/megafile/msa/speccol/sc3500/sc3520/012500/012524/html/12524bio.html on June 10, 2021.

185   *new state constitution:* "Baltimore Celebrates Constitution, Emancipation," *Baltimore Sun*, November 1, 1864, page 1, accessed at https://www.prattlibrary.org/locations/periodicals/civilwar/index.aspx?id=5480 on March 5, 2020; and "A Guide to the History of Slavery in Maryland," 2007, Maryland State Archives and the University of Maryland College Park, page 16, accessed at https://msa.maryland.gov/msa/intromsa/pdf/slavery_pamphlet.pdf on March 5, 2020.

185   *African American Marylanders:* "Black Marylanders 1860: African American Population by County, Status, & Gender," *Legacy of Slavery in Maryland at the Maryland State Archives*, accessed at https://msa.maryland.gov/msa/mdslavery/html/research/census1860.html on March 5, 2020.

185   *former slave market:* Henson, *A Negro Explorer at the North Pole*, 3; Henry Louis Gates Jr. and Cornel West, *The African-American Century: How Black Americans*

Have Shaped Our Country (New York: Simon & Schuster, 2002), 12; and Henry Louis Gates Jr., *Life upon These Shores: Looking at African American History, 1513–2008* (New York: Alfred A. Knopf, 2011), 245.

185 *tobacco plantations:* Christopher I. Sperling, "Agriculture and Slavery in Prince George's County, Maryland," *African Diaspora Archaeology Newsletter* 10, no. 1 (2007): 1, 3.

185 *the predominant crop:* Sperling, "Agriculture and Slavery in Prince George's County, Maryland," 1.

185 *free Black population:* "Black Marylanders 1860: African American Population by County, Status, & Gender," *Legacy of Slavery in Maryland at the Maryland State Archives.*

185 *Henry Louis Gates Jr:* Henry Louis Gates Jr., *Stony the Road: Reconstruction, White Supremacy, and the Rise of Jim Crow* (New York: Penguin Publishing Group, 2019), 17.

185 *District of Columbia:* Henson, *A Negro Explorer at the North Pole,* 2–3.

186 *Henson's first biographers:* Floyd Miller, *Ahdoolo! The Biography of Matthew A. Henson* (New York: E. P. Dutton, 1963), 17.

186 *N Street School:* Henson, *A Negro Explorer at the North Pole,* 3.

186 *twenty-five thousand:* "Ending Slavery in the District of Columbia," DC.gov, accessed at https://emancipation.dc.gov/page/ending-slavery-district-columbia on March 5, 2020.

186 *Henson ran away:* Henson, *A Negro Explorer at the North Pole,* 3; and Garey Reynolds, "Matt Henson—Trek to the Pole," *Baltimore Afro American,* November 18, 1961, page 17.

186 *Jack's thrilling stories:* Reynolds, "Matt Henson—Trek to the Pole," page 17.

186 *As a cabin boy:* Gates and West, *The African-American Century,* 12–13; Henson, *A Negro Explorer at the North Pole,* 3; Miller, *Ahdoolo!,* 16, 18; and Reynolds, "Matt Henson—Trek to the Pole."

186 *he later wrote:* Henson, *A Negro Explorer at the North Pole,* 3.

187 *B. H. Stinemetz:* *Evening Star,* February 5, 1880, page 2.

187 *When Captain Childs died:* Henson, *A Negro Explorer at the North Pole,* 3; and Miller, *Ahdoolo!,* 18.

187 *Peary was born:* William H. Hobbes, "Robert Edwin Peary," *Encyclopedia Arctica 15: Biographies,* Dartmouth College Library, accessed at https://collections.dartmouth.edu/arctica-beta/html/EA15-55.html on June 25, 2021. Subsequent information in this paragraph comes from the same source.

187 *an expedition to Nicaragua:* "RADM Robert E. Peary, CEC, USN: Isthmian Canal surveyor, Arctic explorer, and early air advocate," Naval History and Heritage Command, US Navy Seabee Museum, January 27, 2020, accessed at https://www.history.navy.mil/content/history/museums/seabee/explore/civil-engineer-corps-history/robert-e—peary.html on June 25, 2021; and Miller, *Ahdoolo!,* 18.

187 *unexplored and unmapped:* Hobbes, "Robert Edwin Peary."

187 *second trip to Nicaragua:* William M. Farrell, "Henson Recalls Peary's Staking of Claim to the North Pole 40 Years Ago," *New York Times,* April 7, 1949, pages 31, 58; Henson, *A Negro Explorer at the North Pole,* 3; and Gates, *Life upon These Shores,* 245.

187   *also a valet:* Henson, *A Negro Explorer at the North Pole*, 3; Gates, *Life upon These Shores*, 245; Miller, *Ahdoolo!*, 18; and "Matt Henson: First to the North Pole?" *The News.*, February 10, 1972, page 17.

187   *Henson readily accepted:* Henson, *A Negro Explorer at the North Pole*, 3; Gates, *Life upon These Shores*, 245; and Miller, *Ahdoolo!*, 18.

188   *for two years:* James Mills, "The Legacy of Arctic Explorer Matthew Henson," *National Geographic*, February 28, 2014, accessed at https://www.national geographic.com/adventure/article/the-legacy-of-arctic-explorer-matthew -henson on April 2, 2023.

188   *Together they trekked:* Miller, *Ahdoolo!*, 19–21.

188   *connection they formed:* Miller, *Ahdoolo!*, 18–22.

188   *League Island Navy Yard in:* Reynolds, "Matt Henson—Trek to the Pole"; and Henson, *A Negro Explorer at the North Pole*, 3.

188   *daring new expedition:* Miller, *Ahdoolo!*, 23–25; Henson, *A Negro Explorer at the North Pole*, 3; Hobbes, "Robert Edwin Peary"; and Reynolds, "Matt Henson—Trek to the Pole."

188   *He wagered $100:* "Matt Henson: First to the North Pole?" *News.*

188   *Fridtjof Nansen:* Robin Hanbury-Tenison, *The Oxford Book of Exploration* (New York: Oxford University Press, 2005), 482; "Admiral Robert E. Peary," *American Polar Society*, accessed at https://americanpolar.org/about/polar-luminaries /admiral-robert-e-peary/ on August 4, 2023; and Larson, *To the Edges of the Earth*, 11.

188   *what one journalist:* J. Arthur Bain's interview with Fridtjof Nansen, "Dr. Nansen's Story," *Madison Daily Leader*, January 4, 1897, page 1. The quotation that follows in this paragraph is from the same source and page. Bain believed that "Dr. Nansen will either solve the problem or perish in the attempt."

188   *through global exploration:* Lyle Dick, "Robert Peary's North Polar Narratives and the Making of an American Icon," *American Studies* 45, no. 2 (2004): 6.

189   *global polar expeditions:* Jonathan M. Karpoff, "Public versus Private Initiative in Arctic Exploration: The Effects of Incentives and Organizational Structure," *Journal of Political Economy* 109, no. 1 (2001): 45; and Kelly Lankford, "Arctic Explorer Robert Peary's Other Quest: Money, Science, and the Year 1897," *American Nineteenth-Century History* 9, no. 1 (2008): 43.

189   *new investigative trip:* Miller, *Ahdoolo!*, 25–27.

189   *map the northern coast:* "Peary's Arctic Journey," *Sun*, April 24, 1892, page 5.

189   *there was a dispute:* Larson, *To the Edges of the Earth*, 26.

189   *journey's central purpose:* "To Explore Greenland: The Final Preparations of Lieut. Peary's Party," *Sun*, June 5, 1891, page 7.

189   *from New York City:* "To Explore Greenland"; Josephine Diebitsch Peary, *My Arctic Journal: A Year Among Ice-Fields and Eskimos* (New York: The Contemporary Publishing Company, 1893), 1; and Henson, *A Negro Explorer at the North Pole*, 5.

189   *broke the two bones:* Henson, *A Negro Explorer at the North Pole*, 5; Gilbert Grosvenor, Foreword, in Robert Peary, *The North Pole: Its Discovery in 1909 Under the Auspices of the Peary Arctic Club* (New York: Frederick A. Stokes Company, 1910), xxix; and Hobbes, "Robert Edwin Peary."

189   *they hunkered down:* Larson, *To the Edges of the Earth*, 25.

189   *expedition to:* Larson, *To the Edges of the Earth*, 25; and Hugh J. Lee, "Peary's Transections of North Greenland, 1892–1895," *Proceedings of the American Philosophical Society* 82, no. 5 (1940): 926.

190   *enormous ice cap:* Grosvenor, *The North Pole*, xxix.

190   *"a barren waste":* Grosvenor, *The North Pole*, xxix; and Robert E. Peary, "The Great White Journey," in Josephine Peary, *My Arctic Journal*, 226.

190   *"hissing white torrent":* Robert E. Peary, "The Great White Journey," in Josephine Peary, *My Arctic Journal*, 232–33. The information and quotations in the paragraph that follows come from the same source, pages 226–27.

190   *proof at last:* Larson, *To the Edges of the Earth*, 25.

190   *"gazed from the summit:* Peary, "The Great White Journey," 227.

190   *The exhausted men:* Robert E. Peary, "The Great White Journey," 228.

190   *John Verhoeff:* Henson, *A Negro Explorer at the North Pole*, 5–6.

190   *permanently changed:* Henson, *A Negro Explorer at the North Pole*, 7.

191   *hundreds of years:* "Inuit," *The Canadian Encyclopedia*, accessed at https://www .thecanadianencyclopedia.ca/en/article/inuit on June 28, 2021.

191   *slim wooden canoes:* Series 18: Photographic materials Sub-series C: Loose photographs, Box 9.11, Archives of the Explorers Club.

191   *"dressing in the same":* Henson, *A Negro Explorer at the North Pole*, 7.

191   *"white Sahara":* Josephine Peary, *My Arctic Journal*, 232; Henson, *A Negro Explorer at the North Pole*, 40–41; "Matthew Henson, of North Pole Fame, to Speak Here," *Star*, October 17, 1930, page 1; and Mills, "The Legacy of Arctic Explorer Matthew Henson."

191   *"I have come":* Henson, *A Negro Explorer at the North Pole*, 6–7.

191   *seven subsequent treks:* Henson traveled to the Arctic in the following years: 1891–92; 1893–95; 1896; 1897; 1898–1902; 1905–6; 1908–9. "Matthew Alexander Henson," *Encyclopaedia Britannica*, accessed at https://www.britannica .com/biography/Matthew-Alexander-Henson on June 14, 2021.

191   *"He has shared":* Peary, *The North Pole*, 20.

191   *fathered Inuit children:* "A World Away, Explorers' Kin Meet," *New York Times*, June 7, 1987.

192   *communicating with the Inuits:* Lowell Thomas, "First at the Pole: Lowell Thomas Interviews Matthew Henson," *Lowell Thomas Interviews* (New York: NBC Radio Network, 1939), accessed at https://matthewhenson.com/lowell THOMAS/LowellThomas1939.pdf on June 20, 2021.

192   *even have chosen Henson:* Russell W. Gibbons, letter to the editor, "Matthew Henson: Black Explorer Used and Discarded by Peary," *New York Times*, June 21, 1987, page E24.

192   *most of his toes:* Henson, *A Negro Explorer at the North Pole*, 12; and Hobbes, "Robert Edwin Peary."

192   *meager rations:* "Cache at Cape Sabine," in "Instructions for Matt Henson," likely 1899, Beulah M. Davis Special Collections, Morgan State University; Thomas, "First at the Pole: Lowell Thomas Interviews Matthew Henson"; and

Robert Peary, "The Discovery of the North Pole," *Hampton's Magazine*, published as a book in 1910 by the Columbian-Sterling Publishing Company, 773.

192 *He lectured widely:* Larson, *To the Edges of the Earth*, 33, 114.

192 *Peary Arctic Club:* Statement by President Morris K. Jessup of the Peary Arctic Club, New York, April 1907, PACC Series 13 Box 5, Archives of the Explorers Club; and "1898—The Peary Arctic Club," Peary-MacMillan Arctic Museum, accessed at https://learn.bowdoin.edu/pearys-north-pole-explorations/maps/1898-1902-expeditions/1898-01a-peary-arctic-club.html on November 25, 2022.

192 *sign contracts:* Miller, *Ahdoolo!*, 70; and Bruce Henderson, "Who Discovered the North Pole?" *Smithsonian Magazine*, April 2009, accessed at https://www.smithsonianmag.com/history/who-discovered-the-north-pole-116633746/ on June 14, 2021. Henderson argues that Peary was "one of the last of the imperialistic explorers, chasing fame at any cost and caring for the local people's well-being only to the extent that it might affect their usefulness to him. (In Greenland in 1897, he ordered his men to open the graves of several natives who had died in an epidemic the previous year—then sold their remains to the American Museum of Natural History in New York City as anthropological specimens. He also brought back living natives—two men, a woman and three youngsters—and dropped them off for study at the museum; within a year four of them were dead from a strain of influenza to which they had no resistance.)"

193 Plessy v. Ferguson: *Plessy v. Ferguson*, 163 U.S. 537 (1896), accessed at https://supreme.justia.com/cases/federal/us/163/537/#tab-opinion-1917401 on June 28, 2021; and Gates, *Life upon These Shores*, 203.

193 *Lynching rates:* "Bar Graph of Lynchings of African Americans, 1890–1929," American Social History Project, CUNY Graduate Center, accessed at https://herb.ashp.cuny.edu/items/show/1884 on June 28, 2021.

193 *Ida B. Wells-Barnett:* Ida B. Wells-Barnett, *Southern Horrors: Lynch Law in All Its Phases* (New York: The New York Age, 1892), accessed at https://www.gutenberg.org/files/14975/14975-h/14975-h.htm#THE_OFFENSE on March 3, 2020. The quotation that follows is from the same source.

193 *white mobs:* Gates, *Life upon These Shores*, 242.

194 *African Americans organized:* John Hope Franklin and Evelyn Brooks Higginbotham, *From Slavery to Freedom: A History of African Americans* (New York: McGraw Hill, 2011), 274.

194 *pseudoscientific arguments:* Gates, *Stony the Road*, 69.

194 *Some posited that:* Frederick Ludwig Hoffman, *Race Traits and Tendencies of the American Negro* (New York: Publication for the American Economic Association by the Macmillan Company, 1896), 76; and Rudolph Matas, *The Surgical Peculiarities of the American Negro* (Transactions of the American Surgical Association, vol. 14, 1896), 48.

194 *remembered one conversation:* N. B. Dodson, "Afro-Americans Honor Henson," *Statesman*, October 30, 1909, page 5. The quotation that follows is from the same source and page.

195 *secured $75,000:* Peary, *The North Pole*, 13; lecture notes of Matthew Henson, page 3, Beulah M. Davis Special Collections, Morgan State University.

195 *Peary ensured that:* Peary, *The North Pole*, 19.

195 *the SS* Roosevelt*:* "Peary Arctic Club: Expedition of 1908, S. S. Roosevelt, New York, July 6, 1908," PACC_5_150_0025, Archives of the Explorers Club; Henson, *A Negro Explorer at the North Pole*, 15; and Robert E. Peary and R. A. Harris, "Peary Arctic Club Expedition to the North Pole, 1908–9," *Geographical Journal* 36, no. 2 (1910): 129.

195 *Crowds of people:* "Peary Arctic Club: Expedition of 1908"; Henson, *A Negro Explorer at the North Pole*, 15; and Peary and Harris, "Peary Arctic Club Expedition to the North Pole, 1908–9."

195 *up the East River:* "Particulars of the S. S. Roosevelt," PACC_5_150_0027, Archives of the Explorers Club.

195 *"such a din":* "Peary Off Again for North Pole," *Richmond Times Dispatch,* July 7, 1908, page 1.

195 *a white-duck suit:* "President Pleased: Peary's Ship Roosevelt Inspected and Approved by Its Namesake," *Morning Journal-Courier,* July 8, 1908, page 9.

195 *words of encouragement:* "President Pleased," *Morning Journal-Courier.*

195 *sojourn on Long Island:* Henson, *A Negro Explorer at the North Pole,* 16.

195 *approached the coastline:* Henson, *A Negro Explorer at the North Pole,* 21.

196 *the wrong name:* "Peary Off Again for North Pole," *Richmond Times Dispatch.*

196 Evening Star *called him:* "Peary's Coming Voyage," *Evening Star,* June 23, 1908, page 3.

196 *Cape Sheridan:* Henson, *A Negro Explorer at the North Pole,* 34–35.

196 *spent his time:* Henson, *A Negro Explorer at the North Pole,* 38.

196 *two prized tools:* Matthew Henson's pocketknife and saw, Beulah M. Davis Special Collections, Morgan State University.

196 *His diary notes:* Henson, *A Negro Explorer at the North Pole,* 42. The information in the next two sentences comes from the same source, page 39.

196 *believing, he later said:* Lecture notes of Matthew Henson to the Harriet Tubman Community Club, Hempstead, New York, undated, page 2, Beulah M. Davis Special Collections, Morgan State University.

197 *When winter arrived:* Henson, *A Negro Explorer at the North Pole,* 47–48. The information and quotation in this paragraph come from the same source, pages 47–49.

197 *Working alongside him:* Henson, *A Negro Explorer at the North Pole,* 48–52. The information and quotation in this paragraph come from the same source and pages.

197 *spring's approach signaled:* Henson, *A Negro Explorer at the North Pole,* 52–53.

197 *targeted Cape Columbia:* Henson, *A Negro Explorer at the North Pole,* 53.

197 *413 nautical miles:* Henson, *A Negro Explorer at the North Pole,* 69; and Peary, "The Discovery of the North Pole," 773.

197 *In total, seven:* Peary, "The Discovery of the North Pole," 773.

197 *Henson set off:* Henson, *A Negro Explorer at the North Pole,* 55. Unless otherwise noted, the information and quotations in this paragraph come from the same source, pages 53–59.

198 *would fatally lower:* Henson, *A Negro Explorer at the North Pole,* 119–20.

198 *tried to keep warm:* Henson, *A Negro Explorer at the North Pole*, 59, 65, 131; Arctic gloves, in Beulah M. Davis Special Collections, Morgan State University; and Beth Miller, "North Pole to UD," University of Delaware, accessed at https://www.cas.udel.edu/news/Pages/matthew-henson-mittens.aspx on November 28, 2022.

198 *they joyfully reunited:* Henson, *A Negro Explorer at the North Pole*, 61–62.

198 *As dawn broke:* Peary, *The North Pole*, 192; and Henson, *A Negro Explorer at the North Pole*, 77.

198 *some would break:* Peary, "The Discovery of the North Pole," 178, 780.

198 *the North Pole:* Henson, *A Negro Explorer at the North Pole*, 77.

198 *Henson's team led:* Henson, *A Negro Explorer at the North Pole*, 77–78. The information in this paragraph is from the same source, pages 78–102. The quotation is from page 102.

199 *One hundred and thirty-three:* Thomas, "First at the Pole: Lowell Thomas Interviews Matthew Henson"; and Peary, "The Discovery of the North Pole," 169.

199 *two separate teams:* Peary, "The Discovery of the North Pole," 170.

199 *Henson was joined by:* Henson, *A Negro Explorer at the North Pole*, 100, 137–39.

199 *In Henson's eyes:* Henson, *A Negro Explorer at the North Pole*, 137.

199 *Peary's group included:* Peary, "The Discovery of the North Pole," 169; Peary, *The North Pole*, 7, 69; and Henson, *A Negro Explorer at the North Pole*, 127. The quotations that follow in the next paragraph come from Henson, *A Negro Explorer at the North Pole*, 127–29.

199 *"a memory of toil":* Henson, *A Negro Explorer at the North Pole*, 128–29; and Larson, *To the Edges of the Earth*, 207.

200 *relying on the compasses:* Thomas, "First at the Pole: Lowell Thomas Interviews Matthew Henson."

200 *pocket watch:* "Pocket watch likely carried by Matthew Henson in 1908–1909 Arctic expedition," National Museum of African American History and Culture, accessed at https://nmaahc.si.edu/object/nmaahc_2017.31 on June 22, 2021.

200 *As the days passed:* Henson, *A Negro Explorer at the North Pole*, 131–32. The information and quotations in this paragraph are from the same source and pages.

200 *thirty-five degrees Fahrenheit:* "Arctic Ocean Temperature in April," accessed at https://seatemperature.info/april/arctic-ocean-water-temperature.html on November 29, 2022.

200 *Moments from death:* Henson, *A Negro Explorer at the North Pole*, 131–32. The information and quotations in this paragraph are from the same source and pages.

200 *a matter of dispute:* It is unclear which member of the group was the first to arrive at the North Pole. According to Henson, in an interview with Lowell Thomas, Henson claimed it was he who reached the North Pole first because he was the explorer who broke the trail on the lead sledge, reaching the location of their campsite at the North Pole forty-five minutes before Peary, who was incapacitated by exhaustion and riding in a sled. Peary's version of events does not match Henson's, however, making it difficult to discern what exactly transpired. "Matthew (Matt) Alexander Henson," The State House, Annap-

olis, Maryland, November 18, 1961, in Beulah M. Davis Special Collections, Morgan State University; Thomas, "First at the Pole: Lowell Thomas Interviews Matthew Henson"; Henson, *A Negro Explorer at the North Pole*, 132–33; Peary, *The North Pole*, 7; and Peary, "The Discovery of the North Pole," 175.

200   *confirm their position:* Peary, "The Discovery of the North Pole," 175–76; and Peary, *The North Pole*, 291.

200   *Henson was overjoyed:* Henson, *A Negro Explorer at the North Pole*, 133. The two quotations that follow come from the same source and page.

200   *Peary walked among:* Peary, *The North Pole*, 296.

201   *Morris K. Jessup:* Peary, "The Discovery of the North Pole," 175; and "Northward over the Great Ice: Robert E. Peary and the Quest for the North Pole," Peary-MacMillan Arctic Museum, accessed at https://www.bowdoin.edu/arctic-museum/exhibits/2008/northward-over-the-great-ice.html on June 22, 2021.

201   *the return journey:* Peary, "The Discovery of the North Pole," 175. Unless otherwise noted, the information and quotations in this paragraph are from the same source, pages 175–76.

201   *"a huge paleocrystic":* Henson, *A Negro Explorer at the North Pole*, 136.

201   *a quick return:* Peary, "The Discovery of the North Pole," 283–84.

201   *seventeen-day trek:* Henson, *A Negro Explorer at the North Pole*, 141.

201   *"a horrid nightmare":* Henson, *A Negro Explorer at the North Pole*, 141–42.

201   *encountered strong winds:* Peary, "The Discovery of the North Pole," 286.

201   *rode on a sledge:* Henson, *A Negro Explorer at the North Pole*, 140.

201   *reached the camp:* Henson, *A Negro Explorer at the North Pole*, 141; and Peary, *The North Pole*, 7.

201   *Exhaustion overcame them:* Henson, *A Negro Explorer at the North Pole*, 143; and Peary, "The Discovery of the North Pole," 291.

202   *Henson discreetly assessed:* Henson, *A Negro Explorer at the North Pole*, 143.

202   *he saw in the mirror:* Henson was able to access a mirror aboard the *Roosevelt*. Henson, *A Negro Explorer at the North Pole*, 144; and "Events at Pole Told by Henson," *New York Times*, September 16, 1909, page 1–2.

202   *Even the dogs:* Peary, "The Discovery of the North Pole," 291.

202   *more than ninety miles:* Peary, *The North Pole*, 317; and Henson, *A Negro Explorer at the North Pole*, 146–47.

202   *Ross Marvin:* Peary, "The Discovery of the North Pole," 318–19.

202   *enormous chunks of ice:* Henson, *A Negro Explorer at the North Pole*, 160–62.

202   *they finally departed:* Henson, *A Negro Explorer at the North Pole*, 159–61.

202   *in Indian Harbor:* Peary, "The Discovery of the North Pole," 295.

202   *shocking news:* Peary, "The Discovery of the North Pole," 294.

202   *outrageously claimed:* Larson, *To the Edges of the Earth*, 259–60.

203   *"had not been any":* Henson, *A Negro Explorer at the North Pole*, 177–78.

203   *"the gale increased":* Henson, *A Negro Explorer at the North Pole*, 183–84; and Peary, *The North Pole*, 334.

203   *he wired the tale:* "US Census Bureau History: Robert Peary and the Exploration of the North Pole," US Census Bureau, accessed at https://www.census.gov/history/www/homepage_archive/2019/april_2019.html on June 18, 2021.

203   *the Associated Press:* Henson, *A Negro Explorer at the North Pole,* 184.

203   *public relations blitz:* The full details of the Frederick Cook North Pole controversy are beyond the scope of this chapter, but the National Geographic Society argues that "Cook was unable to provide any navigational records of his achievement . . . and [the] rest of his team later reported that they did not quite reach the pole." "North Pole," National Geographic Society; and "Frederick Albert Cook," *Encyclopaedia Britannica,* accessed at https://www.britannica.com/biography/Frederick-Albert-Cook on June 16, 2021. For a full account, see Darrell Hartman, *Battle of Ink and Ice: A Sensational Story of News Barons, North Pole Explorers, and the Making of Modern Media* (New York: Viking, 2023).

203   *dispel critics' doubts:* In 1989, the National Geographic Society, a sponsor of the Peary expedition, completed its analysis of Peary's data and photographs and concluded, in a 230-page report, that Peary, Henson, and the other members of the team reached a point five miles or fewer from the North Pole. The US Census Bureau reports that "today, many researchers studying polar expedition records conclude that Peary may not have stood at the North Pole, but he was probably as close to that point on the Arctic Ocean's ice as the instruments of the day could accurately record." "US Census Bureau History: Robert Peary and the Exploration of the North Pole"; "First People to Reach the North Pole," *Guinness World Records,* accessed at https://www.guinnessworldrecords.com/world-records/first-people-to-reach-the-north-pole on June 29, 2021; and Warren E. Leary, "Peary Made It to the Pole After All, Study Concludes," *New York Times,* December 12, 1989, page 1.

203   *the expedition's success:* Peary, "The Discovery of the North Pole," 295.

204   *his public telegraph:* Peary, "The Discovery of the North Pole," 295.

204   *Peary took much:* "Gives Peary Medal as Pole Discoverer: National Geographic Society Board Satisfied He Reached Farthest North April 6, 1909," *New York Times,* November 4, 1909, page 4.

204   *framed the discovery:* Henson, *A Negro Explorer at the North Pole,* 136. The quotation that immediately follows is from the same source and page.

204   *Henson emphasized how:* Lecture notes of Matthew Henson to the Harriet Tubman Community Club, Hempstead, New York.

205   *the front page:* "Peary Discovers the North Pole After Eight Trials in 23 Years," *New York Times,* September 7, 1909, page 1.

205   *Nine days later:* "Events at the Pole Told by Henson," *New York Times,* September 16, 1909, page 1.

205   *the* Nashville Banner*:* "Peary's Man Friday Helped Him Find Pole," *Tacoma Times,* September 14, 1909, page 1; and "Ham and Japheth at the Pole," *Nashville Globe,* October 8, 1909, page 1.

205   *one journalist:* "Negro at the North Pole: Matthew Henson, Peary's Valet, Proves That Black Man Can Stand Coldest Weather," *New Era,* November 4, 1909, page 3.

205   *another reporter:* "Ham and Japheth at the Pole," *Nashville Globe.* The *Globe* article critiques an editorial about Henson that was published the prior week in the *Nashville Banner.*

206 *Henson's victory showed:* "Ham and Japheth at the Pole," *Nashville Globe*, pages 1, 5.

206 *Peary's own account:* Peary, "The Discovery of the North Pole," 170. The quotations that follow in this paragraph are from the same source and page.

206 *previous Arctic experience:* "Henson Tells the Story of Dash to the North Pole," *Advocate*, September 23, 1909, page 1; and E. L. Blackshear, "A Tribute to Matthew Henson," *Colored American Magazine* 17, no. 5, page 383.

206 *packed audiences:* "Matthew Henson, Peary's Negro, Coming in Style," *Saint Louis Post Dispatch*, December 12, 1909, Parts 4 and 5, page 7.

206 *ultimately fell out:* Thomas, "First at the Pole: Lowell Thomas Interviews Matthew Henson."

207 *She excoriated Peary:* "Says Peary Has Forgotten His Ally: Wife of Henson, Who Accompanied Explorer to Pole, Tells of Neglect," *San Francisco Chronicle*, March 11, 1910.

207 *a maintenance man:* Jay Scriba, "Matthew A. Henson," *Negro Digest*, December 1963, page 21; reprinted by permission from *Milwaukee Journal*.

207 *collector of customs:* Thomas, "First at the Pole: Lowell Thomas Interviews Matthew Henson"; and Dorothy Dougherty, "New York City U.S. Custom House Employee: Matthew Henson," National Archives accessed at https://prologue .blogs.archives.gov/2020/04/06/new-york-city-u-s-custom-house-employee -matthew-henson/ on June 30, 2021.

207 *The exclusive Explorers Club:* Scriba, "Matthew A. Henson."

207 *Expedition Medal:* Gates and West, *The African-American Century*, 15; and Gates, *Life upon These Shores*, 246.

207 *at the White House:* Senators Jones and Parran, Senate Resolution No. 23, March 9, 1961, in "Unveiling and Dedication of Tablet in Memory of Matthew Alexander Henson," The State House, Annapolis, Maryland, November 18, 1961, in Beulah M. Davis Special Collections, Morgan State University.

207 *Woodlawn Cemetery:* "Matt Henson, Who Reached Pole with Peary in 1909, Dies at 88," *New York Times*, March 10, 1955, page 27; and "Matthew Alexander Henson," Arlington National Cemetery, accessed at https://www.arlingtoncemetery .mil/Explore/Notable-Graves/Explorers/Matthew-Henson on June 30, 2021.

207 *Arlington National Cemetery:* "Matthew Alexander Henson," Arlington National Cemetery; and Scriba, "Matthew A. Henson."

9. AMELIA EARHART: THE AVIATOR

208 *Amelia Earhart had begun:* Amelia Earhart, *The Fun of It: Random Records of My Own Flying and of Women in Aviation* (New York: Brewer, Warren, and Putnam, 1932), 214.

208 *Charles Lindbergh completed:* "World Honors Lindy's Flight," *Brownsville Herald*, May 20, 1932, page 1.

208 *bought her own airplanes:* "Biography," *The Family of Amelia Earhart*, accessed at https://www.ameliaearhart.com/biography/ on July 5, 2022; and Debra Michals, "Amelia Earhart," National Women's History Museum, accessed at https://www.womenshistory.org/education-resources/biographies/amelia -earhart on on July 5, 2022.

209 *Frenchman Louis Blériot:* Allan Janus, "Blériot's Cross-Channel Flight," National Air and Space Museum, accessed at https://airandspace.si.edu/stories/editorial/bl%C3%A9riots-cross-channel-flight on July 5, 2022.

209 *hazards of aviation:* "Bleriot's Flyer a Show," *New York Times*, July 26, 1909, page 2; and "Bleriot XI Monoplane," in R. G. Grant, *Flight: The Complete History of Aviation* (New York: DK Publishing, 2017), 42. The information in the paragraph that follows comes from Grant, *Flight*, page 52.

210 *invested more heavily:* Grant, *Flight*, 54, 58–59. The sentence that follow comes from the same source, page 58.

210 *J. A. Armstrong:* Lieutenant J. A. Armstrong, "The Answer," circa 1918, Gilder Lehrman Institute, accessed at https://www.gilderlehrman.org/sites/default/files/inline-pdfs/T-06570_1918_07_29_30.docx.pdf on June 8, 2021.

210 *German Zeppelin L-3:* Richard P. Hallion, *Taking Flight: Inventing the Aerial Age from Antiquity Through the First World War* (New York: Oxford University Press, 2003), 350.

210 *The surprise raid:* "Flashlight in Sky Followed by Bombs," *New York Times*, January 20, 1915, page 2. The information about this raid and the quotation from the *New York Times* are from the same source and page.

210 Times *of London:* "The Air Raid: Evidence at the Inquests," *Times*, January 22, 1915, page 34.

211 *defended the raid:* "'Fortified Yarmouth': German Official Excuses," *Times*, January 22, 1915, page 34.

211 *"The possibilities":* Roald Amundsen, "The Expedition," in Roald Amundsen and Lincoln Ellsworth, *Our Polar Flight* (New York: Dodd, Mead, and Company, 1925), 3–4. The quotation that follows is from the same source and page.

211 *Ellsworth even remembered:* Lincoln Ellsworth, "The Amundsen-Ellsworth Polar Flight," in Amundsen and Ellsworth, *Our Polar Flight*, 104.

211 *Dornier Wal aircraft:* Amundsen, "The Expedition," in Amundsen and Ellsworth, *Our Polar Flight*, 145.

211 *Amundsen remembered:* Ellsworth, "The Amundsen-Ellsworth Polar Flight," in Amundsen and Ellsworth, *Our Polar Flight*, 107. The information that follows in this paragraph is from the same source, pages 108–10, and the quotation is from page 110.

212 *President Calvin Coolidge:* Photograph, "President Coolidge today conferred the Congressional medal on Commander Byrd and Machinist Bennet," February 19, 1927, Library of Congress, accessed at https://www.loc.gov/resource/cph.3c31298/ on July 5, 2022.

212 *a failed endeavor:* "Richard Byrd and Floyd Bennett Receive Medal of Honor," *Encyclopedia Virginia*, Virginia Humanities, accessed at https://encyclopediavirginia.org/673hpr-e744b949eb785fc on June 9, 2022.

212 *approached the North Pole:* "11–14 May 1926," This Day in Aviation, accessed at https://www.thisdayinaviation.com/11-1926/ on June 8, 2022.

212 *they recorded:* Roald Amundsen and Lincoln Ellsworth, *First Crossing of the Polar Sea* (Garden City: Doubleday, Doran & Company, 1928), 142.

212 *Hjalmar Riiser-Larsen:* Amundsen and Ellsworth, *First Crossing of the Polar Sea*, 141.

212 *Looking down from:* Amundsen and Ellsworth, *First Crossing of the Polar Sea,* 141–42; and "The *Norge*'s Success," *Western Morning News,* May 13, 1926, page 5.

213 *Riiser-Larsen later wrote:* Amundsen and Ellsworth, *First Crossing of the Polar Sea,* 142.

213 *immediate international acclaim:* "First aircraft flight over the North Pole," Guinness World Records, accessed at https://www.guinnessworldrecords.com /world-records/first-aircraft-flight-over-the-north-pole on June 8, 2022.

213 *Coolidge expressed his:* Rosanne Butler, "'It Was a Great Occasion': Calvin Coolidge Comes to Williamsburg," Calvin Coolidge Presidential Foundation, accessed at https://coolidgefoundation.org/resources/essays-papers-addresses-10/ on June 8, 2022.

213 *In Rome:* "Norge I. Safe," *Daily Mail,* May 17, 1926, page 8.

213 *the contemporary theory:* "Amundsen," *Sydney Morning Herald,* May 13, 1926, page 9.

213 *first transatlantic flight:* Nola Taylor Tillman, "Charles Lindbergh and the First Solo Transatlantic Flight," June 11, 2019, Space.com, accessed at https://www .space.com/16677-charles-lindbergh.html on June 13, 2022.

213 *charming neo-Gothic residence:* Site visit, Amelia Earhart's house, Atchison, Kansas, on July 25, 2021.

214 *bold Kaw Indians:* William Clark, September 14, 1806, in *Journals of Lewis and Clark Expedition,* accessed at https://lewisandclarkjournals.unl.edu/item /lc.jrn.1806-09-14#ln40091401 on June 13, 2022.

214 *but, she lamented:* Earhart, *The Fun of It,* 5.

214 *by Earhart's day:* Earhart, *The Fun of It,* 5.

214 *From age three:* Susan Butler, *East to the Dawn: The Life of Amelia Earhart* (Reading: Addison-Wesley, 1997), 29; and "Amelia Earhart," *Kansapedia,* Kansas Historical Society, accessed at https://www.kshs.org/kansapedia /amelia-earhart/12041 on June 14, 2022.

214 *remained in Kansas City:* Butler, *East to the Dawn,* 30.

214 *"exceedingly fond":* Earhart, *The Fun of It,* 5. The information in the sentence that follows is from the same source and page.

214 *In grade school:* Earhart, *The Fun of It,* 8. The quotation that immediately follows is from the same source and page.

214 *With her school friends:* Earhart, *The Fun of It,* 14.

215 *Mark Twain's character:* Mark Twain (Samuel Clemens), *The Adventures of Tom Sawyer* (Hartford: The American Publishing Company, 1884), chapter 29, accessed at https://www.gutenberg.org/files/74/74-h/74-h.htm#c31 on June 14, 2022.

215 *relocated with their parents:* Mary S. Lovell, *The Sound of Wings: The Life of Amelia Earhart* (New York: St. Martin's Press, 1989), 13–15.

215 *forced to move:* Lovell, *The Sound of Wings,* 16–17.

215 *earned high marks:* William J. Scanlan to Jack Pitman, January 18, 1956, Minnesota Historical Society, accessed at https://www.mnhs.org/blog/collections upclose/amelia-earhart-found-in-st-paul on June 15, 2022.

215   *Amelia finished coursework:* Lovell, *The Sound of Wings*, 19–23.

215   *the Ogontz School:* Earhart, *The Fun of It*, 19; and Lovell, *The Sound of Wings*, 23.

215   *"men without arms":* Earhart, *The Fun of It*, 19.

215   *Spadina Military:* Earhart, *The Fun of It*, 19–20; and Lovell, *The Sound of Wings*, 26.

216   *Laboring for ten hours:* Earhart, *The Fun of It*, 19–20.

216   *she was a witness:* Amelia Earhart, *20 Hrs., 40 Min.* (New York: G. P. Putnam's Sons, 1928), 34.

216   *Canada's Royal Flying Corps:* Glenn B. Foulds and Jonathan Scotland, "Royal Flying Corps," *The Canadian Encyclopedia*, accessed at https://www.thecanadian encyclopedia.ca/en/article/royal-flying-corps on July 5, 2022.

216   *found herself drawn:* Earhart, *The Fun of It*, 20.

216   *Alighting on the airfields:* Earhart, *20 Hrs., 40 Min.*, 37.

216   *"no civilian had":* Earhart, *The Fun of It*, 20.

216   *One powerful sensation:* Earhart, *The Fun of It*, 20.

216   *she listened eagerly:* Neta Snook Southern, *I Taught Amelia to Fly* (New York: Vantage Press, 1974), 102; and Lovell, *The Sound of Wings*, 27.

216   *did not yet contemplate:* Earhart, *20 Hrs., 40 Min.*, 39.

216   *women were banned:* Claudia M. Oakes, "United States Women in Aviation Through World War I," *Smithsonian Studies in Air and Space* 2 (Washington, DC: Smithsonian Press, 1978), 17.

216   *Columbia University:* Earhart, *The Fun of It*, 23.

216   *asked her to move:* Lovell, *The Sound of Wings*, 30–31.

217   *factories churned out:* Andrew Rolle and Arthur C. Verge, *California: A History*, 8th ed. (West Sussex, UK: Wiley-Blackwell, 2014), 234.

217   *Glenn Martin Company:* Otto Pastron, "How Much Did California Contribute to Aviation and Naval Assets During WW1?" The United States World War I Centennial Commission, accessed at https://www.worldwar1 centennial.org/index.php/california-in-ww1-then/3979-how-much-did -california-contribute-to-aviation-naval-assets-during-ww1.html on June 16, 2022. The information in the sentence that follows is from the same source and page.

217   *rebuild their marriage:* Butler, *East to the Dawn*, 94; and Lovell, *The Sound of Wings*, 31.

217   *"particularly active":* Earhart, *20 Hrs., 40 Min.*, 44.

217   *a regular attendee:* Earhart, *The Fun of It*, 24, and Butler, *East to the Dawn*, 94.

217   *magnetic pull of flight:* Earhart, *The Fun of It*, 24.

217   *Edwin agreed:* Lovell, *The Sound of Wings*, 32–33.

217   *pilot Frank Hawks:* Earhart, *The Fun of It*, 24–25; and Lovell, *The Sound of Wings*, 32–33. Unless otherwise noted, the quotations and information that follow in this paragraph and the next come from Earhart, *The Fun of It*, 25.

218   *he could not pay:* Earhart, *The Fun of It*, 25; and Lovell, *The Sound of Wings*, 33.

218   *a telephone company:* Earhart, *The Fun of It*, 25.

218   *learning how to fly:* Earhart, *The Fun of It*, 26.

218   *signaled a new trend:* "Women's Bureau: History," US Department of Labor, accessed at https://www.dol.gov/agencies/wb/about/history on June 20, 2022.

218   *Nineteenth Amendment:* "US Constitution: Nineteenth Amendment," *Constitution Annotated,* accessed at https://constitution.congress.gov/constitution/amendment-19/ on June 21, 2022.

218   *she began wearing:* Earhart, *The Fun of It*, 26. The quotations that follow in this paragraph come from the same source and page.

219   *notion of feminism:* Susan Ware, *Still Missing: Amelia Earhart and the Search for Modern Feminism* (New York: W. W. Norton and Company, 1993), 118–19.

219   *Anita Snook:* Earhart, *The Fun of It*, 28; and Lovell, *The Sound of Wings*, 33.

219   *pioneering female aviators:* Oakes, "United States Women in Aviation," 22, 26, 28.

219   *an impressive list:* "Neta Snook Southern," Ames History Museum, accessed at https://ameshistory.org/content/neta-snook on June 21, 2022.

219   *she taught Earhart:* Amelia Earhart to Neta Snook Southern, January 26, 1929, Gilder Lehrman Collection, accessed at https://www.gilderlehrman.org/sites/default/files/content-images/07243.001.jpg on June 21, 2022; and Southern, *I Taught Amelia to Fly*, 102, 122.

219   *a Curtiss Canuck:* Earhart, *The Fun of It*, 28.

219   *Earhart's admitted habit:* Southern, *I Taught Amelia to Fly*, 2, 122. The quotation that follows come from the same source, page 124.

219   *sunny yellow Kinner:* Lovell, *The Sound of Wings*, 39–40.

220   *gaining confidence:* Lovell, *The Sound of Wings*, 39.

220   *at national parks:* Earhart, *The Fun of It*, 49–51; and Lovell, *The Sound of Wings*, 47.

220   *"As thrilling to me":* Earhart, *The Fun of It*, 51.

220   *Transcontinental Airway:* "Transcontinental Service," Smithsonian National Postal Museum, accessed at https://postalmuseum.si.edu/exhibition/airmail-in-america-us-aerial-mail-service-1918%E2%80%931926/transcontinental-service on July 6, 2022.

220   *Katharine Lee Bates's:* Transcontinental Airmail Map, July 1, 1924, Post Office, New York, NY, Classification Section, accessed at https://upload.wikimedia.org/wikipedia/commons/7/75/Transcontinental_Air_Mail_Map_1924.jpg on June 21, 2022; and Katharine Lee Bates, "America the Beautiful," 1893, Gilder Lehrman Collection, accessed at https://www.gilderlehrman.org/history-resources/spotlight-primary-source/america-beautiful-1893 on June 21, 2022.

220   *"It is such things":* Earhart, *The Fun of It*, 51.

220   *earn a living:* Lovell, *The Sound of Wings*, 49.

220   *Denison House:* Earhart, *20 Hrs., 40 Min.*, 97; Lovell, *The Sound of Wings*, 49–52; and Keith O'Brien, *Fly Girls: How Five Daring Women Defied All Odds and Made Aviation History* (New York: Houghton Mifflin Harcourt, 2018), 17–18.

220   *In her spare time:* Earhart, *20 Hrs., 40 Min.*, 90–91.

220   *unexpected phone call:* Earhart, *The Fun of It*, 58; and Lovell, *The Sound of Wings*, 52.

221   *"Hello. You don't know":* Earhart, *The Fun of It*, 59. The information that follows in this paragraph comes from the same source and page.

221   *spark of interest:* Hilton Howell Railey, *Touch'd with Madness* (New York: Carrick and Evans, 1938), 103.

221   *Two men were preparing:* Earhart, *The Fun of It*, 60.

221   *Amy Phipps Guest:* Earhart, *The Fun of It*, 60.

221   *her family refused:* Lovell, *The Sound of Wings*, 98.

221   *Earhart wanted to fly:* Earhart, *The Fun of It*, 61.

221   *"Despite my intentions":* Earhart, *The Fun of It*, 61.

221   *During the weeks:* Earhart, *The Fun of It*, 65. The information that follows in this paragraph comes from the same source, page 63–65.

222   *wrote her will:* Lovell, *The Sound of Wings*, 106.

222   *an exploratory excursion:* Earhart, *20 Hrs., 40 Min.*, 108. The information and quotation that follow in this paragraph come from the same source and page.

222   *Earhart donned clothes:* Earhart, *20 Hrs., 40 Min.*, 111–12.

222   *her determined gray eyes:* Kathleen C. Winters, *Amelia Earhart: The Turbulent Life of an American Icon* (New York: Palgrave MacMillan, 2010), 50; and Earhart, *20 Hrs., 40 Min.*, 112–13.

222   *she chose to bring:* Joseph A. Mussulman, "Army Hygiene," *Discover Lewis and Clark*, accessed at https://lewis-clark.org/a-military-corps/army-regulations/army-hygiene/ on June 23, 2022; "The Uniforms and Equipment of the Lewis and Clark of the Expedition," *Corps of Discovery*, The US Army Center of Military History, accessed at https://history.army.mil/lc/the%20mission/facts/uniforms.htm on June 23, 2022; and Earhart, *20 Hrs., 40 Min.*, 115.

222   *Richard Byrd's book:* Earhart, *20 Hrs., 40 Min.*, 115–16.

222   *"I am sending":* Earhart, *20 Hrs., 40 Min.*, 116.

222   *took off from Boston:* Earhart, *20 Hrs., 40 Min.*, 119. The information and quotations that follow in this paragraph come from the same source, pages 129–30.

223   *their view throughout:* Earhart, *20 Hrs., 40 Min.*, 184.

223   *that violently shook:* Earhart, *20 Hrs., 40 Min.*, 176.

223   *three oranges:* Earhart, *20 Hrs., 40 Min.*, 179.

223   *quickly running out:* Earhart, *20 Hrs., 40 Min.*, 196.

223   *from the gloom:* Earhart, *20 Hrs., 40 Min.*, 196–98. The quotations that follow in this paragraph come from the same source and pages.

224   *The lead story:* "Miss Earhart's Plane Comes Down in Wales: Tanks Almost Empty After Stormy Passage," *Boston Globe*, June 19, 1928, page 1.

224   *In England:* "First Woman to Succeed in Atlantic Flight," *Guardian*, June 19, 1928, page 11.

224   *After traveling to London:* Lovell, *The Sound of Wings*, 126.

224   *Earhart's publicist:* Lovell, *The Sound of Wings*, 130–33.

224   *to file for divorce:* Lovell, *The Sound of Wings*, 153–54; and "Putnam, George Palmer," Special Collections Library, Purdue University, accessed at https://archives.lib.purdue.edu/agents/people/1288 on July 6, 2022.

224 *the passage of time:* Cutting in Amelia Earhart's scrapbook from 1932, Special Collections Library, Purdue University, likely from *Illustrated Love Magazine*, January 1932, 25–27, cited in Lovell, *The Sound of Wings*, 154.

224 *proposing to her:* Doris L. Rich, *Amelia Earhart: A Biography* (Washington, DC: The Smithsonian Institution, 1989), 112; and Amelia Earhart to George Putnam, February 7, 1931, Special Collections, Purdue University Archives, accessed at https://earchives.lib.purdue.edu/digital/collection/earhart/id/2988 on June 27, 2022.

224 *when she wed Putnam:* O'Brien, *Fly Girls*, 125.

224 *a letter to Putnam:* Amelia Earhart to George Putnam, February 7, 1931, Special Collections, Purdue University Archives, accessed at https://earchives.lib .purdue.edu/digital/collection/earhart/id/2988 on June 27, 2022.

225 *Amelia asked:* George Palmer Putnam, *Soaring Wings: A Biography of Amelia Earhart* (New York: Harcourt, Brace and Company, 1939), 99. The quotation that follows in this paragraph comes from the same source and page.

225 *Putnam later reflected:* Putnam, *Soaring Wings*, 98.

225 *in her autobiography:* Earhart, *The Fun of It*, 210. The quotations that follow in this paragraph come from the same source and page.

225 *transcontinental air derby:* Rich, *Amelia Earhart*, 89, 100, 103.

225 *faced growing competition:* Rich, *Amelia Earhart*, 131; Lovell, *The Sound of Wings*, 175; and O'Brien, *Fly Girls*, 153.

226 *who needed assistance:* Lovell, *The Sound of Wings*, 160–61.

226 *Six had died:* Rich, *Amelia Earhart*, 129.

226 *Bernt Balchen:* Putnam, *Soaring Wings*, 100; and Lovell, *The Sound of Wings*, 176.

226 *Balchen oversaw improvements:* Lovell, *The Sound of Wings*, 175–77.

226 *repaired its fuselage:* Earhart, *The Fun of It*, 210–11; "Amelia Earhart Lands in Canada on Sea Hop," *Daily News*, May 20, 1932, page 3; "Woman Ace Plans Solo Ocean Hop," *Daily Times*, May 20, 1932, page 1.

226 *studied weather reports:* Earhart, *The Fun of It*, 212–13.

226 *For his part:* Lovell, *The Sound of Wings*, 176.

227 *Earhart avoided generating:* Putnam, *Soaring Wings*, 102. The quotation that follows in this paragraph comes from the same source and page.

227 *Earhart conferred:* Putnam, *Soaring Wings*, 104; and Rich, *Amelia Earhart*, 132.

227 *The weather report:* Earhart, *The Fun of It*, 212–13. The information and quotation that follow in this paragraph come from the same source and pages.

227 *bid him farewell:* O'Brien, *Fly Girls*, 156.

227 *asked him privately:* Bernt Balchen's diary entry, cited in Lovell, *The Sound of Wings*, 180.

227 *five years to the day:* Earhart, *The Fun of It*, 214. The information and quotation that follow in this paragraph come from the same source and page.

228 *Not long after:* Earhart, *The Fun of It*, 214–15. The information and quotations that follow in this paragraph come from the same source and pages.

228 *had to safely descend:* Earhart, *The Fun of It*, 215.

228   *the exhaust manifold:* Earhart, *The Fun of It*, 216–17.

228   *to keep flying:* Earhart, *The Fun of It*, 216.

229   *"should come down":* Earhart, *The Fun of It*, 217. The information and quotations that follow in this paragraph come from the same source, page 218.

229   *2,026-mile flight:* "First Woman to Fly the Atlantic," *Guardian*, May 22, 1932, accessed at https://www.theguardian.com/theobserver/1932/may/22/life1.life magazine on April 7, 2023; Jean L. Backus, *Letters from Amelia: An Intimate Portrait of Amelia Earhart* (Boston: Beacon Press, 1982), 126; and Lovell, *The Sound of Wings*, 184–85.

229   *sent a cablegram:* "Hoover Voices Nation's Pride; King Likely to Honor Flier," *New York Times*, May 22, 1932, page 1. The quotation that follows in this paragraph comes from the same source and page.

230   *received many awards:* Photograph, "Harriet Chalmers Adams Awards Amelia Earhart the Society's Gold Medal," Earhart, Box I:II, RSWG, accessed at https://worldhistory.columbia.edu/content/transatlantic-rebelles-mapping -first-decade-society-woman-geographers on July 6, 2022.

230   *smashed additional records:* "Earhart Solos the Pacific," National Air and Space Museum, accessed at https://pioneersofflight.si.edu/content/earhart-solos -pacific on July 6, 2022; and "Earhart Was the First Person to Fly Solo from Los Angeles via Mexico City to Newark," Smithsonian National Postal Museum, accessed at https://postalmuseum.si.edu/earhart-was-the-first -person-to-fly-solo-from-los-angeles-via-mexico-city-to-newark on July 6, 2022.

230   *departed from Oakland:* Lovell, *The Sound of Wings*, 240.

230   *to complete their trip:* "Amelia's Route on Attempted World Flight," *Honolulu Star-Bulletin*, July 2, 1937, page 1.

230   *reached a new country:* "Amelia's Route on Attempted World Flight," *Honolulu Star-Bulletin*.

230   *Lae, New Guinea:* "More About Earhart," *Miami News*, June 1, 1937, page 6.

230   *successfully reached Lae:* "Amelia's Route on Attempted World Flight," *Honolulu Star-Bulletin*.

230   *newspaper ominously warned:* "Earhart Is Facing Hazardous Flight," *Kingsport Times*, June 30, 1937, page 7.

231   *was expected to take:* Lovell, *The Sound of Wings*, 272–77.

231   *The skies were calm:* Lovell, *The Sound of Wings*, 277.

231   *briefly broke through:* Lovell, *The Sound of Wings*, 281.

231   *a panicked message:* "Radio, transcripts, Earhart flight," page 43, George Palmer Putnam Collection of Amelia Earhart Papers, Purdue University Libraries, Archives and Special Collections, accessed at https://earchives.lib.purdue.edu /digital/collection/earhart/id/3054/rec/2172 on April 7, 2023.

231   *her last message:* "Radio, transcripts, Earhart flight," page 43, George Palmer Putnam Collection of Amelia Earhart Papers.

231   *the Associated Press:* "Last Word from Plane Indicates Gas Nearly Gone, No Land in Sight," *Honolulu Star-Bulletin*, June 2, 1937, page 1.

231 *last received radio transmission:* "Radio, transcripts, Earhart flight," page 43, George Palmer Putnam Collection of Amelia Earhart Papers.

231 the *New York Times*: "Amelia Earhart," *New York Times*, July 20, 1937, page 22.

10. SALLY RIDE: THE ASTRONAUT

232 *workers busily prepared:* Lynn Sherr, *Sally Ride: America's First Woman in Space* (New York: Simon & Schuster, 2014), 154; and "Space Shuttle Challenger–STS-7 Launch," accessed at https://www.youtube.com/watch?v=DjtXJpdxeew on July 13, 2022.

232 *a white RV:* Cliff Lethbridge, "Launch Complex 39 Fact Sheet," *Spaceline*, accessed at https://www.spaceline.org/cape-canaveral-launch-sites/launch-complex-39-fact-sheet/ on July 14, 2022; and Sherr, *Sally Ride*, 155.

233 *smiled and chatted:* "Space Shuttle Challenger–STS-7 Launch," accessed at https://www.youtube.com/watch?v=DjtXJpdxeew on July 13, 2022.

233 *Howard Hughes:* Roger D. Greene, "Hughes Makes Mother Earth Just a Little Round Apple," *Morning Call*, July 15, 1938, page 12; and photograph of the landing of Howard Hughes's Lockheed 14 aircraft, New York, July 14, 1938, Howard Hughes Public Relations Photograph Collection, 1930–50, PH-00373, Special Collections, University Libraries, University of Nevada, Las Vegas, Nevada, accessed at https://special.library.unlv.edu/ark%3A/62930/d1db7vx42 on November 25, 2023.

233 *Kirke Simpson:* Kirke L. Simpson, "'Round-the-World Flight Raises New Defense Problems," *Morning Call*, July 15, 1938, page 12. The quotation that immediately follows is from the same source and page.

234 *murdering his opponents:* Doug Bandow, "Stalin Died 60 Years Too Late," April 1, 2013, Cato Institute, accessed at https://www.cato.org/commentary/stalin-died-60-years-too-late on July 25, 2022.

234 *fighter planes:* "Invasion of Poland, Fall 1939," *Holocaust Encyclopedia*, accessed at https://encyclopedia.ushmm.org/content/en/article/invasion-of-poland-fall-1939 on July 15, 2022.

234 *a new word:* blitzkrieg: "Invasion of Poland, Fall 1939," *Holocaust Encyclopedia*; and "Luftwaffe," *Encyclopaedia Britannica*, accessed at https://www.britannica.com/topic/Luftwaffe on July 15, 2022.

234 *Floyd Bennett Field:* "Photograph of the landing of Howard Hughes's Lockheed 14 aircraft," New York, July 14, 1938.

234 *seventy-eight million:* Iris Kesternich, Bettina Siflinger, James P. Smith, and Joachim K. Winter, "The Effects of World War II on Economic and Health Outcomes Across Europe," *Review of Economics and Statistics* 96, no. 1 (2014): 103–18, accessed at https://www.ncbi.nlm.nih.gov/pmc/articles/PMC4025972/ on August 19, 2022.

234 *Japanese dive-bombers:* Raymond R. Panko, "Nakajima B5N2 'Kate' Type 97–3 Carrier Attack Aircraft at Pearl Harbor," Pearl Harbor Aviation Museum, August 18, 2017, accessed at https://www.pearlharboraviationmuseum.org/blog/nakajima-b5n2/ on July 15, 2022; and "Attack on Pearl Harbor," *Encyclopaedia Britannica*, accessed at https://cdn.britannica.com/01/192801

-050-99E264FA/World-War-II-Japanese-attack-Pearl-Harbor-Oahu-Hawaii
-December-7-1941.jpg on July 15, 2022.

234 *his somber address:* Franklin D. Roosevelt, Speech by Franklin D. Roosevelt, New York (transcript), December 8, 1941, Library of Congress, accessed at www.loc.gov/item/afccal000483/ on July 15, 2022.

235 *War Production Board:* David Vergun, "During WWII, Industries Transitioned from Peacetime to Wartime Production," Department of Defense, March 27, 2020, accessed at https://www.defense.gov/News/Feature-Stories/story/Article/2128446/during-wwii-industries-transitioned-from-peacetime-to-wartime-production/ on July 15, 2022.

235 *ninety-six thousand aircraft:* "Aerospace Industry," *Encyclopaedia Britannica,* accessed at https://www.britannica.com/technology/aerospace-industry/World-War-II on August 19, 2022; and site visit, the National World War II Museum, New Orleans, November 20, 2021.

235 *Rose Widmer:* Rose Widmer (1923–2016) was the great-aunt of the author.

235 *The first was rocketry:* Douglas Brinkley, *American Moonshot: John F. Kennedy and the Great Space Race* (New York: HarperCollins, 2019), 8–9.

235 *deadly ballistic missile:* Brinkley, *American Moonshot,* 67.

235 *with German scientists:* Some of these German scientists had been members of the Nazi Party in Germany. Lev Golinkin, "Why Do Stanford, Harvard, and NASA Still Honor a Nazi Past?" *New York Times,* December 13, 2022; Brinkley, *American Moonshot,* 74; and "V-2 Rocket," *Encyclopaedia Britannica,* accessed at https://www.britannica.com/technology/V-2-rocket on July 19, 2022.

235 *carry the fuel and oxygen:* Michael Greshko, "Rockets and Rocket Launches, Explained," *National Geographic,* January 4, 2019, accessed at https://www.nationalgeographic.com/science/article/rockets-and-rocket-launches-explained on July 19, 2022.

235 *first atomic bomb:* Brinkley, *American Moonshot,* 77.

235 *a B-29 bomber:* Brinkley, *American Moonshot,* 77.

236 *246,000 people:* Dan Listwa, "Hiroshima and Nagasaki: The Long-Term Health Effects," Columbia Center for Nuclear Studies, accessed at https://k1project.columbia.edu/news/hiroshima-and-nagasaki on July 19, 2022.

236 *When Germany surrendered:* Michael Schaller, Janette Thomas Greenwood, Andrew Kirk, Sarah J. Purcell, Aaron Sheehan-Dean, and Christina Snyder, *American Horizons: US History in a Global Context,* Volume 2, 4th ed. (New York: Oxford University Press, 2021), 865–66.

236 *President Harry Truman:* President Harry S. Truman's Address before a Joint Session of Congress, March 12, 1947, The Avalon Project, Yale Law School, accessed at https://avalon.law.yale.edu/20th_century/trudoc.asp on July 19, 2022; and Schaller, Greenwood, Kirk, et al., *American Horizons,* 870.

236 *$400 million:* Schaller, Greenwood, Kirk, et al. *American Horizons,* 873; and "Truman Doctrine," *Encyclopaedia Britannica,* accessed at https://www.britannica.com/event/Truman-Doctrine on August 25, 2022.

236 *the USSR detonated:* Irek Sabitov and Bill Streifer, "The Shock of 'First Lightning': An Intelligence Failure?" *American Intelligence Journal* 31, no. 1 (2013): 54.

236 *In May 1951:* Sherr, *Sally Ride,* 6.

237 *trained by the Soviet Union:* Mark O'Neill, "Soviet Involvement in the Korean War: A New View from the Soviet-Era Archives," *OAH Magazine of History* 14, no. 3, (2000): 20–21.

237 *the suburbs of Los Angeles:* Sherr, *Sally Ride*, 6. Unless otherwise noted, the information that follows in this paragraph comes from the same source, page 10.

237 *but Dale emphasized:* Undated interview with Sally Ride, quoted in "Sally Ride (1951–2012)," NASA, accessed at https://solarsystem.nasa.gov/people/1760 /sally-ride-1951-2012/ on July 20, 2022.

237 *she felt free:* Michael Ryan, "A Ride in Space," interview with Sally Ride, *People*, June 20, 1983, accessed at https://people.com/archive/cover-story-a-ride-in -space-vol-19-no-24/ on August 10, 2022.

237 *Lynn Sherr:* Sherr, *Sally Ride*, 12. The information that follows in this paragraph comes from the same source and page.

237 *Dunlop Maxply racket:* "Tennis Racquet, Sally Ride," National Air and Space Museum Collection, accessed at https://airandspace.si.edu/collection-objects /tennis-racquet-sally-ride/nasm_A20140271000 on August 2, 2022; and Sherr, *Sally Ride*, 21–23.

237 *Her sister recounted:* Ryan, "A Ride in Space."

238 *she promptly retorted:* Sherr, *Sally Ride*, 20.

238 Scientific American: Susan Okie, "Sally Ride Remains an Elusive Character— Even to a Close Friend," *Baltimore Sun*, May 8, 1983, page E8.

238 *Bushnell Sky Rover:* "Telescope, Bushnell Sky Rover, Sally Ride," National Air and Space Museum Collection, accessed at https://airandspace.si.edu/collection -objects/telescope-bushnell-sky-rover-sally-ride/nasm_A20140273000 on August 2, 2022; and Sherr, *Sally Ride*, 19.

238 *glowing planets:* Sherr, *Sally Ride*, 19.

238 *realized her passion:* Okie, "Sally Ride Remains an Elusive Character"; and Sherr, *Sally Ride*, 22–23.

238 *Susan Okie:* Okie, "Sally Ride Remains an Elusive Character."

238 *At Westlake:* Sherr, *Sally Ride*, 25–26.

238 *Okie remembered:* Okie, "Sally Ride Remains an Elusive Character."

238 *Elizabeth Mommaerts:* Okie, "Sally Ride Remains an Elusive Character."

238 *had unsuccessfully revolted:* Paul Lendvai, *One Day That Shook the Communist World: The 1956 Hungarian Uprising and Its Legacy* (Princeton: Princeton University Press, 2008), 208.

239 *major in astrophysics:* Sherr, *Sally Ride*, 27.

239 *launch of* Sputnik: "*Sputnik* and the Dawn of the Space Age," NASA, accessed at https://history.nasa.gov/sputnik.html on August 22, 2022.

239 *spread the news:* "Pervyi v mire iskusstvennyi sputnik Zemli sozdan v Sovetskoi strane!" and "Triumf sovetskoi nauki i tekhniki," *Pravda*, October 6, 1957, page 1.

239 *"American prestige":* "Reaction to the Soviet Satellite: A Preliminary Evaluation," White House Office of the Staff Research Group, Box 35, Special Projects: Sputnik, Missiles, and Related Matters; NAID #12082706, Eisenhower

Library, accessed at https://www.eisenhowerlibrary.gov/research/online
-documents/sputnik-and-space-race on July 26, 2022.

239 *To make matters worse:* "Space Race Timeline," Royal Museums Greenwich, accessed at https://www.rmg.couk/stories/topics/space-race-timeline on August 22, 2022.

239 *Redstone missile:* Ernst Stuhlinger, "The Story of Explorer 1," Wernher von Braun Memorial Lecture, National Air and Space Museum, Smithsonian Institution, January 31, 1978, 20, accessed at https://www.nasa.gov/sites/default /files/atoms/files/story_of_explorer_1_stuhlinger.pdf on August 1, 2022; and "Explorer 1 Overview," NASA, accessed at https://www.nasa.gov/mission _pages/explorer/explorer-overview.html on August 1, 2022. The quotation that follows is from the same source.

239 *a minor victory:* "US Satellite Now in Orbit," *Logan Daily News,* February 1, 1958, page 1.

240 *Congress passed a law:* Steven J. Dick, ed., *NASA'S First 50 Years: Historical Perspectives* (Washington, DC: US Government Printing Office, 2009), xiii.

240 *Eisenhower publicly expressed:* Dwight Eisenhower, "Statement by the President upon Signing the National Aeronautics and Space Act of 1958," The American Presidency Project, accessed at https://www.presidency.ucsb.edu/documents /statement-the-president-upon-signing-the-national-aeronautics-and-space -act-1958 on July 27, 2022.

240 *After* Sputnik*'s launch:* Statement by the National Science Board in response to Russian satellite, October 1957, [DDE's Records as President, Official File, Box 625, OF 146-F-1 Outer Space, Soviet Satellites-Sputnik; NAID #12060499], Eisenhower Library, accessed at https://www.eisenhowerlibrary .gov/sites/default/files/research/online-documents/sputnik/10-1957-statement .pdf on July 26, 2022.

240 *double federal funding:* Public Law 85–864, accessed at https://www.govinfo .gov/content/pkg/STATUTE-72/pdf/STATUTE-72-Pg1580.pdf on July 26, 2022; and Roger L. Geiger, "What Happened After Sputnik? Shaping University Research in the United States," *Minerva* 35, no. 4 (1997): 355.

240 *Congress's panicked call:* Elton C. Fay, "H-Bombing by Satellite Held Future Peril," and "News Summary," *Los Angeles Times,* April 13, 1961, page 2.

240 *Overton Brooks:* "Congressmen Call for an All-Out Space Program," *Los Angeles Times,* April 13, 1961, page 2.

241 *John F. Kennedy:* "The New Frontier," acceptance speech of Senator John F. Kennedy, Democratic National Convention, July 15, 1960, page 6, John F. Kennedy Presidential Library and Museum, accessed at https://www.jfk library.org/asset-viewer/archives/JFKSEN/0910/JFKSEN-0910-015 on July 28, 2022. The quotation that immediately follows is from the same source, page 5.

241 *address to Congress:* John F. Kennedy, Address to Joint Session of Congress, May 25, 1961, John F. Kennedy Presidential Library and Museum, accessed at https://www.jfklibrary.org/learn/about-jfk/historic-speeches/address-to -joint-session-of-congress-may-25-1961 on July 28, 2022. The quotations and information that immediately follow are from the same source.

241 *Congress ultimately approved:* Kennedy, Address to Joint Session of Congress,

May 25, 1961; and "$5.3 Billion Authorized for NASA; Moon Race Criticized," *Congressional Quarterly Almanac*, July 25, 1963, accessed at https://library.cqpress.com/cqalmanac/document.php?id=cqal63-1317418 on August 23, 2022.

242   *all but disqualified:* Sherr, *Sally Ride*, 67.

242   *A small group:* Amy Foster, "The Gendered Anniversary: The Story of America's Women Astronauts," *Florida Historical Quarterly* 87, no. 2 (2008): 153; and Margaret A. Weitekamp, "Lovelace's Woman in Space Program," NASA, accessed at https://history.nasa.gov/flats.html on July 25, 2023.

242   *W. Randolph Lovelace:* Foster, "The Gendered Anniversary," 151; and Margaret Weitekamp, "NASA's Early Stand on Women Astronauts: 'No Present Plans to Include Women on Space Flights,'" National Air and Space Museum, March 17, 2016, accessed at https://airandspace.si.edu/stories/editorial/nasa%E2%80%99s-early-stand-women-astronauts-%E2%80%9Cno-present-plans-include-women-space-flights%E2%80%9D on August 23, 2022.

242   *decided to terminate the trial:* Mary Robinette Kowal, "To Make It to the Moon, Women Have to Escape Earth's Gender Bias," *New York Times*, July 17, 2019, accessed at https://www.nytimes.com/2019/07/17/science/women-astronauts-nasa.html on April 10, 2023.

242   *now infamous letter:* Weitekamp, "NASA's Early Stand on Women Astronauts."

242   Pravda *crowed that:* "Novyi triumf v osvoennii kosmosa!" *Pravda*, June 17, 1963, page 1.

243   *Nikita Khrushchev reportedly:* "Soviet Orbits Woman Astronaut Near Bykovsky for Dual Flight; They Talk by Radio, Are Put on TV," *New York Times*, June 17, 1963, page 1.

243   *the comment:* "Soviet Orbits Woman Astronaut," *New York Times*.

243   *a Soviet coin:* "Coin, Tereshkova 20th Anniversary, Sally Ride," National Air and Space Museum Collection, accessed at https://airandspace.si.edu/collection-objects/coin-tereshkova-20th-anniversary-sally-ride/nasm_A20140232000 on August 2, 2022.

243   *His background:* "Biographies of Apollo 11 Astronauts," NASA, accessed at https://history.nasa.gov/ap11ann/astrobios.htm on August 3, 2022.

243   *then a sophomore:* Sherr, *Sally Ride*, 31–32, 36.

243   *she later remembered:* Sherr, *Sally Ride*, 36; and "Where Were You When Man First Walked on the Moon?" *USA Today*, weekend edition, July 10–12, 2009, page 7. Unless otherwise noted, the quotations and information that follow in this paragraph are from the same source.

244   *the Apollo astronauts:* Sherr, *Sally Ride*, 36.

244   *She had transferred:* Sherr, *Sally Ride*, 43, 78–79.

244   *a front-page headline:* "NASA to Recruit Women," *Stanford Daily*, January 12, 1977, page 1, in Sherr, *Sally Ride*, 80.

244   *"ad made it clear":* Sally K. Ride, interviewed by Rebecca Wright, San Diego, California, October 22, 2002, accessed on November 19, 2020 at https://historycollection.jsc.nasa.gov/JSCHistoryPortal/history/oral_histories/RideSK/RideSK_10-22-02.htm.

244   *"twenty-year effort":* Sherr, *Sally Ride*, 79.

244   *Intrigued, Ride recalled:* Ride, interviewed by Wright.

244   *undergraduate colloquium:* "Colloquium on Life Science in Space Exploration," 1971, Sally K. Ride Papers, National Air and Space Museum Archives, accessed at https://edan.si.edu/slideshow/viewer/?eadrefid=NASM.2014.0025_ref71 on August 9, 2022.

244   *filled out the paperwork:* Sherr, *Sally Ride*, 83. The information in the rest of the paragraph comes from the same source, pages 78–91.

245   *"couldn't believe it was":* James Hoffman and Mathew Tekulsky, "Adventurers in Space: Six Extraordinary Women Look to the Stars and See Their Futures," *Montgomery Advertiser*, March 19, 1978, page 6.

245   *"I thought maybe":* Ride, interviewed by Wright.

245   *objectives began to shift:* With the Space Race behind them, the United States and the Soviet Union collaborated on the Apollo-Soyuz Test Project in 1975, when an American spacecraft and a Soviet spacecraft docked together. John Uri, "45 Years Ago: Historic Handshake in Space," NASA.gov, July 17, 2020, accessed at https://www.nasa.gov/feature/45-years-ago-historic-handshake-in-space on July 26, 2023.

245   *innovative space shuttle:* "First Shuttle Launch," NASA.gov, April 12, 2013, accessed at https://www.nasa.gov/multimedia/imagegallery/image_feature_2488.html on August 23, 2022.

245   *NASA wanted to construct:* "First Shuttle Launch," NASA.gov.

245   *space station:* William N. Callmers, *Space Policy and Exploration* (Hauppauge, New York: Nova Science Publishers, 2008), 89; and "For All Mankind," *Modesto Bee*, January 17, 1978, page A-13.

245   *Its final purpose:* Wayne Biddle, "Shuttle May Carry Intelligence Satellite," *New York Times*, December 20, 1984, page B12; and Michael Cassutt, "The Secret Space Shuttles," *Air & Space Magazine*, August 2009, accessed at https://www.airspacemag.com/space/secret-space-shuttles-35318554/ on August 23, 2022.

245   *NASA's 1978 class:* Sherr, *Sally Ride*, 94, 99.

245   *NASA's new commitment:* Sherr, *Sally Ride*, 94.

245   *Judith Resnik:* "For All Mankind," *Modesto Bee*.

245   *space shuttle program:* Sherr, *Sally Ride*, 98.

245   *her dissertation:* Bound copy of Ride's PhD dissertation, "The Interaction of X-Rays with the Interstellar Medium," June 1978, Sally K. Ride Papers, National Air and Space Museum Archives, accessed at https://sova.si.edu/details/NASM.2014.0025?s=0&n=10&t=C&q=*%3A*&i=0#ref100 on August 9, 2022; and Sherr, *Sally Ride*, 94–95.

245   *forefront of her mind:* Ride, interviewed by Wright. The quotation that immediately follows is from the same source.

246   *scientific seminars:* Sherr, *Sally Ride*, 101.

246   *supersonic trainer jets:* Sherr, *Sally Ride*, 102–3.

246   *This activity was:* Ride, interviewed by Wright.

246   *she reminded herself:* Ride's handwritten aerobatics notes, page 1, Sally K. Ride Papers, National Air and Space Museum Archives, accessed at https://edan.si.edu/slideshow/viewer/?eadrefid=NASM.2014.0025_ref142 on August 9, 2022.

246   *displays of affection:* Sherr, *Sally Ride*, 9.

246   *"protective of her emotions":* Okie, "Sally Ride Remains an Elusive Character." The quotation that immediately follows is from the same source.

246   *dated different men:* Sherr, *Sally Ride*, 50–59.

246   *Bill Colson:* Sherr, *Sally Ride*, 95.

246   *Steve Hawley:* Sherr, *Sally Ride*, 118–19, 138.

246   *Tam O'Shaughnessy:* Sherr, *Sally Ride*, 193–96.

246   *Ride and Hawley divorced:* Sherr, *Sally Ride*, 228, 237.

247   *Tam would later say:* Sherr, *Sally Ride*, 316.

247   *Perhaps Ride feared:* Sherr, *Sally Ride*, 312.

247   *twentieth anniversary:* Danielle Sempsrott, "Retired Astronaut Bob Crippen on the 40th Anniversary of STS-1 and the Beginning of the Shuttle Program," NASA, April 8, 2021, accessed at https://www.nasa.gov/press-release /retired-astronaut-bob-crippen-on-the-40th-anniversary-of-sts-1-and-the -beginning-of on August 23, 2022.

247   *President Ronald Reagan:* Diary of President Ronald Reagan, June 11, 1985, Ronald Reagan Presidential Foundation, accessed at https://www.reagan foundation.org/ronald-reagan/white-house-diaries/diary-entry-06111985/ on August 10, 2022.

247   *Reagan also:* Ronald Reagan, "Address to the Nation on Defense and National Security," March 23, 1983, Ronald Reagan Presidential Library, accessed at https://www.reaganlibrary.gov/archives/speech/address-nation-defense-and -national-security on August 25, 2022.

247   *NASA's budget growth:* "NASA Spending Since 1958," Office of Management and Budget, linked in "NASA Budgets: US Spending on Space Travel Since 1958 Updated," *Guardian*, accessed at https://www.theguardian.com/news /datablog/2010/feb/01/nasa-budgets-us-spending-space-travel on August 10, 2022.

247   *reusable spacecraft,* Buran: Cathleen Lewis, "The Soviet *Buran* Shuttle: One Flight, Long History," National Air and Space Museum, November 15, 2013, accessed at https://airandspace.si.edu/stories/editorial/soviet-buran-shuttle -one-flight-long-history on August 10, 2022.

247   *George Abbey:* Ride, interviewed by Wright. Unless otherwise noted, the information and quotation that follow in this paragraph are from the same source.

247   *her philosophy:* Todd Halvorson, "Sally Ride Discusses Anniversary of Her Historic Trip to Space," June 18, 2008, accessed at https://www.space.com/5532 -sally-ride-discusses-anniversary-historic-trip-space.html on August 23, 2022.

248   *capsule communicator:* Mark A. Garcia, "Sally Ride—First American Woman in Space," NASA, June 18, 2018, accessed at https://www.nasa.gov/feature/sally -ride-first-american-woman-in-space on August 23, 2022.

248   *serve as a mission specialist:* Halvorson, "Sally Ride Discusses Anniversary of Her Historic Trip."

248   *would act as commander:* Cliff Lethbridge, "STS-7 Fact Sheet," Spaceline.org, accessed at https://www.spaceline.org/united-states-manned-space-flight /space-shuttle-mission-program-fact-sheets/sts-7/ on August 8, 2022.

248   *Four of the five:* Ride, interviewed by Wright. The information in the rest of the paragraph comes from the same source.

248   *Ride would later remember:* Ride, interviewed by Wright.

248   *ninety-six-page notebook:* "Ride's STS-7 Shuttle Training Notebook," Sally K. Ride Papers, National Air and Space Museum Archives, accessed at https://edan.si.edu/slideshow/viewer/?eadrefid=NASM.2014.0025_ref179 on August 23, 2022.

248   *studied the procedures:* "Ride's STS-7 Shuttle Training Notebook," Sally K. Ride Papers.

248   *crew's daily schedule:* "Crew Activity Plans for STS-7 (annotated)," and "STS-7 Flight Schedule, Invitation to Launch and Bus Passes," Sally K. Ride Papers, National Air and Space Museum Archives, accessed at https://edan.si.edu/slideshow/viewer/?eadrefid=NASM.2014.0025_ref180 and https://edan.si.edu/slideshow/viewer/?eadrefid=NASM.2014.0025_ref194 on August 9, 2022.

248   *To ready her body:* Sherr, *Sally Ride*, 112–14.

249   *the host of queries:* Judy Mann, "Dr. Ride's Ride," June 24, 1983, *Washington Post*, accessed at https://www.washingtonpost.com/archive/local/1983/06/24/dr-rides-ride/5cf74ac3-ad4a-4852-be09-1abcf2cd0525/ on August 10, 2022.

249   *the slight when:* Mann, "Dr. Ride's Ride"; and "Interview with Astronaut Sally Ride and STS-7," raw footage for *The Kid Show*, 1980s, Texas Archive of the Moving Image, accessed at https://texasarchive.org/2019_01218 on August 10, 2022.

249   *"remain[ed] calm, unrattled":* Ryan, "A Ride in Space."

249   *ate a large breakfast:* Sherr, *Sally Ride*, 154–55.

249   *in-flight jumpsuit:* "Jacket, In-Flight Suit, Shuttle, Sally Ride, STS-7," National Air and Space Museum Collection, accessed at https://airandspace.si.edu/collection-objects/jacket-flight-suit-shuttle-sally-ride-sts-7/nasm_A19830241000 on August 12, 2022.

249   *video broadcast:* "Space Shuttle Challenger–STS-7 Launch," accessed at https://www.youtube.com/watch?v=DjtXJpdxeew on July 13, 2022.

249   *President Ronald Reagan had:* Ronald Reagan, Press Conference, June 1, 1983, accessed at https://www.youtube.com/watch?v=JSbj2EJnrJc on July 13, 2022.

250   *fear was palpable:* Ride, interviewed by Wright. The information and quotations that follow in this paragraph are from the same source.

250   *a little more at ease:* Ride, interviewed by Wright. The quotations that follow in this paragraph are from the same source.

250   *2.5 million miles:* "Space Shuttle: STS-7," NASA.gov, accessed at https://www.nasa.gov/mission_pages/shuttle/shuttlemissions/archives/sts-7.html on August 23, 2022.

250   *using the robotic arm:* John Noble Wilford, "Shuttle Rockets to Orbit with Five Aboard," *New York Times*, June 19, 1983, accessed at https://archive.nytimes.com/www.nytimes.com/library/national/science/nasa/061983sci-nasa-wilford.html; and Photograph, "Challenger in Space," June 22, 1983, NASA, accessed at https://www.nasa.gov/mission_pages/shuttle/flyout/multimedia/challenger/1983-06-22.html on August 23, 2022.

250 *"Oh, my gosh":* Ride, interviewed by Wright; and Al Rossiter Jr., "Sally Ride and Her Colleagues Aboard the Shuttle Challenger," *United Press International,* June 20, 1983, accessed at https://www.upi.com/Archives/1983/06/20/Sally-Ride-and -her-colleagues-aboard-the-shuttle-Challenger/5742424929600/ on August 16, 2022.

251 *Despite her worries:* William J. Broad, "Cool, Versatile Astronaut: Sally Kristen Ride," *New York Times,* June 19, 1983, page 1. The quotation that immediately follows is from the same source.

251 *Inclement weather thwarted:* Ride, interviewed by Wright. The quotations that immediately follow are from the same source.

251 *A video camera:* "Space Shuttle Challenger Landing," accessed at https://www .youtube.com/watch?v=9P64vO8EQNQ on August 23, 2022.

251 *the massive spacecraft:* Lethbridge, "STS-7 Fact Sheet."

251 *exited the spacecraft:* "Space Shuttle Challenger Landing."

251 *mission was accomplished:* NASA Post Flight Operation Report, August 5, 1983, page 2, NASA, accessed at https://ntrs.nasa.gov/api/citations/19830023440 /downloads/19830023440.pdf on August 16, 2022.

252 *Ride told reporters:* Sally Ride, Press Conference, June 24, 1983, accessed at https://www.youtube.com/watch?v=nxo84aJJvWc on August 23, 2022.

252 *Johnson Space Center:* Sherr, *Sally Ride,* 168–69.

252 *negative news headlines:* Sherr, *Sally Ride,* 170; and "Big Bouquet Spurned by Woman Astronaut," *Spokane Chronicle,* June 25, 1983, page 7.

252 *Ride explained that:* "Astronaut Sally Ride Spurns Bouquet of Roses, Carna- tions," *Lancaster New Era,* June 25, 1983, page 22; and Sherr, *Sally Ride,* 170.

252 *arguing that she had:* "President Telephones to Congratulate Crew," *Miami Her- ald,* June 25, 1983, 18A.

252 *the* Miami Herald*:* Mary Voboril, "Sally Ride Earns Place in History Next to Earhart," *Miami Herald,* June 25, 1983, 18A.

252 *In Earhart's hometown:* "Sally Ride Honored," *Daily Spectrum,* July 26, 1983, page 14.

253 *she said at the event:* "Elizabeth Dole Lauds Female Flying Pioneer," *St. Joseph Gazette,* July 25, 1983, page 6A; and "Sally Ride Honored in Earhart Forest," *Salina Journal,* July 25, 1983, page 7.

253 *Soviet Union's efforts:* Charles Seabrook, "Soviet Union, US Take Different Paths in Space Race, But No Winner Is Apparent," *Atlanta Constitution,* July 3, 1983, page 9B.

253 *their week-long mission:* "STS-41-G," NASA, accessed at https://www.nasa.gov /mission_pages/shuttle/shuttlemissions/archives/sts-41G.html on August 16, 2022.

253 *completed a space walk:* "STS-41-G," NASA.

253 *Ride was devastated:* Sherr, *Sally Ride,* 203.

253 *an upcoming flight*: Sherr, *Sally Ride,* 203.

253 *lent her knowledge:* Sherr, *Sally Ride,* 210, 214.

254 *Sally Ride Science:* Denise Grady, "American Woman Who Shattered Space Ceiling," *New York Times,* July 23, 2012, accessed at https://www.nytimes

.com/2012/07/24/science/space/sally-ride-trailblazing-astronaut-dies-at-61
.html on August 23, 2022.

254   *Reflecting on her life:* Halvorson, "Sally Ride Discusses Anniversary of Her Historic Trip."

EPILOGUE: THE FUTURE OF AMERICAN EXPLORATION

255   *planet's average temperature:* "World of Change: Global Temperatures," NASA, accessed at https://earthobservatory.nasa.gov/world-of-change/global-temperatures on September 27, 2022.

256   *the amplified effects:* Chelsea Harvey, "The Arctic Is Warming Four Times Faster than the Rest of the Planet," *E & E News* and *Scientific American*, August 12, 2022, accessed at https://www.scientificamerican.com/article/the-arctic-is-warming-four-times-faster-than-the-rest-of-the-planet/ on September 27, 2022.

256   *267 billion tons:* Renee Cho, "What Lies Beneath Melting Glaciers and Thawing Permafrost?" Columbia Climate School, September 13, 2022, accessed at https://news.climate.columbia.edu/2022/09/13/what-lies-beneath-melting-glaciers-and-thawing-permafrost/ on September 27, 2022.

256   National Geographic *reports:* Neil Shea, "A Thawing Arctic Is Heating Up a New Cold War," *National Geographic*, August 15, 2019, accessed at https://www.nationalgeographic.com/adventure/2019/08/how-climate-change-is-setting-the-stage-for-the-new-arctic-cold-war-feature/ on October 4, 2022.

256   *"new Cold War":* Shea, "A Thawing Arctic Is Heating Up."

256   *through the Arctic:* Brown University, "Melting Arctic Ice Could Transform International Shipping Routes, Study Finds," *Science Daily*, June 20, 2022, accessed at www.sciencedaily.com/releases/2022/06/220620152119.htm on October 4, 2022.

256   *Tourism to the Arctic:* "Top Arctic Trips and Tours in 2022," accessed at https://www.adventure-life.com/arctic/tours/2022 on September 27, 2022.

256   *a passenger capacity:* Adam Minter, "Rich Tourists Can Actually Preserve the Arctic," *Washington Post*, July 21, 2022, accessed at https://www.washingtonpost.com/business/rich-tourists-can-actually-preserve-the-arctic/2022/07/21/3a4d1e32-08fe-11ed-80b6-43f2bfcc6662_story.html on September 27, 2022.

256   *evidence of human settlement:* Bridget Alex, "These Ice Age Humans Somehow Survived North of the Arctic Circle," *Discover*, November 26, 2019, accessed at https://www.discovermagazine.com/planet-earth/these-ice-age-humans-somehow-survived-north-of-the-arctic-circle on October 4, 2022. The information in the rest of the paragraph comes from the same source.

257   *62 percent less:* "Mars and Beyond: The Road to Making Humanity Multiplanetary," SpaceX, accessed at https://www.spacex.com/human-spaceflight/mars/ on October 4, 2022.

257   *The SpaceX company:* Brett Tingley, "These 2 Private Companies Aim to Beat SpaceX to Mars with 2024 Flight," July 19, 2022, Space.com, accessed at https://www.space.com/relativity-space-private-mars-mission-launching-2024 on October 4, 2022.

257   *SpaceX ultimately hopes:* "Mars and Beyond," SpaceX.

257   *Apollo 17 mission:* "Apollo 17," NASA, April 7, 2011, accessed at https://www
.nasa.gov/mission_pages/apollo/missions/apollo17.html on October 4, 2022.

257   *federal budget dollars:* "NASA Spending Since 1958," Office of Management
and Budget, linked in "Nasa Budgets: US Spending on Space Travel Since
1958 Updated," *Guardian*, accessed at https://www.theguardian.com/news
/datablog/2010/feb/01/nasa-budgets-us-spending-space-travel on August 10,
2022.

257   *new space shuttle program:* Mike Wall, "Presidential Visions for Space Exploration: From Ike to Biden," Space.com, January 20, 2021, accessed at https://
www.space.com/11751-nasa-american-presidential-visions-space-exploration
.html on October 4, 2022.

257   *NASA predicts that:* Jeff Foust, "NASA Weighs Changes to Artemis 3 If Key
Elements Are Delayed," *Space News*, accessed at https://spacenews.com/nasa
-weighs-changes-to-artemis-3-if-key-elements-are-delayed/ on August 10,
2023; Ben Scott, "A Visual Guide of NASA's Plan to Get Back to the Moon,"
National Geographic, August 26, 2022, accessed at https://www.national
geographic.com/magazine/graphics/a-visual-guide-of-nasas-plan-to-get-back
-to-the-moon on October 4, 2022; and Alexandra Witze, "NASA's Artemis Moon Mission Is Set to Launch: Here's the Science on Board," *Nature*,
August 24, 2022, accessed at https://www.nature.com/articles/d41586-022
-02293-8 on October 4, 2022.

257   *new Artemis program:* "Artemis," NASA, accessed at https://www.nasa.gov
/specials/artemis/ on October 4, 2022.

257   *the agency plans to:* "Why the Moon?" video, NASA, accessed at https://www
.nasa.gov/specials/artemis/ on October 4, 2022.

258   *search for water:* Alex Fox, "Four Things We've Learned About NASA's
Planned Base Camp on the Moon," *Smithsonian Magazine*, August 29, 2022,
accessed at https://www.smithsonianmag.com/science-nature/four-things-weve
-learned-about-nasas-planned-base-camp-on-the-moon-180980589 on October 4, 2022.

258   *NASA proclaims:* "Artemis," NASA.

258   *John Raymond:* John W. Raymond, "How We're Building a 21st-Century Space
Force," *Atlantic*, December 20, 2020, accessed at https://www.theatlantic.com
/ideas/archive/2020/12/building-21st-century-space-force/617434/ on October 4, 2022.

259   *250 million years:* Laurence Tognetti, "The Record for the Farthest Galaxy Just
Got Broken Again, Now Just 250 Million Years After the Big Bang," *Universe
Today*, August 2, 2022, accessed at https://www.universetoday.com/156987
/the-record-for-the-farthest-galaxy-just-got-broken-again-now-just-250
-million-years-after-the-big-bang/ on October 4, 2022.

259   *an Ariane 5 rocket:* "James Webb Space Telescope Launch Timeline As It Happened," *SciTech Daily*, December 25, 2021, accessed at https://scitechdaily.com
/james-webb-space-telescope-launch-timeline-as-it-happened/ on October 4,
2022; and Joseph Biden, "Remarks by President Biden and Vice President Harris in a Briefing to Preview the First Images from the James Webb Space Telescope," July 11, 2022, White House, accessed at https://www.whitehouse.gov

/briefing-room/statements-releases/2022/07/11/remarks-by-president-biden
-and-vice-president-harris-in-a-briefing-to-preview-the-first-images-from
-the-james-webb-space-telescope/ on October 4, 2022.

259   *president Joseph Biden:* Biden, Remarks by President Biden and Vice President
Harris. The quotation that immediately follows comes from the same source.

INDEX

Mississippi River, 17, 43, 44, 46
Mississippi River Valley, 45
Missouri River, 17–22, 27–32, 36, 43,
    65, 213–14
Misti, El, 178–79
Mommaerts, Elizabeth, 238–39
Monterey, CA, 55
moon, 7, 226, 241–45, 257–58
Morgan, John, 141, 142
Morrison, William, 159
mountain men, 40, 43, 46
    see also Beckwourth, James
Mount Rainier National Park, 109
Muir, Daniel, 89–92
Muir, John, 87–110, 120, 128, 129
    accident of, 94–95
    childhood of, 88–92, 96
    emigration to America, 88–91, 96
    environmental protection and, 88,
        98–100, 104–10, 132, 166
    as inventor, 92, 95, 105
    Johnson and, 106–8
    Merriams and, 113–14, 119
    Native Americans and, 91–92, 106, 109
    as shepherd, 97, 98
    Sierra Club cofounded by, 109
    in Sierra Nevada, 96, 100, 105, 106, 166
    at University of Wisconsin, 92, 94
    writings of, 98–100, 105–7, 109–10
    in Yosemite Valley, 87–88, 93, 94,
        96–100, 106–9, 113–14
Muir, Louisa Strentzel, 99
Murderer's Bar, 56–57
Mushenge, 154
Musk, Elon, 257
Mussolini, Benito, 233
My First Summer in the Sierra (Muir), 98

N

NACA (National Advisory Committee
    for Aeronautics), 240, 243

Nagasaki, 235–36
Nansen, Fridtjof, 188
Narkeeta, 195
NASA (National Aeronautics and Space
    Administration), 232, 233, 240,
    242–43, 247, 253–54, 257, 258
    women recruited by, 244–45
    see also space exploration; space
        shuttle program
Nash, Roderick, 98, 105
Nashville Banner, 205
Nashville Globe, 205–6
National Aeronautic Association, 220
National Defense Education Act, 240
National Era, 62
National Geographic, 180–81, 256
National Geographic Society, 180, 189,
    204, 230
national parks, 87, 107–9, 220
National Science Board, 240
National Security Act, 236
National Security Council, 236
Native Americans, 1–6, 9, 17–19, 40,
    43–45, 85, 95, 113, 164, 214, 220
    in Battle of Little Bighorn, 69
    Beckwourth and, 43, 46–48, 50, 51,
        56, 62, 91
    bison and, 100, 101
    Boone and, 1–2, 4
    in California, 41, 50–51, 56, 58, 63,
        93–94, 106, 165
    Cheyenne, 50, 62, 68–69, 100
    citizenship for, 102
    Civil War and, 67
    Crow, 47–48, 62
    Dawes Act and, 102
    diseases and, 14, 21, 40, 45, 51
    Ghost Dance religion of, 102
    Hidatsa, 14–16, 20, 22–25, 27, 31, 33
    lead mined by, 46, 91
    Lewis and Clark Expedition and,
        19–25, 30–38